Land-use Change Impacts on Soil Processes

Tropical and Savannah Ecosystems

FSC
www.fsc.org
MIX
Paper from
responsible sources
FSC® C013604

Land-use Change Impacts on Soil Processes

Tropical and Savannah Ecosystems

Edited by

Francis Q. Brearley

Manchester Metropolitan University

and

Andrew D. Thomas

Aberystwyth University

www.cabi.org

CABI is a trading name of CAB International

CABI
Nosworthy Way
Wallingford
Oxfordshire OX10 8DE
UK

Tel: +44 (0)1491 832111
Fax: +44 (0)1491 833508
E-mail: info@cabi.org
Website: www.cabi.org

CABI
745 Atlantic Avenue
8th Floor
Boston, MA 02111
USA

Tel: +1 617 682 9015
E-mail: cabi-nao@cabi.org

A catalogue record for this book is available from the British Library, London, UK.

Library of Congress Cataloging-in-Publication Data

Land-use change impacts on soil processes : tropical and savannah ecosystems / edited by Francis Q. Brearley, Manchester Metropolitan University; and Andrew D. Thomas, Aberystwyth University.
pages cm
Includes bibliographical references and index.
ISBN 978-1-78064-210-9 (hbk : alk. paper) 1. Soils--Environmental aspects--Case studies. 2. Land use--Environmental aspects--Case studies. 3. Soil ecology--Case studies. I. Brearley, Francis Q., editor. II. Thomas, Andrew D. (Andrew David), 1970- editor. III. Title: Land use change impacts on soil processes.

S596.L36 2015
577.5'7--dc23

2015021721

ISBN-13: 978 1 78064 210 9

Commissioning editors: Vicki Bonham and Nicki Dennis
Editorial assistant: Emma McCann
Production editors: Tracy Head and Emma Ross

Typeset by SPi, Pondicherry, India.
Printed and bound in the UK by CPI Group (UK) Ltd, Croydon, CR0 4YY.

Contents

Contributors

Philip Antwi-Agyei, Sustainability Research Institute, School of Earth and Environment, University of Leeds, Leeds LS2 9JT, UK; and Department of Environmental Science, College of Science, Kwame Nkrumah University of Science and Technology, Kumasi, Ghana (p.antwi-agyei@leeds.ac.uk)

Justin M. Becknell, Department of Ecology, Evolution, and Behavior, University of Minnesota, St Paul, MN 55108, USA (justinbecknell@gmail.com)

Luke Beesley, Environmental and Biochemical Sciences, The James Hutton Institute, Craigiebuckler, Aberdeen, AB15 8QH, UK (luke.beesley@hutton.ac.uk)

Clóvis D. Borges, Cell and Molecular Biology Laboratory, Center for Nuclear Energy in Agriculture CENA, University of São Paulo USP, Piracicaba, São Paulo, CEP 13.416-000, Brazil (clovisdb@usp.br)

Francis Q. Brearley, School of Science and the Environment, Manchester Metropolitan University, Chester Street, Manchester, M1 5GD, UK (f.q.brearley@mmu.ac.uk)

Chikondi P. Chabvuta, Centre for Community Organization and Development, PO Box 2109, Lilongwe, Malawi (chikonatie@gmail.com)

Lovemore Chipungu, School of the Built Environment and Development Studies, University of KwaZulu-Natal, Level 8 Shepstone Building, Howard College, Durban 4001, South Africa (chipungu@ukzn.ac.za)

Marife D. Corre, Büsgen Institute – Soil Science of Tropical and Subtropical Ecosystems, Georg-August University Göttingen, Büsgenweg 2, D-37077 Göttingen, Germany (mcorre@gwdg.de)

Heather D'Angelo, Department of Ecology, Evolution, and Environmental Biology, Columbia University, 116th Street and Broadway, New York, NY 10027, USA (hd2144@columbia.edu)

Free de Koning, Conservation International Ecuador, Catalina Aldáz N34-181 y Portugal, Quito, Ecuador (fdekoning@conservation.org)

Raghavan Dinesh,Indian Institute of Spices Research, Marikunnu PO, Calicut – 673012, Kerala, India (rdinesh2005@gmail.com, rdinesh@spices.res.in)

Andrew J. Dougill, Sustainability Research Institute, School of Earth and Environment, University of Leeds, Leeds, LS2 9JT, UK (a.dougill@see.leeds.ac.uk)

David R. Elliott, School of Science and the Environment, Manchester Metropolitan University, Chester Street, Manchester, M1 5GD, UK (d.elliott@mmu.ac.uk)

Kazumichi Fujii, Forestry and Forest Products Research Institute, Tsukuba 305-8687, Japan (fjkazumichi@gmail.com)

Shinya Funakawa, Graduate School of Agriculture, Kyoto University, Kyoto 606-8502, Japan (funakawa@kais.kyoto-u.ac.jp)

Arkalgud Ganeshamurthy, Indian Institute of Horticultural Research, Hessargahatta Lake PO, Bengaluru – 560089, Karnataka, India (angmurthy@gmail.com)

Subrata Ghoshal Chaudhuri, National Bureau of Soil Survey and Land Use Planning (Regional Centre), Salt Lake City, Bidhan Nagar, Kolkata – 700091, West Bengal, India (sgc2004@gmail.com)

Caitlyn Gillikin, Department of Biology, Barnard College, Columbia University, 3009 Broadway, New York, NY 10027, USA (caitlyngillikin@gmail.com)

Tasmin N.L. Griffith, School of Science and the Environment, Manchester Metropolitan University, Chester Street, Manchester, M1 5GD, UK (tamsinlouise1210@hotmail.com)

Catherine Grimaldi, INRA, Institut National de la Recherche Agronomique, UMR 1069, Soil Agro and hydroSystem, F-35000 Rennes, France (catherine.grimaldi@rennes.inra.fr)

Michel Grimaldi, IRD, Institut de Recherche pour le Développement, iEES-Paris, Institute of Ecology and Environmental Sciences of Paris, UMR 242, F-93143 Bondy, France (michel.grimaldi@ird.fr)

Stéphane Guédron, ISTerre, Institute of Earth Science, Université Grenoble 1, IRD – UMR 5559 (IRD/UJF/CNRS) – BP 53, F-38041 Grenoble, France (stephane.guedron@ujf-grenoble.fr)

Michael Hardman, School of Environment and Life Sciences, University of Salford, Salford Crescent, Salford, M5 4WT, UK (m.hardman@salford.ac.uk)

Chie Hayakawa, The University of Tokyo, Graduate School of Agricultural and Life Sciences, Tokyo 113-8657, Japan (a-hykw@mail.ecc.u-tokyo.ac.jp)

Simon Kaiser, Tropical Silviculture and Forest Ecology, Georg-August University Göttingen, Büsgenweg 1, 37077 Göttingen, Germany (simon-kaiser@gmx.de)

Takashi Kosaki, Department of Tourism Science, Tokyo Metropolitan University, Tokyo 192-0364, Japan (kosakit8@tmu.ac.jp)

Eiko E. Kuramae, Microbial Ecology Department, Netherlands Institute of Ecology NIOO-KNAW, Wageningen, the Netherlands (e.kuramae@nioo.knaw.nl)

Magdalena López-Ulloa, Environmental Engineering, Universidad de las Americas, Avenida de los Granados E 12-41 y Colimes Esq., Quito, Ecuador (maggielopez62@gmail.com)

Hangwelani H. Magidimisha, Human Science Research Council, Demography Governance and Service Delivery, Private Bag X07, Dalbridge, Durban, 4014, South Africa (hmagidimisha@hsrc.ac.za)

Helen Mairs, School of Science and the Environment, Manchester Metropolitan University, Chester Street, Manchester, M1 5GD, UK (helenmairs@gmail.com)

David W.P. Manning, Odum School of Ecology, University of Georgia, Athens, GA 30602, USA (manningd@uga.edu)

Krista L. McGuire, Department of Biology, Barnard College, Columbia University, 3009 Broadway, New York, NY 10027, USA; and

Department of Ecology, Evolution, and Environmental Biology, Columbia University, 1200 Amsterdam Ave, New York, NY 10027, USA (kmcguire@barnard.edu)

Lucas W. Mendes, Cell and Molecular Biology Laboratory, Center for Nuclear Energy in Agriculture CENA, University of São Paulo USP, Piracicaba, São Paulo, CEP 13.416-000, Brazil (lwmendes@cena.usp.br)

Dina C. Merrer, Department of Chemistry, Barnard College, Columbia University, 3009 Broadway, New York, NY 10027, USA (dmerrer@barnard.edu)

David D. Mkwambisi, Lilongwe University of Agriculture and Natural Resources, Bunda College Campus, PO Box 219, Lilongwe, Malawi (david.mkwambisi@bunda.luanar.mw)

Acácio A. Navarrete, Cell and Molecular Biology Laboratory, Center for Nuclear Energy in Agriculture CENA, University of São Paulo USP, Piracicaba, São Paulo, CEP 13.416-000, Brazil (navarrete@cena.usp.br)

L. Ronald Ng Cheong, Mauritius Sugarcane Industry Research Institute, Moka Road, Réduit, Mauritius (ronald.ngcheong@msiri.mu)

Jennifer S. Powers, Department of Ecology, Evolution, and Behavior, University of Minnesota, St Paul, MN 55108, USA; and Department of Plant Biology, University of Minnesota, St Paul, MN 551008, USA (powers@umn.edu)

Jeroen M. Schoorl, Soil Geography and Landscape, Wageningen University, PO Box 47, 6700 AA Wageningen, Netherlands (jeroen.schoorl@wur.nl)

Luitgard Schwendenmann, School of Environment, The University of Auckland, Private Bag 92019, Auckland 1142, New Zealand (l.schwendenmann@auckland.ac.nz)

Andrew D. Thomas, Department of Geography and Earth Sciences, Aberystwyth University, Penglais, Aberystwyth, SY23 3DB, UK (ant23@aber.ac.uk)

Siu Mui Tsai, Cell and Molecular Biology Laboratory, Center for Nuclear Energy in Agriculture CENA, University of São Paulo USP, Piracicaba, São Paulo, CEP 13.416-000, Brazil (tsai@cena.usp.br)

Gunshiam Umrit, Mauritius Sugarcane Industry Research Institute, Moka Road, Réduit, Mauritius (gunshiam.umrit@msiri.mu)

Edzo Veldkamp, Büsgen Institute – Soil Science of Tropical and Subtropical Ecosystems, Georg-August University Göttingen, Büsgenweg 2, D-37077 Göttingen, Germany (eveldka@gwdg.de)

Acknowledgements

We thank the authors for their contributions and the reviewers who helped improve them. We are also grateful to the publishers and our families for their support and patience while we completed this book.

1 Land-use Change Impacts on Soil Processes in Tropical and Savannah Ecosystems: An Introduction

Francis Q. Brearley[1]* and Andrew D. Thomas[2]
[1]*School of Science and the Environment, Manchester Metropolitan University, Manchester, UK;* [2]*Department of Geography and Earth Sciences, Aberystwyth University, Aberystwyth, UK*

1.1 Introduction

> For most of history, few things have mattered more to human communities than their relations with soil.
> (McNeill and Winiwarter, 2004, p. 1627)

Soils are the thinnest, outermost layer of the Earth's land surface: a complex, heterogeneous combination of weathered parent material, living and dead organic matter, water and gases upon which humans are wholly dependent. Soils take thousands of years to fully develop; yet poor management can lead to rapid and ultimately detrimental changes in their physical, chemical and biological characteristics. The disparity between the time taken to form and the speed with which soils can degrade means they are inherently fragile and thus require prudent management. There is, however, widespread evidence to suggest that careful and sustainable soil management is not the norm, and that the global soil resource is being depleted, threatening the numerous ecosystem services they provide (Banwart, 2011; Richter *et al.*, 2011; UNCCD, 2012; Koch *et al.*, 2013). In compiling and editing this book we have brought together a collection of case studies from around the globe that illustrate the impact of a range of land-use changes on the physical, chemical and biological characteristics of soils and provide a snapshot of the challenges we face in ensuring sustainable soil management in tropical and savannah environments.

1.2 The Importance of Soil to Human Well-being

There are numerous ways in which soils support human well-being.

First, the sustained fertility of soil is essential for food, fibre, animal feed, timber and, increasingly, biofuel production. With the world's population projected to increase to 9 billion people by 2050, the sustainable and efficient use of soils will be central to global food security (Stocking, 2003; Godfray *et al.*, 2010) and the diversification of energy sources through biomass production (Tilman *et al.*, 2009). The demands we place on soils are likely to increase in the future, thus ensuring soils are managed sustainably and are not degraded further will become ever more important.

Second, soils are integral to the water cycle (Falkenmark and Rockström, 1996), making clean water available to humans and helping to regulate stream and river flows. Degraded and

* Corresponding author: f.q.brearley@mmu.ac.uk

polluted soils within a catchment will typically result in degraded and polluted stream water. Eroded and compacted soils are more likely to lead to overland flow generation, leading to *in situ* reductions in soil moisture and greater instances of crop drought as well as increased problems linked to flooding and sediment deposition. Effective management of soils therefore underlies effective catchment management.

Third, the organic carbon (C) stored in soils is estimated to be around 1500 Pg to a depth of 1 m globally (Post *et al.*, 1982; Eswaran *et al.*, 1993; Batjes, 1996) (a Petagram is equivalent to 10^{15} g or 1 billion tonnes). This is twice the mass of C in the atmosphere as CO_2 and three times the C in aboveground vegetation. Soil respiration (the flux of plant- and microbially generated CO_2 from soils to the atmosphere) is the second largest terrestrial C flux (Bond-Lamberty and Thomson, 2010). Land-use changes, particularly forest clearance and grazing intensification can significantly increase soil respiration C fluxes markedly with likely positive feedback to future climate warming (Grace *et al.*, 2006; Thomas, 2012). There is, however, great potential for increasing the soil C stock through adoption of land management practices that increase soil organic matter (SOM) content, the win–win scenario of increased C storage and soil fertility advocated by Lal (2004) and others. Careful consideration is, however, needed to account for the complete C balance to ensure that well-meaning land management interventions do not lead to unintentional consequences for the fluxes of other greenhouse gases (Powlson *et al.*, 2011). A clearer understanding, and quantification, of the amount of C stored in soils under different land uses is therefore essential for an improved understanding of the global C cycle and how this will be altered in the near future by human activities.

Finally, soils are host to an enormous reservoir of microbial biodiversity; thousands of species can be found in just a gram of soil, most of which we know very little about and which are overlooked by many scientists. However, understanding the diversity and functioning of soil microbial communities (archaea, bacteria and fungi in particular) is of major importance due to their key roles in regulating the biogeochemical cycles of C, nitrogen and other elements (Falkowski *et al.*, 2008; Singh *et al.*, 2010). In addition, the potential for utilizing microorganisms in the search for the next generation of antibiotics and other pharmaceuticals is only just being realized as limitations associated with the inability to culture most microbes are being overcome through rapid advances in molecular techniques (Daniel, 2004; Singh and Macdonald, 2010; Lewis, 2012).

In order to manage soils sustainably and to ensure we continue to benefit from the essential range of ecosystem services they provide, we need a much better understanding and awareness of when and how changes in soil characteristics become long-term degradation. A particular challenge is to predict the soil response to land-use changes that may be non-linear or characterized by thresholds (Nikolaidis, 2011). This is a priority in tropical and savannah environments with high population densities because: (i) soils typically contain highly weathered clay minerals with limited capacity to form complexes with organic matter (Feller and Beare, 1997; Zech *et al.*, 1997; Rasche and Cadisch, 2013); (ii) persistently high soil temperatures and moisture facilitate rapid oxidation of SOM (Joergensen, 2010; Hayakawa *et al.*, 2014); and (iii) high rainfall intensities equate to high potential energy and erosion potential as loss of vegetation cover and aggregate stability can result in very large increases in soil erosion through overland flow and landslides (Lal, 1990; Sidle *et al.*, 2006).

A loss of soil 'quality' often leads to reduced yields and accelerated erosion. In some cases, particularly in drier areas affected by dust storms, this can have immediate and adverse impacts on human health (Griffin *et al.*, 2001). For example, cases of valley fever, or *Coccidioidomycosis*, in the human population are prevalent in agricultural areas experiencing wind erosion. Cases of valley fever have increased in the south-west USA over the last decade (Centers for Disease Control and Prevention, 2013) and may be associated with a deterioration of soils and an increase in dust storm events.

It is encouraging to see that there is an increasing number of international and national organizations, policies, agreements and initiatives to protect and raise awareness of our soil resource. Examples include the International Soil Reference and Information Centre (ISRIC), the United Nations Convention to Combat Desertification (UNCCD), the Food and Agriculture Organization of the United Nations (UN-FAO) Global Soil Partnership and World Soils Week among others. Although 2015 is the International Year of Soils, it is dispiriting to report that the European Union

(EU) Soil Framework Directive has recently been withdrawn, despite the estimated annual cost of soil degradation across the EU of €38 billion per year (European Commission, 2006, in Hartemink and McBratney, 2008). There are clearly still many challenges ahead.

1.3 Land-use Change and Soil Properties

Changing land use, particularly the removal of forest cover for agriculture and the intensification of grazing, has a major and widespread impact on soils (Ellis and Ramankutty, 2008; Don *et al.*, 2011). Over half of the world's soils are now cultivated, grazed, subject to logging disturbance or have been built upon (Richter, 2007; Ellis, 2011). This has resulted in major changes to soil properties as well as the soil-related resources available to us (Yaalon, 2007). The major land-use changes affecting soils in the tropics are complex and numerous and it is misleading to generalize too much; however, the following are all notable drivers of change in various parts of the tropics.

The expansion of agriculture into areas of native vegetation across tropical regions is widespread (Gibbs *et al.*, 2010). Agricultural expansion may be small scale (agroforestry or swidden agriculture) but large-scale expansion of monoculture crops is likely to have a more pervasive effect on soil processes and functions due to the much larger disturbance to the soil through ploughing and addition of nutrients and other chemicals. Conversion to agriculture is arguably the most prevalent land-use transition in tropical regions leading to a number of alterations to soil including: (i) loss of SOM and C; (ii) periodic but rapid additions of fertilizers and pesticides; (iii) changes to soil hydrology, both directly through irrigation and indirectly through changes to evapotranspiration rates; and (iv) impacts upon soil organisms that will feed back to influence soil properties and functioning.

In sub-tropical and dry sub-humid savannahs, pastoralism is the dominant agricultural land use. In many areas, a reduction in the area of communal grazing land available to those without land tenure due to privatization is leading to increased animal stocking densities (Thomas, 2012). This leads to changes in the natural vegetation cover, particularly in the balance between trees, shrubs and grasses that, in turn, affects the fire return period (Grace *et al.*, 2006; Eldridge *et al.*, 2011). Soils are affected indirectly by all these changes, but they are also affected directly by the trampling action of livestock (Thomas, 2012).

The number of people living in urban areas is now more than that in rural areas, with the rate of increase being most rapid in tropical countries (Seto *et al.*, 2011). This is important as urbanization has an obvious and dramatic effect on soils (Marcotullio *et al.*, 2008), although it has received far less attention when compared to the role that agriculture and forestry can have on soils and soil productivity. Renewed interest in urban agriculture in tropical countries (Hamilton *et al.*, 2013; Orsini *et al.*, 2013) has increased the importance of improving our understanding of soils in urban environments and the potential for urban-based agriculture to address food needs. With this, however, comes a new set of sustainability related challenges such as urban planning policy, the utilization of waste products for fertilizers and the potential for soil contamination. This is a particularly important research area for contemporary soil science.

Soils are also commonly affected by chemical and biological contaminants from a variety of sources, including organic manures (van der Perk, 2013). These are a threat to soil productivity and health, particularly as pollutants may be subsequently ingested if food grown on contaminated land is consumed. Deep weathering of soils, typical of much of the tropics, can often exacerbate heavy metal additions through mining and cause widespread pollution. For example, mercury is used in gold mining to form a mercury–gold amalgam and help in the recovery of gold. Many areas of the tropics have land used for artisanal small-scale gold mining that releases around 1000 t of mercury a year to the environment (Telmer and Veiga, 2009) where it leads to pollution of watercourses, soils, sediments and higher trophic levels.

Changes to soil properties and processes are not necessarily wholly undesirable, and a certain amount of change should be viewed as an unavoidable consequence of utilizing soils for the wider human good. Indeed, there are examples of soil properties improving as a consequence of land-use changes, such as the *terra preta do índio* (or Amazonian dark earths) created by pre-Colombian inhabitants of South America prior to the arrival of Europeans (Glaser *et al.*, 2001; Marris, 2006). These are some of the most fertile tropical soils and the conditions that led to their creation are still being elucidated; indeed, much of the focus on biochar (e.g. Sohi *et al.*, 2010) in current research stems from the discovery of the inclusion of charred remains in *terra preta* soil. On the Indonesian island of Java, there has been an island-wide increase in soil C since the 1970s following a prior decline from the 1930s onwards (Minasny *et al.*, 2011). This has been attributed to increased application of crop residues and animal manure to improve soil fertility following government intervention. Other examples of restoration of soil processes following disturbance include the re-growth of plantations on old pastures or recovery of secondary forests on old shifting cultivation sites that can both lead to recovery of SOM and a subsequent increase in soil fertility (e.g. Don *et al.*, 2011). In addition, a number of agricultural management practices such as reduction of tillage, the use of cover crops and the use of organic amendments can all induce C storage in soils (Lal, 2004).

1.4 Themes and Scope of this Book

This book is not a comprehensive account of all the drivers of soil change in the tropics, nor does it cover all areas of the tropics. It does, however, take a broad view of the tropics, with the inclusion of studies from South Africa and the dry sub-humid tropics of the Kalahari as well as the humid tropics. Tropics are drawn from a wide geographical area including South and Central America, South-east Asia, India and Africa, and the chapters have been contributed by authors from all of these areas as well as Europe, North America, Australasia and Japan. It thus provides a snapshot of a range of factors affecting soils across the globe. From this, we shed light on emerging topics that can be can be grouped into the following themes.

1.4.1 The effects of land-use change on soil microbial populations

Advances in molecular and biochemical techniques now mean that the enumeration of soil microbial biomass, diversity and functionality is much more straightforward than it was even a few years ago. This has allowed the extent to which clearing natural vegetation for crops alters the soil microbial populations and the functions they perform to be elucidated from many sites. Studies from the Andaman Islands of India (Dinesh *et al.*, Chapter 2), Malaysia (D'Angelo *et al.*, Chapter 3; Brearley, Chapter 4) and the Amazon (Mendes *et al.*, Chapter 5) all report consistent findings on the impact of the clearance of natural vegetation cover for agriculture. Between them, the studies show a reduction in SOM, microbial biomass or activity and changes to microbial community composition. These changes then have clear knock-on effects on other key soil processes and functions such as nitrogen cycling, moisture retention, erodibility and acidification (Fujii *et al.*, Chapter 6). The message is clear that in nearly all circumstances, clearance of natural vegetation for agriculture will lead to detrimental changes to soil properties.

1.4.2 Urban soils, agriculture and soil contamination

Two chapters report on the neglected topic of urban soils and agriculture in Kenya, South Africa, Zimbabwe (Chipungu *et al.*, Chapter 7) and Malawi (Mkwambisi *et al.*, Chapter 8). Together, both studies show that, irrespective of the country or city, there are similar challenges and opportunities to improving access to land for crops in urban areas. In nearly all cases, soils are on marginal land and farming is undertaken on an informal basis without the permission of local authorities. Although the utilization of waste from a variety of sources can help improve soil fertility, urban farmers regularly run the risk of soil contamination. The theme of contamination is also explored by Grimaldi *et al.* (Chapter 9), who demonstrate the clear impact gold mining in French Guiana can have on mercury stocks and fluxes in soils.

1.4.3 Land-use effects on soil carbon and soil organic matter

The preservation of SOM is integral to many soil conservation strategies as well as to global efforts to increase terrestrial C sequestration. Studies from a broad range of environments demonstrate how land-use change affects soil C stocks and the rate at which organic C is respired. Corre *et al.* (Chapter 10) show how soil C is unevenly distributed around catchments in Ecuador and how land-use changes can lead to large losses of C through erosional processes. Following on, Thomas *et al.* (Chapter 11) reveal how sensitive the C stock in the Kalahari is to grazing intensity, while Ng Cheong and Umrit (Chapter 12) show how clearance of natural vegetation in Mauritius leads to a loss of the original soil organic C and replacement with that derived from sugar cane cropping. In contrast, Powers *et al.* (Chapter 13) show how soil C stocks can be restored during the process of secondary succession in dry tropical forests of Costa Rica, but Schwendenman and Kaiser (Chapter 14) demonstrate that afforestation with teak may not lead to an increase in soil C stocks in Panama, at least over the timeframe of their study.

In the concluding chapter, we briefly summarize key points from the book, bring together a number of the chapter authors to consider areas that are in need of further research and then conclude with suggestions for a sustainable way forward. This is clearly important given the essential role played by soils in maintaining human civilizations (Hillel, 1992; McNeill and Winiwarter, 2004; Montgomery, 2007) and understanding, balancing and mitigating potential conflicts between the multiple demands placed on soils by humans is paramount to the sustainable utilization of this key resource now and into the future.

References

Banwart, S. (2011) Save our soils. *Nature* 474, 151–152.

Batjes, N.H. (1996) Total carbon and nitrogen in the soils of the world. *European Journal of Soil Science* 47, 151–163.

Bond-Lamberty, B. and Thomson, A. (2010) Temperature-associated increases in the global soil respiration record. *Nature* 464, 579–582.

Centers for Disease Control and Prevention (2013) Increase in reported Coccidioidomycosis – United States, 1998–2011. *Morbidity and Mortality Weekly Report (MMWR)* 62, 217–221.

Daniel, R. (2004) The soil metagenome – a rich resource for the diversity of novel natural products. *Current Opinion in Biotechnology* 15, 199–204.

Don, A., Schumacher, J. and Freibauer, A. (2011) Impact of tropical land-use change on soil organic carbon stocks – a meta-analysis. *Global Change Biology* 17, 1658–1670.

Eldridge, D.J., Bowker, M.A., Maestre, F.T., Roger, E., Reynolds, J.F. and Whitford, W.G. (2011) Impacts of shrub encroachment on ecosystem structure and functioning: Towards a global synthesis. *Ecology Letters* 14, 709–722.

Ellis, E.C. (2011) Anthropogenic transformation of the terrestrial biosphere. *Philosophical Transactions of the Royal Society Series A – Mathematical, Physical and Engineering Sciences* 369, 1010–1035.

Ellis, E.C. and Ramankutty, N. (2008) Putting people in the map: Anthropogenic biomes of the world. *Frontiers in Ecology and the Environment* 6, 439–447.

Eswaran, H., Van Den Berg, E. and Reich, P. (1993) Organic carbon in soils of the world. *Soil Science Society of America Journal* 57, 192–194.

Falkenmark, M. and Rockström, J. (1996) The new blue and green water paradigm: Breaking new ground for water resource planning and management. *Journal of Water Resources Planning and Management* 132, 129–132.

Falkowski, P.G., Fenchel, T. and Delong, E.F. (2008) The microbial engines that drive Earth's biogeochemical cycles. *Science* 320, 1034–1039.

Feller, C. and Beare, M.H. (1997) Physical control of soil organic matter dynamics in the tropics. *Geoderma* 79, 69–116.

Gibbs, H.K., Ruesch, A.S., Achard, F., Clayton, M.K., Holmgren, P., Ramankutty, N. and Foley, J.A. (2010) Tropical forests were the primary source of new agricultural land in the 1980s and 1990s. *Proceedings of the National Academy of Sciences of the USA* 107, 16732–16737.

Glaser, B., Haumaier, L., Guggenberger, G. and Zech, W. (2001) The 'Terra Preta' phenomenon: A model for sustainable agriculture in the humid tropics. *Naturwissenschaften* 88, 37–41.

Godfray, H.C.J., Beddington, J.R., Crute, I.R., Haddad, L., Lawrence, D., Muir, J.F., Pretty, J., Robinson, S., Thomas, S.M. and Toulmin, C. (2010) Food security: The challenge of feeding 9 billion people. *Science* 327, 812–818.

Grace, J., San José, J.J., Meir, P., Miranda, H.S. and Montes, R.A. (2006) Productivity and carbon fluxes of tropical savannas. *Journal of Biogeography* 33, 387–400.

Griffin, D.W., Kellog, C.A. and Shinn. E.A. (2001) Dust in the wind: Long range transport of dust in the atmosphere and its implications for global public and ecosystem health. *Global Change and Human Health* 2, 20–33.

Hamilton, A.J., Burry, K., Mok, H.-F., Barker, S.F., Grove, J.R. and Williamson, V.G. (2014) Give peas a chance? Urban agriculture in developing countries. A review. *Agronomy for Sustainable Development* 34, 45–73.

Hartemink, A.E. and McBratney, A.B. (2008) A soil science renaissance. *Geoderma* 148, 123–129.

Hayakawa, C., Funakawa, S., Fujii, K., Kadono, A. and Kosaki, T. (2014) Effects of climatic and soil properties on cellulose decomposition rates in temperate and tropical forests. *Biology and Fertility of* Soils 50, 633–643.

Hillel, D. (1992) *Out of the Earth*. Aurum Press, London.

Joergensen, R.G. (2010) Organic matter and micro-organisms in tropical soils. In: Dion, P. (ed.) *Soil Biology and Agriculture in the Tropics, Soil Biology* 21, Springer-Verlag, Berlin, pp. 17–44.

Koch, A., McBratney, A.B., Adams, M., Field, D., Hill, R., Lal, R. *et al.* (2013) Soil security: Solving the global soil crisis. *Global Policy Journal* 4, 434–441.

Lal, R. (1990) *Soil Erosion in the Tropics: Principles and Management*. McGraw-Hill, New York.

Lal, R. (2004) Soil carbon sequestration impacts on global climate change and food security. *Science* 304, 1623–1627.

Lewis, K. (2012) Recover the lost art of drug discovery. *Nature* 485, 439–440.

McNeill, J.R. and Winiwarter, V. (2004) Breaking the sod: Humankind, history, and soil. *Science* 304, 1627–1629.

Marcotullio, P.J., Braimoh, A.K. and Onishi, T. (2008) The impact of urbanization on soils. In: Braimoh, A.K. and Vlek, P.L.G. (eds) *Land Use and Soil Resources*. Springer-Verlag, Heidelberg, Germany, pp. 201–250.

Marris, E. (2006) Black is the new green. *Nature* 442, 624–626.

Minasny, B., Sulaeman, Y. and McBratney, A.B. (2011) Is soil carbon disappearing? The dynamics of soil carbon in Java. *Global Change Biology* 17, 1917–1924.

Montgomery, D.R. (2007) *Dirt: The Erosion of Civilizations*. University of California Press, Berkeley, California.

Nikolaidis, N.P. (2011) Human impacts on soils: Tipping points and knowledge gaps. *Applied Geochemistry* 26, S230–S233.

Orsini, F., Kahane, R., Nono-Womdim, R. and Gianquinto, G. (2013) Urban agriculture in the developing world: A review. *Agronomy for Sustainable Development* 33, 695–720.

Post, W.M., Emanuel, W.R., Zinke, P.J. and Stangenberger, A.G. (1982) Soil carbon pools and world life zones. *Nature* 298, 156–159.

Powlson, D.S., Whitmore, A.P. and Goulding, K.W.T. (2011) Soil carbon sequestration to mitigate climate change: A critical re-examination to identify the true and the false. *European Journal of Soil Science* 62, 42–55.

Rasche, F. and Cadisch, G. (2013) The molecular microbial perspective of organic matter turnover and nutrient cycling in tropical agroecosystems – What do we know? *Biology and Fertility of Soils* 49, 251–262.

Richter, D. deB. (2007) Humanity's transformation of Earth's soil: Pedology's new frontier. *Soil Science* 172, 957–967.

Richter, D. deB., Bacon, A.R., Mobley, M.L., Richardson, C.J., Andrews, S.S., West, L. *et al.* (2011) Human-soil relations are changing rapidly: Proposals from SSSA's cross-divisional soil change working group. *Soil Science Society of America Journal* 75, 2079–2084.

Seto, K.C., Güneralp, B. and Hutyra, L.R. (2012) Global forecasts of urban expansion to 2030 and direct impacts on biodiversity and carbon pools. *Proceedings of the National Academy of Sciences of the USA* 109, 16083–16088.

Sidle, R.C., Ziegler, A.D., Negishi, J.N., Nik, A.R., Siew, R. and Turkelboom, F. (2006) Erosion processes in steep terrain – truths, myths, and uncertainties related to forest management in Southeast Asia. *Forest Ecology and Management* 224, 199–225.

Singh, B.K. and Macdonald, C.A. (2010) Drug discovery from uncultivable microorganisms. *Drug Discovery Today* 15, 792–799.

Singh, B.K., Bardgett, R.D., Smith, P. and Reay, D. (2010) Microorganisms and climate change: Feedbacks and mitigation options. *Nature Reviews Microbiology* 8, 779–790.

Sohi, S.P., Krull, E., Lopez-Capel, E. and Bol, R. (2010) A review of biochar and its use and function in soil. *Advances in Agronomy* 105, 47–82.

Stocking, M.A. (2003) Tropical soils and food security: The next 50 years. *Science* 302, 1356–1359.

Telmer, K. and Veiga, M.M. (2009) World emissions of mercury from artisanal and small scale gold mining. In: Pirrone, N. and Mason, R. (eds) *Mercury Fate and Transport in the Global Atmosphere Emissions, Measurements and Models*. Springer-Verlag, Heidelberg, Germany, pp. 131–172.

Thomas, A.D. (2012) Impact of grazing intensity on seasonal variations of soil organic carbon and soil CO_2 efflux in two semi-arid grasslands in southern Botswana. *Philosophical Transactions of the Royal Society Series B – Biological Sciences B* 367, 3076–3086.

Tilman, D., Socolow, R., Foley, J.A., Hill, J., Larson, E., Lynd, L. *et al.* (2009) Beneficial biofuels – the food, energy, and environment trilemma. *Science* 325, 270–271.

United Nations Convention to Combat Desertification (UNCCD) (2012) *Zero Net Land Degradation*, UNCCD Secretariat Policy Brief. Bonn, Germany.

Van der Perk, M. (2013) *Soil and Water Contamination* (2nd edition). CRC Press, Boca Raton, Florida.

Yaalon, D.H. (2007) Human-induced ecosystem and landscape processes always involve soil change. *BioScience* 57, 918–919.

Zech, W., Senesi, N., Guggenberger, G., Kaiser, K., Lehmann, J., Miano, et al. (1997) Factors controlling humification and mineralization of soil organic matter in the tropics. *Geoderma* 79, 117–161.

2 Effects of Land-use Changes on Biochemical and Microbial Parameters in Soils of the Andaman Islands, India

Raghavan Dinesh,[1]* Arkalgud Ganeshamurthy[2] and Subrata Ghoshal Chaudhuri[3]
[1]ICAR-Indian Institute of Spices Research, Calicut, Kerala, India; [2]ICAR-Indian Institute of Horticultural Research, Bengaluru, Karnataka, India; [3]ICAR-National Bureau of Soil Survey and Land Use Planning (Regional Centre), Salt Lake City, Bidhan Nagar, Kolkata, West Bengal, India

2.1 Introduction

Large areas of forest in the tropics are presently undergoing deforestation due to anthropogenic influences such as forest clearance, human settlement and conversion for agriculture (Gibbs *et al.*, 2010; Villoria *et al.*, 2014). It is well known that forest clearance for agriculture, an increasingly prevalent situation in the tropics, removes natural vegetation, reduces biodiversity, and simplifies the landscape and ecosystem structure (Li *et al.*, 2005; Rosa *et al.*, 2014). The effects of these changes include reductions in productivity because of increasing losses of nutrients and soil; downstream impacts, such as reductions in water quality through increased sedimentation and changes in water yield; and widespread reductions in biodiversity and the supply of various ecological services (Lamb and Gilmour, 2003; Ahrends *et al.*, 2010; Kessler *et al.*, 2012; Rosa *et al.*, 2014).

In the case of soils, the long-term negative impacts of forest clearance and cultivation often include the degradation and subsequent loss of organic carbon (C) and nitrogen (N) (An *et al.*, 2008; Kessler *et al.*, 2012; Brearley, Chapter 4, this volume; Ng Cheong and Umrit, Chapter 12, this volume). However, assessing the long-term impact of deforestation on soil environments is difficult because parameters such as tree growth or soil organic matter (SOM) are slow in revealing changes (Dick, 1994). In the short term, deforestation may have detrimental effects on soil biological and biochemical properties, since forest clearance modifies the microclimatic conditions at the soil surface and the rates and quality of SOM input to the soil (Lemenih *et al.*, 2005).

There are many studies that focus on soil microbial and biochemical properties as these reflect the size and activity of microbial processes (Cheng *et al.*, 2013; Dinesh and Chaudhuri, 2013). This is because biologically mediated processes in soils play a key role in the mineralization of organic C and in nutrient cycling, and tend to react most rapidly to changes in the external environment, such as land use and management, than do most physical and chemical parameters (Sparling, 1992; Trasar-Cepeda *et al.*, 2008). In addition, these properties are more sensitive to environmental stress, play a major role in SOM decomposition and provide rapid information on soil quality (García *et al.*, 2000). The biochemical

* Corresponding author: rdinesh2005@gmail.com, rdinesh@spices.res.in

parameters include variables directly related to microbial activity and the activities of extracellular hydrolytic enzymes involved in the C, N, phosphorus and sulfur cycles in soil. These soil biochemical and microbiological parameters are considered potential indicators of management impacts on soil quality (Gil-Sotres *et al.*, 2005) and could illustrate the effects of anthropogenic activities and disturbance on the soil (Kennedy and Papendick, 1995).

Microbial indices, and their relationships, have been extensively studied in various forest systems of the temperate (Leirós *et al.*, 2000; Díaz-Raviña *et al.*, 2005; Gamboa and Galicia, 2011) and tropical (Salamanca *et al.*, 2002; Dinesh *et al.*, 2003, 2004; Tripathi and Singh, 2013, Mendes *et al.*, Chapter 5, this volume) regions. Soil characteristics often differ between temperate and tropical soils: higher temperatures in tropical regions lead to faster turnover rates of microbial biomass and SOM in comparison to temperate climatic conditions (Joergensen, 2010). Consequently, there will be large differences in soil biological properties between tropical and temperate soil types. Information on microbial indices of forest soils subjected to long-term disturbance, especially those in the tropics, will provide valuable insights into the extent of soil deterioration and help design management strategies that may contribute to the recovery of soil quality. In this chapter, we highlight the findings of our investigations on microbial and biochemical parameters in soils under undisturbed humid tropical forests in the Andaman Islands of India and adjacent plantations of coconut, arecanut, rubber, padauk and teak established by clear-felling portions of these forests. The major objective was to study the long-term changes in biochemical/microbial indicators of soil quality due to clear-felling of tropical forests for establishment of plantations. Besides soil microbial biomass (SMB)-C, and -N and soil respiration, we also determined ergosterol (a fungal biomarker), adenylates (adenosine tri- (ATP), mono- (AMP), di- (ADP) phosphate) and the ratios of these parameters in these soils. We hypothesized that such land-use change in the tropical forests would markedly influence soil organic substrate and nutrient levels, which would subsequently alter the soil microbial biomass, respiration rates and microbial community over the long term.

2.2 Materials and Methods

2.2.1 Study site

The study sites were located in South Andaman, India (10° 30′–13° 42′ N and 92° 14′–94° 16′ E). The climate is wet tropical with an annual rainfall of 3100 mm mainly spread over 7 months from May to November. The temperature ranges from a minimum of 18.6°C to a maximum of 33.0°C. The wet tropical forests of South Andaman are classified into evergreen (EG), semi-evergreen (SE) and moist-deciduous (MD). The characteristics of the study sites are given in Table 2.1. Portions of these forests were cleared manually to establish commercial plantations of coconut (*Cocos nucifera*; Arecaceae), arecanut (*Areca catechu*; Arecaceae), rubber (*Hevea brasiliensis*; Euphorbiaceae), padauk (*Pterocarpus dalbergioides*; Leguminosae) and teak (*Tectona grandis*; Lamiaceae).

2.2.2 Soil sampling

Three undisturbed forest types (EG, SE, MD) and adjacent plantations of coconut, arecanut, rubber, padauk and teak were selected and soil cores (0–30 cm) taken from 20 randomly selected points distributed uniformly over an area of 10 ha from each, using an Edelmann corer (7 cm diameter, 30 cm length) immediately after the end of the monsoon (in December 2000–January 2001). The soil samples were cleared of leaf litter and organic material and transferred for storage in sealed plastic bags. The soil samples were then sieved (<2 mm), analysed for their moisture content on the day of collection and stored at 4°C. Subsamples for the determination of organic C and total N were sieved to pass through a 0.5-mm mesh. The soils, classified as Fluventic Sulfaquents (Entisols), were derived from alluvial and marine deposits. At 0–30 cm the soil is a light brown (7.5YR 5/6), sandy clay loam with medium permeability, firm and sticky.

2.2.3 Soil biochemical/microbial parameters

The soil biochemical parameters measured were soil microbial biomass (SMB), soil microbial

Table 2.1. Site characteristics of the study areas to examine soil properties in undisturbed forests and plantations in the Andaman Islands (India).

Forest type	Major tree species: I Storey	II Storey	III Storey	Ground storey	Adjacent plantation
Evergreen	*Artocarpus chaplasha* (Moraceae) *Calophyllum soulattari* (Clusiaceae) *Dipterocarpus alatus* (Dipterocarpaceae) *Dipterocarpus gracilis* (Dipterocarpaceae)	*Amoora wallichii* (Meliaceae) *Artocarpus gomezianus* (Moraceae) *Pterocarpus dalbergioides* (Leguminosae) *Pterocymbium tinctorium* (Sterculiaceae)	*Aglaia andamanica* (Meliaceae) *Artocarpus lakoocha* (Moraceae) *Baccaurea sapida* (Euphorbiaceae) *Sterculia villosa* (Sterculiaceae)	*Sterculia villosa* (Sterculiaceae) *Artocarpus lakoocha* (Moraceae). *Baccaurea sapida* (Euphorbiaceae) *Aglaia andamanica* (Meliaceae)	Rubber (*Hevea brasiliensis*; Euphorbiaceae) (20–27)[a]
Semi-evergreen	*Artocarpus chaplasha* (Moraceae) *Dipterocarpus pilosus* (Dipterocarpaceae). *Planchonia andamanica* (Lecythidaceae) *Pterygota alata* (Sterculiaceae).	*Amoora wallichii* (Meliceae) *Lannea coromandelica* (Anacardiaceae) *Podocarpus neriifolius* (Podocarpaceae) *Pterocarpus dalbergioides* (Leguminosae)	*Lagerstroemia hypoleuca* (Lythraceae) *Dillenia pentagyna* (Dilleniaceae) *Lannea coromandelica* (Anacardiaceae) *Albizia lebbeck* (Leguminosae)	*Baccaurea sapida* (Euphorbiaceae) *Myristica* sp. (Myristicaceae) *Saprosma* ternata (Rubinaceae)	Arecanut (*Areca catechu*; Arecaceae) (24–25)[a] Teak (*Tectona grandis*; Lamiaceae) (45–53)[a]
Moist deciduous	*Canarium euphyllum* (Burseraceae) *Ganophyllum falcatum* (Sapindaceae) *Salmalia insignis* (Bombacaceae) *Tetrameles nudiflora* (Datiscaceae)	*Diospyros marmorata* (Ebenaceae) *Sageraea elliptica* (Annonaceae) *Amoora wallichii* (Meliaceae) *Terminalia bialata* (Combretaceae)	*Nauclea gageana* (Rubiaceae) *Adenanthera pavonina* (Leguminosae) *Pterocymbium tinctorium* (Sterculiaceae) *Anthocephalus cadamba* (Rubiaceae)	*Podocarpus neriifolius* (Podocarpaceae) *Artocarpus gomeziana* (Moraceae). *Ventilago madraspatana* (Rhamnaceae) *Byttneria andamanensis* (Sterculiaceae)	Coconut (*Cocos nucifera*; Arecaceae) (31–33)[a] Padauk (*Pterocarpus dalbergioides*; Leguminosae) (35–41)[a]

Note: [a]Stand age.

biomass-C (SMB-C) and -N (SMB-N). Soil microbial biomass is considered one of the most promising indicators of soil quality because it responds promptly to environmental changes, often much more rapidly than physical and chemical parameters and even crop productivity (Dinesh *et al.*, 2010, 2012; Kaschuk *et al.*, 2010). Soil microbial biomass can act as a major source or sink for soil C and nutrients and potentially influence the amounts of organic C and N retained within SOM.

The other important parameter measured was soil respiration, which is one of the most frequently used indicators for measuring soil organism activity. When soil is disturbed, a change in soil respiration can be observed due to more rapid growth and greater mineralization rates of the microorganisms (Joergensen, 2010). The changes in levels of adenylates (ATP, ADP, AMP) due to deforestation and cultivation were also studied. The ATP content is considered a useful indicator of microbial biomass in soil (Contin *et al.*, 2001) as ATP and the sum of adenylates are closely related to SMB-C in a wide range of sites (Dyckmans *et al.*, 2003). In numerous studies, the sum of adenylates has been found to give good estimates of the SMB (Schinner *et al.*, 1995; Dyckmans and Raubuch, 1997), and the adenylate energy charge (AEC) is an important index for the energetic state of the soil microbial community. Ergosterol, a reliable and relatively inexpensive biochemical indicator of fungal biomass in soil ecosystems (Engelking *et al.*, 2008; Joergensen and Wichern, 2008), was also determined.

SMB-C and SMB-N were measured on fresh soil samples using the fumigation-extraction method (Vance *et al.*, 1987). The C and N in the extracts were determined by the Walkley–Black wet combustion method (Nelson and Sommers, 1982) and micro-Kjeldahl digestion procedure (Bremner and Mulvaney, 1982), respectively, and we subsequently employed k_{EC} of 0.45 (Joergensen, 1996) and k_{EN} of 0.54 (Joergensen and Mueller, 1996). For measuring basal soil respiration, 100 g (oven-dry basis) of moist sample was adjusted to 55% of its water-holding capacity, placed in a 1-l stoppered glass jar and pre-incubated for 3 days at 20°C in the dark. The CO_2 production was then measured for another 3 days. The CO_2 produced was trapped in 20 ml 0.05 M NaOH, which was then measured by titration of the excess NaOH with 0.05 M HCl.

The metabolic quotient (qCO_2) was calculated as follows:

$$[(\text{mg } CO_2\text{–C evolved in 3 days g}^{-1}\text{ soil})/(\text{mg biomass C g}^{-1}\text{ soil})/3\text{ days} \times 1000] = [\text{mg } CO_2\text{–C mg}^{-1}\text{ biomass C per day}] \quad (2.1)$$

The results are expressed in mg CO_2–C g^{-1} biomass C per day (Salamanca *et al.*, 2002). Adenylates were determined using the procedure of Bai *et al.* (1989) as described by Dyckmans and Raubuch (1997). We used the Dimethylsulfoxide (DMSO), Na_3PO_4 (10 mM) buffer + EDTA (20 mM) at pH12 and Celsis nucleotide-releasing buffer for microbial ATP as extractants. After derivatization with chloracetaldehyde, the adenine nucleotides were determined by high performance liquid chromatography (HPLC) as described by Joergensen and Castillo (2001). Ergosterol was measured in moist samples of soil extracted with ethanol by oscillating shaking (Djajakirana *et al.*, 1996), followed by reversed-phase HPLC with 100% methanol as the mobile phase and detection at 282 nm.

2.2.4 Statistical analyses

Analyses for various soil properties were performed on all samples and mean values determined; the values are expressed on an oven dry soil basis (24 h at 105°C). The significance of treatment effects was determined by one-way analysis of variance (ANOVA). Where the F-values were significant, *post hoc* comparisons were made using the Tukey–Kramer honestly significant difference (HSD) test.

2.3 Results

2.3.1 Soil microbial biomass

There was a 55% difference in soil organic carbon (SOC) between forests (mean 17.7 ± 3.4 mg g^{-1}) and plantations (mean 8.0 ± 1.3 mg g^{-1}). Similarly, mean soil total N was 1.69 ± 0.28 mg g^{-1} in the forests, while it was 0.89 ± 0.14 mg g^{-1} in the plantations, a difference of 47%. The mean SMB-C in the forest sites was

511 ± 22 µg g^{-1} (Table 2.2), which was lower by 71% in the plantation sites (174 ± 9 µg g^{-1}). SMB-N also showed a similar pattern; the mean SMB-N in the forest sites was 43.4 ± 4.5 µg g^{-1}, while under the plantations it was 14.9 ± 3.4 µg g^{-1}, a difference of 66%. The mean SMB-C/SOC ratio (*q*MIC) was 2.9 ± 0.7% in the forested sites and 2.1 ± 0.6% in the plantations (Table 2.3). In contrast to *q*MIC, the mean SMB-C/SMB-N ratio was 11.6 ± 1.4 in the forest soils – almost identical to the ratio (11.7 ± 1.6) observed under the plantations (Table 2.3). SMB-C was positively correlated with SOC ($r^2 = 0.64$; $P < 0.0001$; Fig. 2.1a).

2.3.2 Soil respiration

Mean soil respiration under the forests was 17.6 ± 2.0 µg CO_2 g^{-1} per day, which declined about six-fold to 2.9 ± 0.5 µg CO_2 g^{-1} per day in the plantations, the lower values corresponding to sites with lower SMB-C (Fig. 2.1a; $r^2 = 0.61$; $P < 0.0001$). In contrast, the metabolic quotient (qCO_2; i.e. soil respiration per unit of microbial biomass) under the plantations showed an increase (mean 77.1 ± 4.7 mg CO_2 g^{-1} SMB-C per day) compared to the forest soils (mean 41.5 ± 4.7 mg CO_2 g^{-1} SMB-C per day; Table 2.3). A significant negative relationship ($r^2 = -0.64$; $P < 0.0001$) existed between qCO_2 and the SMB-C/SOC ratio (Fig. 2.1b).

2.3.3 Adenylates and adenylate energy charge

The data on adenylates (Table 2.4) indicated greater concentrations of ATP (mean 2.79 ± 0.51 nmol g^{-1}), AMP (mean 0.28 ± 0.05 nmol g^{-1}) and ADP (mean 0.50 ± 0.11 nmol g^{-1}) under the forest sites compared to the plantations (mean 0.75 ± 0.16, 0.15 ± 0.04, 0.24 ± 0.06 nmol g^{-1}, respectively), leading to the sum of the adenylates under the forests (3.57 ± 0.57 nmol g^{-1}) being greater than under the plantations (1.15 ± 0.36 nmol g^{-1}; Table 2.4). Greater concentrations of adenylates resulted in a greater AEC in the forest soils (mean 0.85 ± 0.13) relative to plantation soils (mean 0.76 ± 0.13). Greater ATP concentrations also led to higher ATP/SMB-C ratios in the forest soils (Table 2.3), on average, the ATP/SMB-C ratio in the forest sites was 5.6 ± 1.3 µmol g^{-1}; significantly greater than that observed under the plantations (4.3 ± 1.3 µmol g^{-1}). We observed a negative, but non-significant, relationship between the ATP/SMB-C ratio and SMB-C/SMB-N ratio in the forest soils ($r^2 = -0.12$; n.s.) but a significant negative relationship ($r^2 = -0.54$; $P < 0.0001$) was observed in the plantation sites.

Table 2.2 Soil organic carbon, total N, soil microbial biomass-C, soil microbial biomass-N and soil respiration rates (mean ± SE) in soils under forests and adjacent plantations in the Andaman Islands (India).

	Soil organic C (mg g^{-1})	Total N (mg g^{-1})	SOC/total N	Soil microbial biomass (µg g^{-1})		Soil respiration (µg CO_2–C g^{-1} per day)
				C	N	
Forest sites						
Evergreen	14.4 ± 2.3^{bc}	1.41 ± 0.30^{bc}	10.2 ± 1.7^{a}	422 ± 26^{bc}	34.8 ± 4.2^{b}	11.9 ± 2.1^{c}
Semi-evergreen	16.4 ± 1.9^{b}	1.54 ± 0.29^{b}	10.6 ± 2.2^{a}	494 ± 18^{b}	45.9 ± 4.3^{a}	14.4 ± 1.8^{b}
Moist deciduous	22.4 ± 2.6^{a}	2.13 ± 0.26^{a}	10.5 ± 2.12^{a}	617 ± 23^{a}	49.4 ± 5.1^{a}	26.6 ± 2.2^{a}
Mean	17.7 ± 3.4	1.69 ± 0.28	10.5 ± 1.9	511 ± 22	43.4 ± 4.5	17.6 ± 2.0
Plantation sites						
Arecanut	7.6 ± 1.3^{de}	0.86 ± 0.14^{de}	8.8 ± 1.2^{b}	148 ± 11^{e}	16.3 ± 3.6^{cd}	2.3 ± 0.4^{e}
Coconut	6.9 ± 1.2^{de}	0.83 ± 0.16^{de}	8.3 ± 1.4^{b}	145 ± 8^{e}	15.2 ± 3.3^{cd}	2.2 ± 0.6^{e}
Rubber	10.8 ± 1.3^{cd}	1.05 ± 0.12^{cd}	10.3 ± 1.6^{a}	270 ± 14^{d}	21.2 ± 3.1^{c}	5.2 ± 0.7^{d}
Teak	7.6 ± 1.4^{de}	0.86 ± 0.15^{de}	8.8 ± 1.7^{b}	154 ± 8^{e}	11.0 ± 3.0^{de}	2.3 ± 0.4^{e}
Padauk	6.9 ± 1.5^{de}	0.83 ± 0.14^{de}	8.3 ± 1.4^{b}	151 ± 6^{e}	11.0 ± 4.0^{de}	2.4 ± 0.3^{e}
Mean	8.0 ± 1.3	0.89 ± 0.14	8.9 ± 1.5	174 ± 9	14.9 ± 3.4	2.9 ± 0.5

Note: Values in a row followed by the same letter are not significantly different at $P < 0.05$.

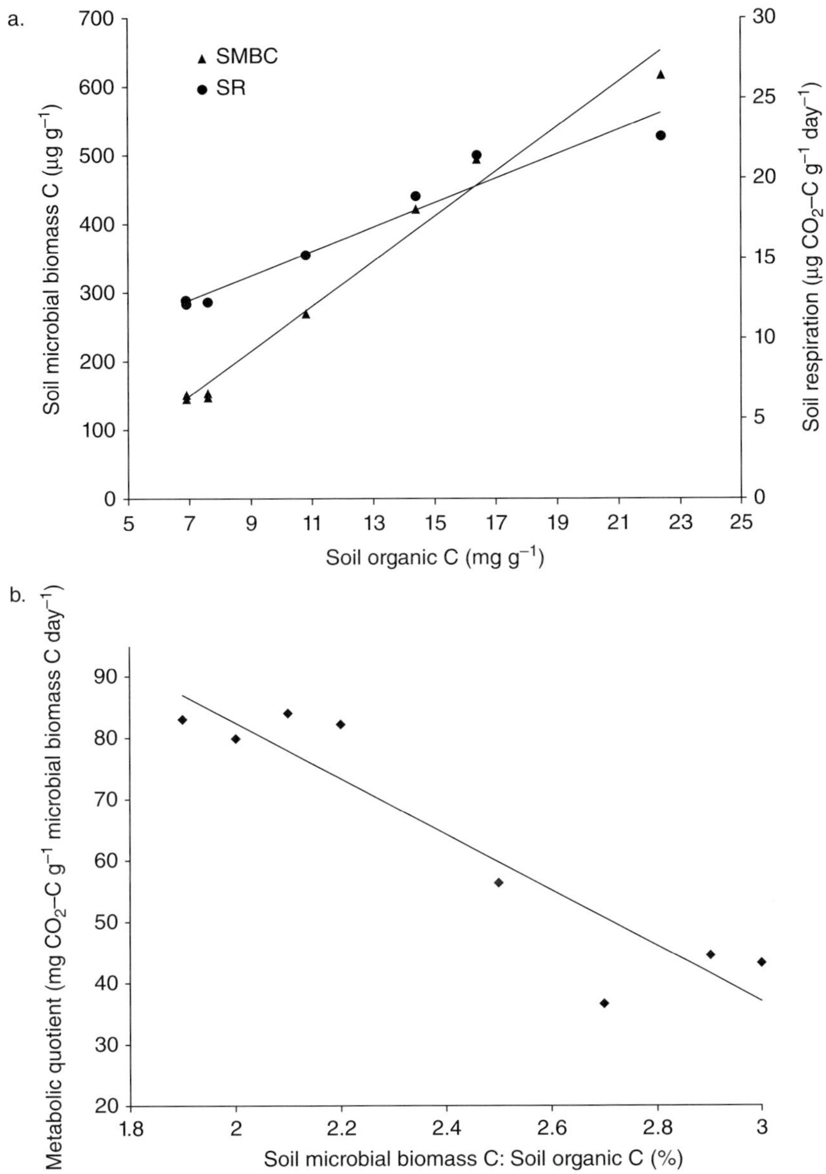

Fig. 2.1. (a) Relationship between soil organic C and soil microbial biomass C (SMBC) and soil respiration (SR); and (b) relationship between soil microbial biomass C/soil organic C ratio and metabolic quotient (qCO_2) in soils under forests and adjacent plantations in the Andaman Islands (India).

2.3.4 Ergosterol

In the forest sites, the mean ergosterol concentration was 3.30 ± 0.36 µg g^{-1} (Table 2.4), while in the plantation soils it was notably lower (0.54 ± 0.11 µg g^{-1}). Likewise, the mean ergosterol/SMB-C ratio in the forest soils (0.64 ± 0.12) was approximately double the ratio in the plantation soils (0.31 ± 0.11). By recalculating ergosterol into fungal biomass by multiplication by 90 (proposed by Djajakirana *et al.*, 1996), we found that fungi represented 58% of

Table 2.3. Ratios (mean ± SE) of soil microbial indices in soils under forests and adjacent plantations in the Andaman Islands (India).

	SMB-C/SOC (qMIC) (%)	SMB-C/SMB-N	qCO$_2$ (mg CO$_2$-C g^{-1} SMB-C per day)	ATP/SMB-C (µmol g^{-1})	Ergosterol/ SMB-C (%)
Forest sites					
Evergreen	2.9 ± 0.6^{ab}	12.1 ± 1.6^{bc}	44.6 ± 4.6^{c}	6.4 ± 1.3^{a}	0.62 ± 0.12^{ab}
Semi-evergreen	3.0 ± 0.6^{a}	10.2 ± 1.5^{cd}	43.3 ± 5.1^{c}	5.8 ± 1.3^{ab}	0.65 ± 0.12^{ab}
Moist deciduous	2.7 ± 0.8^{ab}	12.5 ± 1.2^{ab}	36.6 ± 4.4^{c}	4.6 ± 1.4^{c}	0.66 ± 0.11^{a}
Mean	2.9 ± 0.7	11.6 ± 1.4	41.5 ± 4.7	5.6 ± 1.3	0.64 ± 0.12
Plantation sites					
Coconut	2.1 ± 0.5^{de}	9.50 ± 1.4^{de}	84.1 ± 5.6^{a}	4.4 ± 1.1^{c}	0.30 ± 0.10^{c}
Arecanut	1.9 ± 0.5^{def}	9.10 ± 1.4^{de}	83.1 ± 4.9^{a}	4.3 ± 1.4^{c}	0.30 ± 0.12^{c}
Rubber	2.5 ± 0.7^{bc}	12.7 ± 1.6^{ab}	56.3 ± 3.9^{b}	4.5 ± 1.5^{c}	0.30 ± 0.09^{c}
Teak	2.0 ± 0.5^{de}	13.6 ± 1.5^{a}	79.9 ± 4.3^{a}	4.2 ± 1.1^{c}	0.35 ± 0.11^{c}
Padauk	2.2 ± 0.6^{cd}	13.6 ± 1.9^{a}	82.1 ± 4.8^{a}	4.1 ± 1.3^{c}	0.31 ± 0.12^{c}
Mean	2.1 ± 0.6	11.7 ± 1.6	77.1 ± 4.7	4.3 ± 1.3	0.31 ± 0.11

Note: SMB-C – soil microbial biomass C; SOC – soil organic C; SMB-N – soil microbial biomass N; qCO$_2$ – metabolic quotient (soil respiration/SMB-C); ATP – adenosine tri-phosphate.
Values in a row followed by the same letter are not significantly different at $P < 0.05$.

Table 2.4. Concentrations (mean ± SE) of adenylates (adenosine tri- (ATP), mono- (AMP), di- (ADP) phosphate), adenylate energy charge (AEC) and ergosterol in soils under forests and adjacent plantations in the Andaman Islands (India).

	Adenylates (nmol g^{-1})					
	ATP	AMP	ADP	Total	AEC	Ergosterol (µg g^{-1})
Forest sites						
Evergreen	2.71 ± 0.43^{a}	0.28 ± 0.06^{a}	0.42 ± 0.11^{bc}	3.41 ± 0.67^{a}	0.86 ± 0.13^{a}	2.61 ± 0.33^{bc}
Semi-evergreen	2.86 ± 0.54^{a}	0.28 ± 0.04^{a}	0.56 ± 0.13^{a}	3.70 ± 0.59^{a}	0.85 ± 0.11^{a}	3.20 ± 0.29^{b}
Moist deciduous	2.81 ± 0.57^{a}	0.29 ± 0.06^{a}	0.51 ± 0.10^{ab}	3.61 ± 0.44^{a}	0.85 ± 0.14^{a}	4.10 ± 0.46^{a}
Mean	2.79 ± 0.51	0.28 ± 0.05	0.50 ± 0.11	3.57 ± 0.57	0.85 ± 0.13	3.30 ± 0.36
Plantation sites						
Coconut	0.64 ± 0.13^{c}	0.13 ± 0.04^{c}	0.21 ± 0.07^{e}	0.98 ± 0.21^{b}	0.76 ± 0.14^{b}	0.43 ± 0.09^{d}
Arecanut	0.64 ± 0.10^{c}	0.15 ± 0.04^{c}	0.22 ± 0.07^{e}	1.01 ± 0.36^{b}	0.74 ± 0.13^{b}	0.44 ± 0.10^{d}
Rubber	1.21 ± 0.21^{b}	0.22 ± 0.06^{b}	0.34 ± 0.06^{cd}	1.77 ± 0.46^{b}	0.78 ± 0.14^{b}	0.80 ± 0.17^{d}
Teak	0.64 ± 0.18^{c}	0.12 ± 0.03^{c}	0.21 ± 0.06^{e}	0.97 ± 0.34^{b}	0.77 ± 0.14^{b}	0.54 ± 0.09^{d}
Padauk	0.62 ± 0.16^{c}	0.15 ± 0.04^{c}	0.23 ± 0.05^{e}	1.00 ± 0.41^{b}	0.74 ± 0.11^{b}	0.47 ± 0.11^{d}
Mean	0.75 ± 0.16	0.15 ± 0.04	0.24 ± 0.06	1.15 ± 0.36	0.76 ± 0.13	0.54 ± 0.11

Note: Values in a row followed by the same letter are not significantly different at $P < 0.05$.

the total SMB-C in forest sites and 28% of the SMB-C in the plantation sites. We obtained a significant negative relationships between qCO$_2$ and the ergosterol/SMB-C ratio ($r^2 = -0.71$; $P < 0.0001$) in the forest sites but the relationship was not significant in the plantation sites ($r^2 = 0.32$; n.s.).

2.4 Discussion

Our results revealed markedly lower SOC, SMB-C and SMB-N in the plantation sites, showing a decrease due to deforestation and replacement with agriculture that can be attributed, at least partly, to a reduced input of plant residues due to the absence of fresh overstorey litter. Since SMB in forest sites is largely governed by litterfall, greater SMB-C in the forest sites reflects greater accumulation of plant residues and SOC that are substrates for soil microorganisms. A positive relationship existed between SOC and SMB-C indicating that SMB-C follows a trend parallel to that of SOC when forest is converted to pasture or cropland (Verchot, 2010). While the values of SMB-C observed in this study in the forests (422–617 $\mu g\ g^{-1}$) fall well within the range observed in temperate forest soils (Zhong and Makeschin, 2006), the mean SMB-C in the forests (511 $\mu g\ g^{-1}$) was 55% higher than the worldwide average of 330 $\mu g\ g^{-1}$ soil reported by Wardle (1998). This is not surprising considering that the decomposition of plant residues and the turnover of SOM are higher under tropical conditions (Diels *et al.*, 2004). In contrast to the forest sites, SMB-C in the plantations was 47% lower compared to the worldwide average and 37% lower compared to the mean (281 $\mu g\ g^{-1}$) reported by Joergensen (2010) from a range of plantations in the tropics, thereby suggesting substantial losses in SOM due to conversion to plantations. Several studies have indicated that the removal of natural vegetation followed by cultivation can cause major changes to physical, chemical and biological properties of soil and that these changes are often associated with reductions in SOM, deterioration of soil structure, and decreases in SMB and microbial activity (Dinesh *et al.*, 2004; Bossio *et al.*, 2005; Nourbakhsh, 2007; D'Angelo *et al.*, Chapter 3, this volume). Conversely, SMB-C in soils previously under arable use was found to increase during forest succession, apparently due to increasing inputs and accumulation of SOC (Susyan *et al.*, 2011). In contrast to SMB-C, SMB-N values in the forests (range 34.8–49.4 $\mu g\ g^{-1}$) were an order of magnitude lower than those (213–591 $\mu g\ g^{-1}$) observed in temperate humid forest soils (Gómez-Luna *et al.* 2009).

The mean SMB-C/SOC ratio (*q*MIC), which is an important index for SOM availability to the soil microbial community (Anderson and Domsch, 1989), also showed a decrease in the plantation sites. *q*MIC has been used as an indicator of the changes in SOM status due to alterations in soil conditions (Anderson and Domsch, 2010; Heinze *et al.*, 2010) – values below 2% being a signal of SOM depletion (Anderson, 2003). Therefore, greater *q*MIC values in the forest sites suggested greater availability of SOM than in the plantations. Joergensen (2010) reported that the mean *q*MIC across forests, plantations, pasture and arable soils of the tropics was 1.7% (range 0.1–8.2%) and, similar to our study, it was lower in the plantation soils than the forest soils. However, the *q*MIC has been reported to be lower in the tropics than the humid temperate regions, where values ranging from 2.3 to 2.6% have been reported from arable and grassland soils (Anderson and Domsch, 1989; Joergensen 1995). In addition, extremely low values (mean 0.6%) have been reported in Andisols (Zunino *et al.*, 1982) and, similar to our findings in the plantations, *q*MIC values ranging from 2.2 to 2.5% have been reported in Vertisols and Inceptisols (Joergensen, 2010).

The SMB-C/SMB-N ratio did not show significant changes due to altered land use. This suggests that SMB-C/SMB-N ratio is not a significant indicator of shifts in microbial community structure as observed by Joergensen and Scheu (1999) in temperate soils. However, the SMB-C/SMB-N ratios (mean 11.6 ± 1.4 in the forests and 11.7 ± 1.6 in the plantations) consistently exceeded the corresponding SOC/total N ratios (mean 10.5 ± 1.9 in the forests and 8.9 ± 1.5 in the plantations), possibly due to low N availability and relatively higher SOM availability to soil microbes (Salamanca *et al.*, 2002).

Similarly to SMB-C and SMB-N, soil respiration also declined significantly in the plantations compared to the forests; greater soil respiration rates under the forest sites can be attributed to greater amounts of SOC, which has been found to account for 75% to 81% of the variation in CO_2 production from soils (Wang *et al.*, 2003). Lower soil respiration under the plantations is presumably due to depletion of readily decomposable substrates for microorganisms (Wang *et al.*, 2003) and could also result from the suppression of the decomposition of native SOC (Ding *et al.*, 2010) due to decreases in SMB and activity (Lee and Jose, 2003). Lower soil respiration due to altered land use is also due to the lower substrate quality (lower *q*MIC), resulting in a reduced C use efficiency (higher metabolic quotient, qCO_2). In contrast, under forests, increased C availability via a broader range of substrates (litterfall) and good yield efficiency (higher *q*MIC, low qCO_2) increased soil respiration. Similar to our study, a decrease in soil respiration was observed in degraded lands compared to native vegetation in north-east Brazil (Nunes *et al.*, 2012). Nevertheless, the mean rate of soil respiration in the forests in this study (17.6 μg CO_2 g^{-1} per day) is comparable to that found in 62 tropical forest sites (20 μg CO_2 g^{-1} per day), while the mean rate observed in the plantations (2.9 μg CO_2 g^{-1} per day) is markedly lower than the rate (10 μg CO_2 g^{-1} per day) observed in 22 plantation sites of the tropics (Joergensen, 2010).

In contrast to soil respiration, the metabolic quotient (qCO_2) was markedly lower under the forests compared to the plantations. The metabolic quotient reflects the maintenance energy requirement of soil microbes: levels above 2 g CO_2-C h^{-1} kg SMB-C being the critical threshold for the baseline performance of microbial communities (Anderson, 2003). It is considered the most straightforward index used to evaluate ecosystem development, disturbance or system maturity (Bastida *et al.*, 2008). The qCO_2 has also been found to be affected by a shift in the composition of the microbial populations, for instance the fungal-to-bacterial biomass ratio (Heinze *et al.*, 2010) and has found wide applicability in the assessment of forest ecosystems and different land uses. Lower qCO_2 indicated a decrease in the microbial community maintenance energy requirement in the forest soils and that a smaller microbial community respired at a greater rate in the plantation soils. Subsequently, this could have decreased the substrate use efficiency, i.e. more substrate is catabolized to CO_2 and less substrate is incorporated into the SMB (Anderson and Joergensen, 1997), which suggested that the conversion of total C into microbial C is less efficient (Frazão *et al.*, 2010). In ecological terms, however, a high qCO_2 reflects a high maintenance C demand, and if the soil system cannot replenish the C that is lost through respiration, microbial biomass will decline (Anderson and Domsch, 2010). In soils under different land uses, Joergensen (2010) found that specific qCO_2 values were in the order of pasture > arable > forest > plantation, and that a significant negative relationship between qCO_2 and the SMB-C/SOC ratio existed in these tropical soils. The mean qCO_2 value of the forest soils (41.5 mg CO_2 g^{-1} microbial biomass C per day) is similar to the mean (48 mg CO_2 g^{-1} microbial biomass C per day) observed in 62 tropical forest sites and was almost twice as high as the mean qCO_2 value of 21 mg CO_2 g^{-1} microbial biomass C per day in 99 German soils under different land uses (Anderson and Domsch, 1990; Joergensen, 1995).

The concentrations of adenylates (ATP, AMP and ADP) and the sum of adenylates were markedly greater under the forest sites compared to the plantations. Higher SMB in the forest sites most likely led to higher ATP and adenylate concentrations compared to plantation sites. The mean ATP concentration in forest soils (2.8 nmol g^{-1}) is higher than the mean (2.1 nmol g^{-1}) recorded in 35 tropical forest soils (Joergensen, 2010). In contrast, the mean concentration observed in the plantations (0.75 nmol g^{-1}) is lower than the mean (1.0 nmol g^{-1}) observed in 18 tropical plantation soils (Joergensen, 2010). Greater concentrations of adenylates resulted in greater AEC levels in the forest soils relative to plantation soils (mean 0.76). The values are slightly lower (0.840.88) than those reported by Dinesh *et al.* (2006) and Dinesh and Chaudhuri (2013) in soils under mangrove forests of the Andaman Islands, India, and slightly higher than those (0.67–0.74) reported by Salamanca *et al.* (2002) in soils under secondary tropical forest sites of the Philippines.

We observed negative relationships between the ATP/SMB-C ratio and SMB-C/SMB-N ratio in the forest and plantation soils. Similar significant negative relationships between the ATP/SMB-C ratio and SMB-C/SMB-N ratios were reported by Joergensen and Raubuch (2002) and Salamanca *et al.* (2002). Salamanca *et al.* (2006) also reported that the increase in the ATP/SMB-C ratio and the decrease in the SMB-C/SMB-N ratio were related to N concentration in the leaf litter. They deduced that reduced C input to plantation soils may temporarily restrict the access of soil microorganisms to N-containing organic compounds especially in the dry season.

The ergosterol concentration and the ergosterol/SMB-C ratio, which are indicators of fungal biomass (Joergensen and Wichern, 2008), were lower in the plantation soils compared to the forest soils. Fungal biomass, recalculated using the ergosterol data, indicated a shift in the microbial community structure owing to change in land use, with fungi dominating the microbial community over bacteria in the forest soils. Joergensen (2010) found that the ergosterol content increased significantly in the order forest > plantation > arable. He also found that fungi represented only 25% of total microbial biomass C in tropical soils. However, other reports indicate that fungi dominate the microbial biomass in tropical soils by approximately 50 to 90% (Henrot and Robertson, 1994; Sahani and Behera, 2001; Li *et al.*, 2006a,b) where the worldwide average of 75% was reported for all forms of land-use management (Joergensen and Wichern, 2008). Greater contributions of fungi to microbial biomass in forest soils could be explained by the lack of soil disturbance leading to undisturbed development of fungal hyphae (Pennanen *et al.*, 2004; Appuhn *et al.*, 2006). Apparently, the environment for decomposition favours fungi where soil disturbance is low and both lignin inputs and soil C/N ratios are high as encountered in forest ecosystems (Grandy *et al.*, 2009), as the majority of lignin decomposition is known to occur via fungal rather than bacterial pathways (Lopez *et al.*, 2006). We obtained different patterns in the relationships between qCO_2 and the ergosterol/SMB-C ratio in the forest and plantation sites. Salamanca *et al.* (2006) and Scheller and Joergensen (2008) obtained a positive relationship between qCO_2 and the ergosterol/SMB-C ratio that suggested a negative relationship between substrate use efficiency and the presence of fungi. However, in our forest sites, a negative relationship between qCO_2 and the ergosterol/SMB-C ratio supports the view of Sakamoto and Oba (1994) and Jastrow *et al.* (2007) that fungi are more efficient in the use of organic substrates.

2.5 Conclusions

The study provides clear evidence that soil functioning is significantly impacted due to shifts in land use from forested ecosystems to agricultural plantation systems. This was apparent from the marked changes in SOC and soil microbial properties suggesting that SMB-C, soil respiration and SOC are reliable indices of soil quality in tropical regions. Such soils were characterized by low SMB-C levels, low soil respiration rates and lower SMB-C/SOC ratios and high qCO_2 values, thereby indicating the negative impacts of changed land use on soil quality. Since these microbial parameters are indices of biological stability that ultimately control nutrient cycling and transformation in soils, any decrease in the levels of these indices would result in less 'stable' soils, which may subsequently lead to soil degradation over the medium term. The altered land use also influenced the degree of association between various microbial parameters due to variations in both quantity and quality of SOM. However, it may be possible to partially restore these soils through long-term soil management involving zero/minimum tillage, organic/green manure incorporation, cover cropping, mulching, etc. that would not only sequester SOC thereby enhancing soil quality, but would also prevent soil erosion during high rainfall.

References

Ahrends, A., Burgess, N.D., Milledge, S.A., Bulling, M.T., Fisher, B., Smart, J.C. *et al.* (2010) Predictable waves of sequential forest degradation and biodiversity loss spreading from an African city. *Proceedings of the National Academy of Sciences USA* 107, 14556–14561.

An, S., Zheng, F., Zhang, F., Pelt, S.V., Hamer, U. and Makeschin, F. (2008) Soil quality degradation processes along a deforestation chronosequence in the Ziwuling area, China. *CATENA* 75, 248–256.

Anderson, T.-H. (2003) Microbial eco-physiological indicators to assess soil quality. *Agriculture, Ecosystems Environment* 98, 285–293.

Anderson, T.-H. and Domsch, K.H. (1989) Ratios of microbial biomass C to total organic carbon in arable soils. *Soil Biology and Biochemistry* 21, 471–479.

Anderson, T.-H. and Domsch, K.H. (1990) Application of eco-physiological quotients (qCO_2 and qD) on microbial biomasses from soils of different cropping histories. *Soil Biology and Biochemistry* 22, 251–255.

Anderson, T.-H. and Domsch, K.H. (2010) Soil microbial biomass: The eco-physiological approach. *Soil Biology and Biochemistry* 42, 2039–2043.

Anderson, T.-H. and Joergensen, R.G. (1997) Relationship between SIR and FE estimates of microbial biomass C in deciduous forest soils at different pH. *Soil Biology and Biochemistry* 29, 1033–1042.

Appuhn, A., Scheller, E. and Joergensen, R.G. (2006) Relationships between microbial indices in roots and silt loam soils forming a gradient in soil organic matter. *Soil Biology and Biochemistry* 38, 2557–2564.

Bai, Q.Y., Zelles, L., Scheunert, I. and Korte, F. (1989) Determination of adenine nucleotides in soil by ion-paired reverse-phase high-performance liquid chromatography. *Journal of Microbiological Methods* 9, 345–351.

Bastida, F., Zsolnay, A., Hernández, T. and García C. (2008) Past, present and future of soil quality indices: A biological perspective. *Geoderma* 147, 159–171.

Bossio, D.A., Girvan, M.S., Verchot, L., Bullimore, J., Borelli, T., Albrecht, A. *et al.* (2005) Soil microbial community response to land use change in an agricultural landscape of Western Kenya. *Microbial Ecology* 49, 50–62.

Bremner, J.M. and Mulvaney, C.S. (1982) Nitrogen-total. In: Page, A.L., Miller, R.H. and Keeney, D.R. (eds) *Methods of Soil Analysis*, Part 2, *Chemical and Microbiological Properties*, ASA, SSSA, Madison, Wisconsin, pp. 595–624.

Cheng, F., Peng, X., Zhao, P., Yuan, J., Zhong, C., Cheng,Y., et al.*et al.* (2013) Soil microbial biomass, basal respiration and enzyme activity of main forest types in the Qinling mountains. *PLoS ONE* 8, e67353.

Contin M, Todd, A. and Brookes, P.C. (2001) The ATP concentration in the soil microbial biomass. *Soil Biology and Biochemistry* 33, 701–704.

Díaz-Raviña, M., Bueno, J., González-Prieto, S.J. and Carballas T. (2005) Cultivation effects on biochemical properties, C storage and ^{15}N natural abundance in the 0–5 cm layer of an acidic soil from temperate humid zone. *Soil and Tillage Research* 84, 216–221.

Dick, R.P. (1994) Soil enzyme activity as indicators of soil quality. In: Doran, J.W., Coleman, D.C., Bezdicek, D.F. and Stewart, B.A. (eds) *Defining Soil Quality for Sustainable Environment*. SSSA special publication no. 35, pp. 107–124.

Diels, J., Vanlauwe, B., van der Meersch, M.K., Sanginga, N. and Merckx, R. (2004) Long-term soil organic carbon dynamics in a subhumid tropical climate: ^{13}C data in mixed C3/C4 cropping and modeling with ROTHC. *Soil Biology and Biochemistry* 36, 1739–1750.

Dinesh, R. and Chaudhuri, S.G. (2013) Soil biochemical/microbial indices as ecological indicators of land use change in mangrove forests. *Ecological Indicators* 32, 253–258.

Dinesh, R., Ghoshal Chaudhuri, S., Ganeshamurthy, A.N. and Chanchal Dey (2003) Changes in soil microbial indices and their relationships following deforestation and cultivation in wet tropical forests. *Applied Soil Ecology* 24, 17–26.

Dinesh, R., Ghoshal Chaudhuri, S. and Sheeja, T.E. (2004) Soil biochemical and microbial indices in wet tropical forests: Effects of deforestation and cultivation. *Journal of Plant Nutrition and Soil Science* 167, 24–32.

Dinesh, R., Ghoshal Chaudhuri, S. and Sheeja, T.E. (2006) ATP levels and adenylate energy charge in soils of the mangroves of Andamans. *Current Science* 90, 1258–1263.

Dinesh, R., Srinivasan, V., Hamza, S. and Manjusha, A. (2010) Short-term incorporation of organic manures and biofertilizers influences biochemical and microbial characteristics of soils under an annual crop [Turmeric (*Curcuma longa* L.)]. *Bioresource Technology* 101, 4697–4702.

Dinesh, R., Srinivasan, V., Hamza, S., Manjusha A. and Sanjay Kumar, P. (2012) Short-term effects of nutrient management regimes on biochemical and microbial properties in soils under rainfed ginger (*Zingiber officinale* Rosc.). *Geoderma* 173–174, 192–198.

Ding, W., Yu, H., Cai, Z., Han, F., Xu, Z. (2010) Responses of soil respiration to N fertilization in a loamy soil under maize cultivation. *Geoderma* 155, 381–389.

Djajakirana, G., Joergensen, R.G. and Meyer, B. (1996) Ergosterol and microbial biomass relationship in soil. *Biology and Fertility of Soils* 22, 299–304.

Dyckmans, J., Chander, K., Joergensen, R.G., Priess, J., Raubuch, M. and Sehy, U. (2003) Adenylates as an estimate of microbial biomass C in different soil groups. *Soil Biology and Biochemistry* 35, 1485–1491.

Dyckmans, J. and Raubuch, M. (1997) A modification of a method to determine adenosine nucleotides in forest organic layers and mineral soils by ion-paired reversed-phase high-performance liquid chromatography. *Journal of Microbiological Methods* 30, 13–20.

Engelking, B., Flessa, H. and Joergensen, R.G. (2008) Formation and use of microbial residues after adding sugarcane sucrose to a heated soil devoid of soil organic matter. *Soil Biology and Biochemistry* 40, 97–105.

Frazão, L.A., Piccolo, M.C., Feigl, B.J., Cerri, C.C. and Cerri, C.E.P. (2010) Inorganic nitrogen, microbial biomass and microbial activity of a sandy Brazilian Cerrado soil under different land uses. *Agriculture, Ecosystems and Environment* 135, 161–167.

Gamboa, A.M. and Galicia, L. (2011) Differential influence of land use/cover change on topsoil carbon and microbial activity in low-latitude temperate forests. *Agriculture Ecosystem and Environment* 142, 280–290.

García, C., Hernandez, T., Pascual, J.A., Moreno, J.L. and Ros, M. (2000) Microbial activity in soils of SE Spain exposed to degradation processes. Strategies for their rehabilitation. In: García, C. and Hernandez, T. (eds) *Research and Perspectives of Soil Enzymology in Spain*. Consejo Superìor de Investìgacìones Cìentificas, Madrid, pp. 93–143.

Gibbs, H.K., Ruesch, A.S., Achard, F., Clayton, M.K., Holmgren, P., Ramankutty, N. and Foley, J.A. (2010) Tropical forests were the primary sources of new agricultural land in the 1980s and 1990s. *Proceedings of the National Academy of Sciences USA* 107, 16732–16737.

Gil-Sotres, F., Trasar-Cepeda, C., Leirós, M.C. and Seoane, S. (2005) Different approaches to evaluating soil quality using biochemical properties. *Agriculture, Ecosystems and Environment* 37, 877–887.

Gómez-Luna, B.E., Rivera-Mosqueda, M.C., Dendooven, L., Vázquez-Marrufo, G. and Olalde-Portugal, V. (2009) Charcoal production at kiln sites affects C and N dynamics and associated soil microorganisms in *Quercus* spp. temperate forests of central Mexico. *Applied Soil Ecology* 41, 50–58.

Grandy, A.S., Strickland, M.S., Lauber, C.L., Bradford, M.A. and Fierer, N. (2009) The influence of microbial communities, management, and soil texture on soil organic matter chemistry. *Geoderma* 150, 278–286.

Heinze, S., Rauber, R. and Joergensen, R.G. (2010) Influence of mouldboard plough and rotary harrow tillage on micobial biomass and nutrient stocks in two long-term experiments on loess derived Luvisols. *Applied Soil Ecology* 46, 405–412.

Henrot, J. and Robertson, G.P. (1994) Vegetation removal in two soils of the humid tropics: effect on microbial biomass. *Soil Biology and Biochemistry* 26, 111–116.

Jastrow, J.D., Amonette, J.E. and Bailey, V.L. (2007) Mechanisms controlling soil carbon turnover and their potential application for enhancing carbon sequestration. *Climate Change* 80, 5–23.

Joergensen R.G. (1995) Die quantitative Bestimmung der mikrobiellen Biomasse in Böden mit der Chloroform-Fumigations-Extraktions-Methode. *Göttinger Bodenkundl Ber* 104, 1–229.

Joergensen, R.G. (1996) The fumigation-extraction method to estimate soil microbial biomass: Calibration of the k_{EC} value. *Soil Biology and Biochemistry* 28, 25–31.

Joergensen, R.G. (2010) Organic matter and micro-organisms in tropical soils. In: Dion, P. (ed.) *Soil Biology and Agriculture in the Tropics, Soil Biology 21*, Springer-Verlag, Berlin and Heidelberg, pp. 17–44.

Joergensen, R.G. and Castillo, X. (2001) Interrelationships between microbial and soil properties in young volcanic ash soils of Nicaragua. *Soil Biology and Biochemistry* 33, 1581–1589.

Joergensen, R.G. and Mueller, T. (1996) The fumigation-extraction method to estimate soil microbial biomass: Calibration of the k_{EN} value. *Soil Biology and Biochemistry* 28, 33–37.

Joergensen, R.G. and Raubuch, M. (2002) Adenylate energy charge of a glucose-treated soil without adding a nitrogen source. *Soil Biology and Biochemistry* 34, 1317–1324.

Joergensen, R.G. and Scheu, S. (1999) Depth gradients of microbial and chemical properties in moder soils under beech and spruce. *Pedobiologia* 43, 134–144.

Joergensen, R.G. and Wichern, F. (2008) Quantitative assessment of the fungal contribution to microbial tissue in soil. *Soil Biology and Biochemistry* 40, 2977–2991.

Kaschuk, G., Alberton O. and Hungria, M. (2010) Three decades of soil microbial biomass studies in Brazilian ecosystems: Lessons learned about soil quality and indications for improving sustainability. *Soil Biology and Biochemistry* 42, 1–13.

Kennedy, A.C. and Papendick, R.I. (1995) Microbial characteristics of soil quality. *Journal of Soil and Water Conservation* 50, 243–248.

Kessler, M., Hertel, D., Jungkunst, H.F., Kluge, J.,Abrahamczyk, S., Bos, M, *et al.* (2012) Can joint carbon and biodiversity management in tropical agroforestry landscapes be optimized? *PLoS ONE* 7, e47192.

Lamb, D. and Gilmour, D. (2003) *Rehabilitation and Restoration of Degraded Forests*. IUCN, Gland, Switzerland and Cambridge, UK and WWF, Gland, Switzerland.

Lee, K.H. and Jose, S. (2003) Soil respiration, fine root production and microbial biomass in cottonwood and loblolly pine plantations along a nitrogen fertilization gradient. *Forest Ecology and Management* 185, 263–273.

Leirós, M.C., Trasar-Cepeda, C., Seoane, S. and Gil-Sotres, F. (2000) Biochemical properties of acid soils under climax vegetation (Atlantic oakwood) in an area of the European temperate-humid zone (Galicia, NW Spain): General parameters. *Soil Biology and Biochemistry* 32, 733–745.

Lemenih, M., Karltun, E. and Olsson, M. (2005) Assessing soil chemical and physical property responses to deforestation and subsequent cultivation in smallholders farming system in Ethiopia. *Agriculture, Ecosystems and Environment* 105, 373–386.

Li, Y., Xu, M. and Zou, X. (2006a) Effects of nutrient additions on ecosystem carbon cycle in a Puerto Rican tropical wet forest. *Global Change Biology* 12, 284–293.

Li, Y., Xu, M. and Zou, X. (2006b) Heterotrophic soil respiration in relation to environmental factors and microbial biomass in two wet tropical forests. *Plant and Soil* 281, 193–201.

Li, Y.Y., Shao, M.A., Zheng, J.Y. and Zhang, X.C. (2005) Spatial–temporal changes of soil organic carbon during vegetation recovery at Ziwuling, China. *Pedosphere* 15, 601–610.

Lopez, M.J., Vargas-Garcia, M.D.C., Suarez-Estrella, F. and Moreno, J. (2006) Biodelignification and humification of horticultural plant residues by fungi. *International Biodeterioration and Biodegradation* 57, 24–30.

Nelson, D.W. and Sommers, L.E. (1982) Total carbon, organic carbon and organic matter. In: Page, A.L., Miller, R.H. and Keeney, D.R. (eds) *Methods of Soil Analysis*, Part 2, *Chemical and Microbiological Properties*. ASA, SSSA, Madison, Wisconsin, pp. 539–579.

Nourbakhsh, F. (2007) Decoupling of soil biological properties by deforestation. *Agriculture, Ecosystems and Environment* 121, 435–438.

Nunes J.S., Araujo A.S.F., Nunes L.A.P.L, Lima L.M., Carneiro R.F.V., Salviano, A.A.C. and Tsai, S.M. (2012) Impact of land degradation on soil microbial biomass and activity in northeast Brazil. *Pedosphere* 22, 88–95.

Pennanen, T., Caul, S., Daniell, T.J., Griffiths, B.S., Ritz, K. and Wheatley, R.E. (2004) Community-level responses of metabolic-active soil microorganisms to the quantity and quality of substrate inputs. *Soil Biology and Biochemistry* 36, 841–848.

Rosa, I.M., Ahmed, S.E. and Ewers, R.M. (2014) The transparency, reliability and utility of tropical rainforest land-use and land-cover change models. *Global Change Biology* 20, 1702–1722.

Sahani, U. and Behera, N. (2001) Impact of deforestation on soil physicochemical characteristics, microbial biomass and microbial activity of tropical soil. *Land Degradation and Development* 12, 93–105.

Sakamoto, K. and Oba, Y. (1994) Effect of fungal to bacterial biomass ratio on the relationship between CO_2 evolution and total soil microbial biomass. *Biology and Fertility of Soils* 17, 39–44.

Salamanca, E.F., Raubuch, M. and Joergensen, R.G. (2002) Relationships between soil microbial indices in secondary tropical forest soils. *Applied Soil Ecology* 21, 211–219.

Salamanca, E.F., Raubuch, M. and Joergensen, R.G. (2006) Microbial reaction of secondary tropical forest soils to the addition of leaf litter. *Applied Soil Ecology* 31, 53–61.

Scheller, E. and Joergensen, R.G. (2008) Decomposition of wheat straw differing in N content in soils under conventional and organic farming management. *Journal of Plant Nutrition and Soil Science* 171, 886–892.

Schinner, F., Öhlinger, R., Kandeler, E. and Margesin, R. (1995) *Methods in Soil Biology*. Springer-Verlag, Berlin.

Sparling, G.P. (1992) Ratio of microbial biomass carbon to soil organic carbon as a sensitive indicator of changes in soil organic matter. *Australian Journal of Soil Research* 39, 195–207.

Susyan, E.A., Wirth, S., Ananyeva, N.D. and Stolnikova, E.V. (2011) Forest succession on abandoned arable soils in European Russia – impacts on microbial biomass, fungal-bacterial ratio, and basal CO_2 respiration activity. *European Journal of Soil Biology* 47, 169–174.

Trasar-Cepeda, C., Leirós, M.C., Seoane, S. and Gil-Sotres, F. (2008) Biochemical properties of soils under crop rotation. *Applied Soil Ecology* 39, 133–143.

Tripathi, N. and Singh, R.S. (2013) Cultivation impacts soil microbial dynamics in dry tropical forest ecosystem in India *Acta Ecologica Sinica* 33, 344–353.

Vance, E.D., Brookes, P.C., Jenkinson, D.S. (1987) An extraction method for measuring soil microbial biomass C. *Soil Biology and Biochemistry* 19, 703–707.

Verchot, L.V. (2010) Impacts of forest conversion to agriculture on microbial communities and microbial function. In: Dion, P. (ed.) *Soil Biology and Agriculture in the Tropics*, Soil Biology 21. Springer-Verlag, Berlin and Heidelberg, pp. 45–63.

Villoria, N.B., Byerlee, D. and Stevenson, J. (2014) The effects of agricultural technological progress on deforestation: What do we really know? *Applied Economic Perspectives and Policy* 36, 211–237.

Wang, W.J., Dalal, R.C., Moody, P.W. and Smith, C.J. (2003) Relationships of soil respiration to microbial biomass, substrate availability and clay content. *Soil Biology and Biochemistry* 35, 273–284.

Wardle, D.A. (1998) Controls of temporal variability of the soil microbial biomass: A global scale synthesis. *Soil Biology and Biochemistry* 30, 1627–1637.

Zhong, Z.K. and Makeschin, F. (2006) Differences of soil microbial biomass and nitrogen transformation under two forest types in central Germany. *Plant Soil* 283, 287–297.

Zunino, H., Borie, F., Aguielera, S., Martin, J.P. and Haider, K. (1982) Decomposition of ^{14}C-labeled glucose, plant and microbial products and phenols in volcanic ash-derived soils of Chile. *Soil Biology and Biochemistry* 14, 37–43.

3 Evaluating the Impact of Oil Palm Agriculture and Logging on Soil Microbial Communities in South-east Asia

Heather D'Angelo,[1]* Krista L. McGuire,[1,2] Caitlyn Gillikin,[2] Francis Q. Brearley[3] and Dina C. Merrer[4]

[1]*Department of Ecology, Evolution, and Environmental Biology, Columbia University, New York, USA;* [2]*Department of Biology, Barnard College, Columbia University, New York, USA;* [3]*School of Science and the Environment, Manchester Metropolitan University, Manchester, UK;* [4]*Department of Chemistry, Barnard College, Columbia University, New York, USA*

3.1 Introduction

Since the mid-1900s, South-east Asia's lowland rain forests have been subjected to intense, large-scale deforestation, driven mainly by selective logging and, more recently, agricultural expansion of oil palm plantations (Flint, 1994; Sodhi *et al.*, 2004; Wilcove and Koh, 2010). As a result, South-east Asia now has the highest rate of tropical deforestation in the world, accounting for nearly half of total global forest cover loss (Hansen *et al.*, 2013). These human disturbances have created mosaics of old-growth forest, regenerating forest and oil palm monocultures across the terrestrial landscape, which have resulted in dramatic biodiversity losses (Okuda *et al.*, 2003; Sodhi *et al.*, 2004). Tropical forests contain approximately half of all terrestrial biomass (Sala *et al.* 2000; Pan *et al.*, 2011); thus, soil microbial processes in these ecosystems can be of global importance in regulating transformations of soil organic matter (SOM) and soil carbon (C) storage (Bardgett *et al.*, 2008; Wood *et al.*, 2012). Although land-use changes have been shown to alter soil physicochemical properties, it is still not fully understood how these factors then influence soil microbial communities (Bossio *et al.*, 2005; Adachi *et al.*, 2006). In order to comprehensively evaluate the impact that deforestation is having in this region and predict future ecosystem C balances, a more thorough understanding of these relationships is therefore needed (Allison and Martiny, 2008).

Although selective logging has been an important driver of deforestation in South-east Asia, the expansion of oil palm (*Elaeis guineensis*; Arecaceae) plantations has been the most salient consideration in recent years. Palm oil, the commercial commodity extracted from oil palm fruits and kernels, is the most productive and profitable vegetable oil in the world; and the versatility of this oil in food and non-food commodities has made it a crucial source of income in this region (Koh and Wilcove, 2007). In Malaysia alone, primary forests represented nearly 60% of the total forests converted to oil palm plantations between the years 1990 and 2005, and further deforestation is expected in order to meet

* Corresponding author: hd2144@columbia.edu

increasing future demands for cooking oil and biofuel (Sodhi *et al.*, 2004; Laurance *et al.*, 2010; Wilcove and Koh, 2010; Wicke *et al.*, 2011). Numerous studies have demonstrated that the conversion of tropical forest to oil palm plantations results in biodiversity loss for macrofauna such as primates, butterflies, birds and insects (Peh *et al.*, 2005, 2006; Koh, 2008; Pfeiffer *et al.*, 2008). However, the impacts of these perturbations on soil microbial communities are virtually unexplored.

In addition to understanding the impacts of forest conversion on soil microbial communities, it is also important to understand their composition in forests regenerating from human disturbances such as logging. In the Asian tropics, the majority of forests are in various states of regeneration, and intact, old-growth forests are becoming less reflective of the forested landscape (Asner *et al.*, 2009). There has been much debate regarding the value of degraded tropical forests (Peh *et al.*, 2006; Barlow *et al.*, 2007; Chazdon *et al.* 2009), especially within the context of C payment programmes and conservation initiatives such as Reducing Emissions from Deforestation and Forest Degradation (REDD)+ (Putz and Redford, 2009). Many studies have demonstrated that regenerating tropical forests can recover or continue to retain high biodiversity value even if they are structurally degraded (Dunn, 2004; Wilcove and Koh, 2010; Woodcock *et al.*, 2011), although, again, few studies have evaluated soil microbial communities within this context.

Soil microbial biomass (SMB) is one indicator used to approximate the health and functionality of soils (Kaur *et al.*, 2005; Dinesh *et al.*, Chapter 2, this volume). Microbes conduct several important functions in soil including aggregation, soil water retention, and nutrient capture and cycling. Soil microbial biomass also tracks SOM concentrations (Fierer *et al.*, 2009), which is another indicator of the condition of soil biological, chemical and physical properties (Carter, 2002). One advantage of evaluating SMB is that because microbes respond quickly to disturbance, changes in their composition can be readily assessed in the laboratory. The technique of phospholipid fatty acid analysis (PLFA) for quantifying SMB has the added advantage of providing a coarse community profile of bacteria and fungi. While DNA-based molecular techniques are needed to obtain detailed taxonomic information about microbial composition, PLFA is a more rapid and cost-effective technique for assessing total biomass and community structure of the live pool of soil microbes.

In this chapter we test the following hypotheses: (i) microbial communities would be distinct for each land-use type (primary forest, regenerating forest, and oil palm plantation); (ii) microbial communities from forest sites (primary forest and regenerating forest) would be more similar to each other than those found in the oil palm plantation site; (iii) microbial communities would be distinct at different soil horizon depths; and (iv) SMB would be greater in forest sites than in the oil palm plantation sites. To test these hypotheses, we analysed soil samples collected from an intact tropical rain forest, a tropical rain forest recovering from logging 50 years previously and an oil palm plantation in active production located in and around the Pasoh Forest Reserve in Peninsular Malaysia. We conducted this study in order to gain a preliminary understanding of how measures of SMB compared across land-use type as a proxy for overall soil health and functionality. We expected that fungal biomass would be lower in the oil palm plantation than the forests given that fungi are more sensitive to disturbance than bacteria. However, a decrease in SMB or altered microbial communities could indicate a change in biogeochemical cycling, which has broader implications for global nutrient cycling given the critical role played by soil microbial communities.

3.2 Methods

3.2.1 Study site

This study was conducted in the Negeri Sembilan region of peninsular Malaysia (2°58′N, 102°18′E) in July and August 2010. We collected soils from Pasoh Forest Reserve, which is located within a core remnant of lowland mixed dipterocarp primary forest (Manokaran and LaFrankie, 1990; Condit *et al.*, 1996; Manokaran *et al.*, 2004). Oil palm plantations border the reserve to the east, south and west. This region is aseasonal and receives an annual mean of 2054 mm of rain, ranging from 1728 to 3112 mm with average minimum and maximum temperatures between 22.7°C and 33.2°C. Ultisols are found at higher elevations, while sandy Entisols predominate at lower elevations. Samples were collected from the south-west area of Pasoh, which is dominated by tree communities typical of sandy alluvium, and all of which belong to the Dipterocarpaceae: *Shorea maxwelliana*, *Hopea mengarawan* and *Dipterocarpus crinitus* (Manokaran *et al.*, 2004).

3.2.2 Sampling scheme

In addition to choosing the primary forest at Pasoh Forest Reserve for our first sampling site, we chose a forest remnant that was selectively logged in 1958 and then left to regenerate for our second sampling site (Okuda *et al.*, 2003). For our third sampling site, we selected a nearby 25-year-old oil palm plantation. We collected five soil cores from three replicate plots (20 × 20 m) at each site, totalling nine plots. In order to evaluate the variation in microbial community composition across soil horizons, one 20-cm soil core was taken at each of the four corners and the centre of each plot. Each soil core was divided into three horizons: 0–2, 2–10 and 10–20 cm, which were separated with a sterilized knife before being placed into sample bags. The soils were frozen and then returned to the USA for analysis.

3.2.3 Soil and microbial analyses

In the laboratory, the five soil cores from each plot were composited by soil depth (0–2, 2–10 and 10–20 cm) and homogenized with a sterile 2-mm sieve. To assess SMB, microbial community structure, and bacterial-to-fungal ratios, we used PLFA on 1–2 g of soil from each sample with two analytical replicates for each (White *et al.*, 1996). Lipids were extracted from the soils with a single-phase, phosphate-buffered, $CHCl_3$–CH_3OH solvent mixture (2:1) and separated from neutral and glycolipid fractions by column chromatography on silica gel. Phospholipids were dried under N_2 at 40°C, transesterified to fatty acid methyl esters and quantified by mass spectrometry (Bligh and Dyer, 1959; Zelles *et al.*, 1992) using an Agilent 6980N gas chromatography system equipped with an autosampler and flame ionization detector (FID; Agilent Technologies, Santa Clara, California). PLFA peaks were identified using Agilent ChemStation software and compared to bacterial fatty acid standards (Supelco, Bellefonte, Pennsylvania) and fungal fatty acid standards (Matreya, Pleasant Gap, Pennsylvania). Based on these known microbial biomarkers, we quantified 33 compounds per sample. The abundance of PLFA markers (nmol g^{-1} of dry weight of soil sample) were calculated from the integrals of the compounds in the gas chromatograms and used for further statistical analysis. PLFA values of these 33 peaks were summed to estimate the total SMB for each site and at each depth. Bacterial biomass was estimated using the biomarkers 10Me16:0, 10Me17:0, i15:0, a15:0, 15:0, i17:0, 17:0, cis17:0, cis19:0, 16:1ω7t, 18:1ω7c, and fungal biomass was quantified using 18:2ω6,9 (Frostegård and Bååth, 1996; Whalen and Sampedro, 2010; Frostegård *et al.*, 2011).

3.2.4 Statistical analyses

Analysis of variance (ANOVA) with Tukey's post-hoc tests were used to test for significant differences in SMB across sites using SPSS 19 (Chicago, Illinois). Similarity between samples was calculated using the Bray–Curtis coefficient (Bray and Curtis, 1957; Clarke, 1993) following presence–absence transformation of the PLFA data. To visualize the similarity in microbial community composition across sites, non-metric multi-dimensional scaling (NMDS) was used (Clarke, 1993). To test the significance of the clustering of microbial community composition by site, Analysis of Similarity (ANOSIM) was employed, which is a non-parametric (randomization-based) method of multivariate analysis used to compare the variation in species abundance and composition across samples. NMDS and ANOSIM analyses were done using Primer-E software (Plymouth, UK).

3.3 Results

Total SMB estimated by summing PLFA compounds from each soil depth revealed that SMB was significantly higher in the forest sites than in the oil palm plantation ($P < 0.001$; Fig. 3.1). Across soil depths (0–2, 2–10 and 10–20 cm), the greatest magnitude of difference in SMB across sites was observed within the 10–20 cm profile (Fig. 3.2). At this depth, both forest sites had

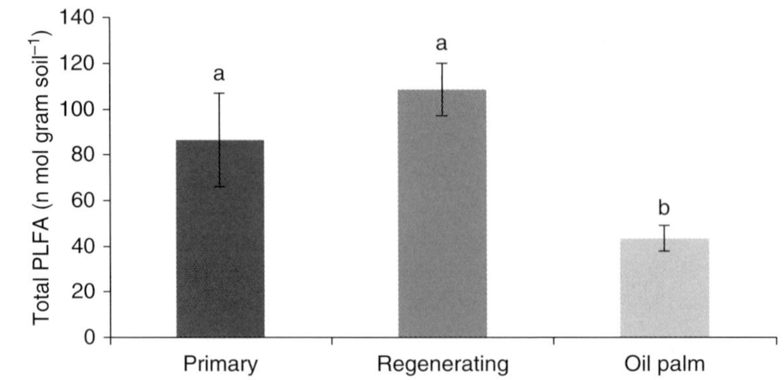

Fig. 3.1. Total PLFAs in primary and regenerating forests, and oil palm soils at Pasoh in Peninsular Malaysia. Different letters above the bars indicate significance at $P < 0.05$.

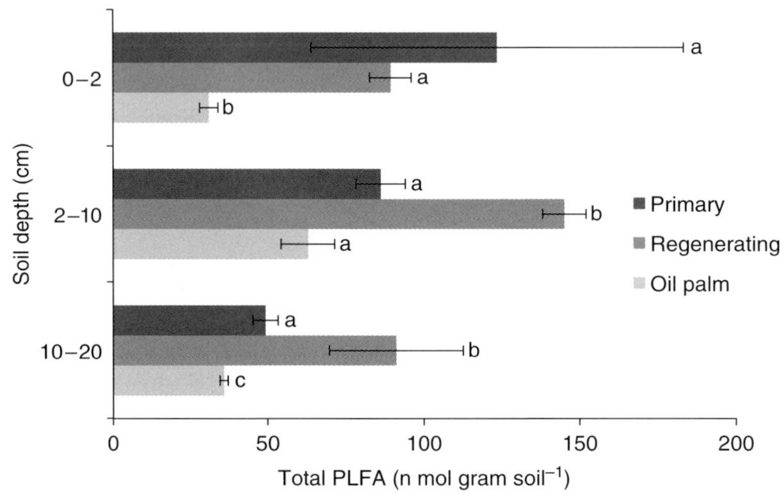

Fig. 3.2. Total PLFAs in primary and regenerating forests, and oil palm soils across three soil depths: 0–2 cm, 2–10 cm and 10–20 cm at Pasoh in Peninsular Malaysia. Comparisons of microbial biomass were made within each soil depth across sites. Different letters next to the bars indicate significance at $P < 0.05$ within a depth category.

greater SMB than the oil palm plantation and between the forest sites, the regenerating forest had greater SMB than the primary forest ($P < 0.05$). In the 0–2 cm depth, both forest sites had greater SMB than the oil palm plantation ($P < 0.05$). Last, in the 2–10 cm depth, there was no significant difference between the primary forest and the oil palm plantation ($P = 0.06$) but the regenerating forest had greater SMB than both the primary forest and the oil palm plantation ($P < 0.001$). At all sites, bacterial-to-fungal ratios declined with sampling depth (Fig. 3.3). Bacterial-to-fungal ratios in the oil palm plantation were significantly lower than in the forest sites and this was most notable in the lower soil horizons ($P < 0.01$; Fig. 3.3), but there was no significant difference in bacterial-to-fungal ratios between the primary or regenerating forest. ANOSIM of lipid composition across samples (all horizons summed for each sample) showed that the microbial community composition was distinct across all three sites ($P < 0.01$ for all contrasts; Fig. 3.4).

3.4 Discussion

Conversion of lowland tropical rain forest to oil palm agriculture results in reduced diversity of

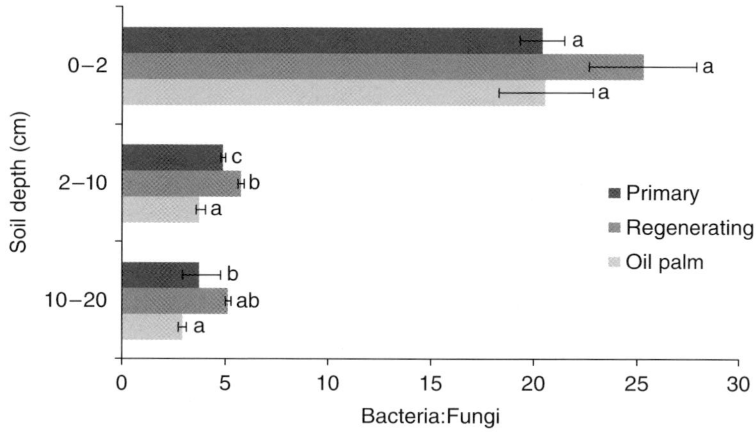

Fig. 3.3. Bacterial-to-fungal ratios in primary and regenerating forests, and oil palm soils at three soil depths at Pasoh in Peninsular Malaysia. Different letters next to the bars indicate significance at $P < 0.05$.

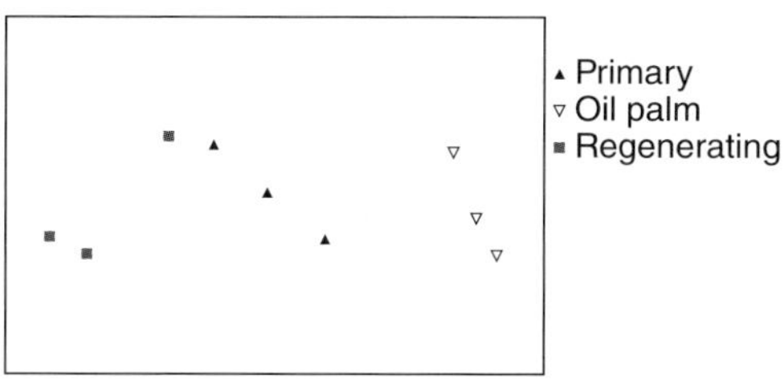

Fig. 3.4. Non-metric multidimensional scaling plot of soil microbial community composition for primary and regenerating forests, and oil palm plantations at Pasoh in Peninsular Malaysia.

numerous macroscopic taxa (Fitzherbert *et al.*, 2008; Koh and Wilcove, 2008; Sodhi *et al.*, 2004). This study found analogous reductions in SMB and alterations in microbial community composition. Our results corroborate previous studies that have reported reductions in SMB following land-use change from forest to plantations (Henrot and Robertson, 1994; Sahani and Behera, 2000; Waldrop *et al.*, 2000; Bossio *et al.*, 2005; Dinesh *et al.*, Chapter 2, this volume); however, this study is one of the few to investigate microbial communities specifically in oil palm plantations in South-east Asia.

Our finding that forest soils had greater SMB than oil palm plantation soils could be due to a number of differences in biotic and abiotic factors across sites. One major difference across sites is that the oil palm plantation has less leaf litter in the upper horizons of soil (personal observations), probably due to the fact that oil palm plantations have a similar tree age structure, a uniform canopy and sparse understorey vegetation (Danielsen *et al.*, 2009). A recent study by Luskin and Potts (2011) evaluating habitat characteristics of forest sites at Pasoh Forest Reserve and nearby oil palm plantations found that, within the plantations, the area covered by leaf litter was reduced and more patchy compared to forest sites. Because leaf litter sustains microbial growth for decomposition and nutrient cycling, and the quantity of SOM is correlated with SMB (Fierer *et al.*, 2009; Dinesh *et al.*, Chapter 2, this volume), we expected to find greater SMB in forest soils based on this observation alone. Oil palm plantations also lack much coarse woody debris, which serves as a C source for decomposer microbes. As would be predicted from this observation, a recent study found that the diversity of wood-inhabiting fungi was significantly reduced in oil palm plantations relative to nearby forest sites (Hattori *et al.*, 2012).

In addition to differences in leaf litter quantity, differences in the chemical constituents of the litter can also influence the size of the SMB. The chemical composition of leaf litter is determined by plant community composition, which is much more heterogeneous in forest habitats compared to monoculture oil palm plantations. Managed plantations also receive external inputs of chemical fertilizers that can alter the SMB (Treseder, 2008; Phosri *et al.*, 2010). Oil palm plantation soils also had notable differences in soil physicochemical characteristics such as a less acidic soil and a greater soil calcium concentration (McGuire *et al.*, 2015), which could further alter SMB and microbial community composition.

Another difference between forest and plantation soils likely to affect SMB may be soil surface temperatures and solar irradiance (Fowler *et al.*, 2011; Luskin and Potts, 2011). Significant temporal variability in soil moisture was reported previously from this site where oil palm plantation soils were hotter and drier during the daytime than their forest counterparts (Luskin and Potts, 2011). Drying and wetting can alter soil microbial communities because bacteria and fungi are inherently different in their abilities to resist and acclimate to stress (Schimel *et al.*, 2007; Brearley, Chapter 4, this volume). Similarly, a decrease in soil moisture can increase rates of cell lysis due to desiccation, which leads to reductions in SMB (Wardle and Parkinson, 1990).

Bacterial-to-fungal ratios were significantly lower in the oil palm plantation, which was counter to expectations, since agricultural sites have been found to have greater bacterial dominance and forests have been found to be more fungal dominated (Fierer *et al.*, 2009). An increase in fungal-to-bacterial abundance has been correlated with more acidic soil pH and higher soil C:N ratios (Fierer *et al.*, 2009; Nilsson *et al.*, 2011). In a recent study, we found higher C:N ratios and more acidic pH in the primary forest soils (McGuire *et al.*, 2015), but this did not reflect a greater relative abundance of fungi relative to bacteria. The reason for higher bacterial-to-fungal ratios in the forest sites remains elusive, but may be related to oil palm plantation management techniques that were not included as covariates in our analyses. The differences in pH between the forest sites and the oil palm plantation may also not have been extreme enough to elicit changes in bacterial-to-fungal ratios. In a recent study that evaluated fungal and bacterial biomass across a pH gradient in the UK, PLFA-based ratios of bacteria-to-fungi did not vary significantly for a pH range of 4.5 to 7.4 (Rousk *et al.*, 2011). Because the pH values of all the sites in this study fall within this range, it is possible that the disparity was not great enough to alter the biomass ratios.

Across the tropics, many degraded forests appear to be transitioning to a state of regeneration (Meyfroidt and Lambin, 2011). Regenerating forests have many characteristics that differ from primary forests aboveground, such as lower canopies and more diverse tree stands (Hamer *et al.*, 2003; Okuda *et al.*, 2003). In contrast, our results show that these forests share many similarities belowground, as indicated by their total SMB and fungal-to-bacterial ratios. More importantly, in almost all of the analyses, this study shows that regenerating forests had more in common with primary forests than they did with oil palm plantations. This observation is a critical point, given that regenerating forests are likely candidates for conversion to oil palm plantations because their conservation value is still widely viewed as uncertain (Berry *et al.*, 2010), and has generated a need for more research comparing a broad range of ecological parameters between regenerating and primary forests.

Since the ecological value of a secondary forest is expected to increase over time (Martin *et al.*, 2013), it is possible that as regenerating forests recover aboveground similarities to their old-growth counterparts, the belowground similarities may increase as well. For example, we observed that the regenerating forest that had been abandoned for 50 years since logging had already regained similar SMB compared to the primary forest. However, the exact course of the trajectory within the last 50 years is unclear and this highlights a gap in our understanding of the mechanisms underlying microbial succession.

Numerous studies have shown that changes in land use have a significant effect on microbial community composition (Waldrop *et al.*, 2000; Carney *et al.*, 2004; Bossio *et al.*, 2005), and our results show that microbial community composition was distinct across all sites. In the context of forest restoration, these results provide a snapshot of how variations due to land-use history alter microbial communities, thus bringing into question whether microbially mediated processes remain functionally equivalent across sites. While it is known that microbial dynamics are the foundation on which higher trophic levels are built, the influence of these differences on succession in an abandoned oil palm plantation remains unclear.

Although this study does not address the taxonomic differences responsible for the patterns we have observed, the marked distinction between the microbial communities of the oil palm plantation and the primary forest is worth noting because of the unique relationship between trees and fungi in this region. In the Malay

Peninsula, primary forests are dominated by large stands of trees in the Dipterocarpaceae. These trees form symbiotic associations with ectomycorrhizal fungi that are important for their growth and survival. Ectomycorrhizal fungi form extensive belowground networks that provide the roots of their plant hosts with essential nutrients such as water, nitrogen and phosphorus in exchange for fixed C (Smith and Read, 2008). Given that the soil in tropical rain forests is often nutrient-poor, the survival and growth of the Dipterocarpaceae depend upon their ability to form ectomycorrhizal associations (Alexander *et al.*, 1992; Brearley, 2012). Studies have shown that fungi are sensitive to stress from intensive agricultural management practices such as fertilizer application and tillage (Munyanziza *et al.*, 1997; Treseder 2004); however, little is known about the response of fungi to oil palm plantation cultivation (but see McGuire *et al.*, 2015). Exploration of this topic should be a key aim of future research in order to recommend better plantation management practices.

Since there were distinct differences in microbial biomarkers across sites, there are likely differences in microbial communities at finer levels of taxonomic resolution, and an initial study by McGuire *et al.* (2015) found a major shift in fungal community composition and a significant reduction in fungal species diversity, especially among ectomycorrhizal fungi, in oil palm plantation soils compared to forest soils. A major limitation of PLFA is that it does not allow for resolution to the species level, so future work should prioritize sequencing the DNA of soil microbes to resolve exactly which taxa have different responses across habitats. However, this preliminary study of microbial communities within forest sites and an oil palm plantation is a useful starting point for further analysis.

3.5 Conclusions

In conclusion, our results that oil palm ecosystems have lower SMB and altered microbial communities compared to forest sites indicate that plantations may experience altered biogeochemical cycling compared to forest sites. Because tropical forests contain the equivalent of 37% of global terrestrial C pools (Dixon *et al.*, 1994), changes in biogeochemical cycling in this region will have global implications. Oil palm plantations are predicted to continue expanding in South-east Asia until at least 2020 (Koh and Ghazoul, 2010). Therefore, efforts to understand the effects of shifts in microbial composition and biomass on nutrient cycling in these ecosystems are of major importance.

Acknowledgements

We acknowledge the National Science Foundation (CHE-0234660 to D.C.M.), which funded the purchase of the gas chromatograph and Marina Orman for technical assistance. Nina Yang, Rosie Gradoville, Carling Bateman and Patahayah Mansor assisted with sample collection. The Forest Research Institute of Malaysia provided invaluable logistic support and infrastructure. Research permits were granted through the Malaysian Economic Planning Unit and the Pasoh Research Committee.

References

Adachi, M., Bekku, Y.S., Rashidah, W., Okuda, T. and Koizumi, H. (2006) Differences in soil respiration between different tropical ecosystems. *Applied Soil Ecology* 34, 258–265.

Alexander, I., Ahmad, N. and See, L.S. (1992) The role of mycorrhizas in the regeneration of some Malaysian forest trees. *Philosophical Transactions of the Royal Society of London Series B: Biological Sciences* 335, 379–388.

Allison, S.D. and Martiny, J.B.H. (2008) Resistance, resilience, and redundancy in microbial communities. *Proceedings of the National Academy of Sciences of the USA* 105, 11512–11519.

Asner, G.P., Rudel, T.K., Aide, T.M., DeFries, R. and Emerson, R. (2009) A contemporary assessment of change in humid tropical forests. *Conservation Biology* 23, 1386–1395.

Bardgett, R.D., Freeman, C. and Ostle, N.J. (2008) Microbial contributions to climate change through carbon cycle feedbacks. *The ISME Journal* 2, 805–814.

Barlow, J., Mestre, L.A.M., Gardner, T.A. and Peres, C.A. (2007) The value of primary, secondary and plantation forests for Amazonian birds. *Biological Conservation* 136, 212–231.

Berry, N.J., Phillips, O.L., Lewis, S.L., Hill, J.K., Edwards, D.P., Tawatao, N.B. *et al.* (2010) The high value of logged tropical forests: Lessons from northern Borneo. *Biodiversity and Conservation* 19, 985–997.

Bligh, E.G. and Dyer, W.J. (1959) A rapid method of total lipid extraction and purification. *Biochemistry and Cell Biology* 37, 911–917.

Bossio, D.A., Girvan, M.S., Verchot, L., Bullimore, J., Borelli, T., Albrecht, A. *et al.* (2005) Soil microbial community response to land use change in an agricultural landscape of western Kenya. *Microbial Ecology* 49, 50–62.

Bray, J.R. and Curtis, J.T. (1957) An ordination of the upland forest communities of southern Wisconsin. *Ecological Monographs* 27, 325.

Brearley, F.Q. (2012) Ectomycorrhizal associations of the Dipterocarpaceae. *Biotropica* 44, 637–648.

Carney, K.M., Matson, P.A. and Bohannan, B.J.M. (2004) Diversity and composition of tropical soil nitrifiers across a plant diversity gradient and among land-use types. *Ecology Letters* 7, 684–694.

Carter, M.R (2002) Soil quality for sustainable land management: Organic matter and aggregation interactions that maintain soil functions. *Agronomy Journal* 94, 38–47.

Chazdon, R.L., Peres, C.A., Dent, D., Sheil, D., Lugo, A.E., Lamb, D. *et al.* (2009) The potential for species conservation in tropical secondary forests. *Conservation Biology* 23, 1406–1417.

Clarke, K. (1993) Non-parametric multivariate analyses of changes in community structure. *Australian Journal of Ecology* 18, 117–143.

Condit, R., Hubbell, S.P., LaFrankie, J.V. and Sukumar, R. (1996) Species-area and species-individual relationships for tropical trees: A comparison of three 50-ha plots. *Journal of Ecology* 84, 549–562.

Danielsen, F., Beukema, H., Burgess, N.D., Parish, F., Brühl, C.A., Donald, P.F. *et al.* (2009) Biofuel plantations on forested lands: Double jeopardy for biodiversity and climate. *Conservation Biology* 23, 348–358.

Dixon, R.K., Solomon, A.M., Brown, S., Houghton, R.A., Trexier, M.C. and Wisniewski, J. (1994) Carbon pools and flux of global forest ecosystems. *Science* 263, 185–190.

Dunn, R.R. (2004) Managing the tropical landscape: A comparison of the effects of logging and forest conversion to agriculture on ants, birds, and lepidoptera. *Forest Ecology and Management* 191, 215–224.

Fierer, N., Strickland, M.S., Liptzin, D., Bradford, M.A. and Cleveland, C.C. (2009) Global patterns in belowground communities. *Ecology Letters* 12, 1238–1249.

Fitzherbert, E.B., Struebig, M.J. and Morel, A. (2008) How will oil palm expansion affect biodiversity? *Trends in Ecology and Evolution* 33, 538–545.

Flint, E.P (1994) Changes in land use in South and Southeast Asia from 1880 to 1980: A data base prepared as part of a coordinated research program on carbon fluxes in the tropics. *Chemosphere* 29, 1015–1062.

Fowler, D., Nemitz, E., Misztal, P., Di Marco, C., Skiba, U., Ryder, J. *et al.* (2011) Effects of land use on surface-atmosphere exchanges of trace gases and energy in Borneo: Comparing fluxes over oil palm plantations and a rainforest. *Philosophical Transactions of the Royal Society B: Biological Sciences* 366, 3196–3209.

Frostegård, Å. and Bååth, E. (1996) The use of phospholipid fatty acid analysis to estimate bacterial and fungal biomass in soil. *Biology and Fertility of Soils* 22, 59–65.

Frostegård, Å., Tunlid, A. and Bååth, E. (2011) Use and misuse of PLFA measurements in soils. *Soil Biology and Biochemistry* 43, 1621–1625.

Hamer, K.C., Hill, J.K., Benedick, S., Mustaffa, N., Sherratt, T.N., Maryati, M. t. et al. (2003) Ecology of butterflies in natural and selectively logged forests of northern Borneo: The importance of habitat heterogeneity. *Journal of Applied Ecology* 40, 150–162.

Hansen, M.C., Potapov, P.V., Moore, R., Hancher, M., Turubanova, S.A., Tyukavina, A. *et al.* (2013) High-resolution global maps of 21st-century forest cover change. *Science* 342, 850–853.

Hattori, T., Yamashita, S. and Lee, S.S. (2012) Diversity and conservation of wood-inhabiting polypores and other aphyllophoraceous fungi in Malaysia. *Biodiversity and Conservation* 21, 2375–2396.

Henrot, J. and Robertson, P.G. (1994) Vegetation removal in two soils of the humid tropics: Effect on microbial biomass. *Soil Biology Biochemistry* 26(1), 111–116.

Kaur, A., Chaudhary, A., Kaur, A. and Choudhary, R. (2005) Phospholipid fatty acid – a bioindicator of environment monitoring and assessment in soil ecosystem. *Current Science* 89, 1103–1112.

Koh, L.P. (2008) Can oil palm plantations be made more hospitable for forest butterflies and birds? *Journal of Applied Ecology* 45, 1002–1009.

Koh, L.P. and Ghazoul, J. (2010) Reply to Sloan and Stork: Spatially explicit scenario analysis for reconciling agricultural expansion, forest protection, and carbon conservation in Indonesia. *Procedures of the National Academy of Sciences USA* 107, 11140–11144.

Koh, L.P. and Wilcove, D.S. (2007) Cashing in palm oil for conservation. *Nature* 448, 993–994.

Koh, L.P. and Wilcove, D.S. (2008) Is oil palm agriculture really destroying tropical biodiversity? *Conservation Letters* 1, 60–64.

Laurance, W.F., Koh, L.P., Butler, R., Sodhi, N.S., Bradshaw, C.J.A., Neidel, J.D. *et al.* (2010) Improving the performance of the Roundtable on Sustainable Palm Oil for nature conservation. *Conservation Biology* 24, 377–381.

Luskin, M.S. and Potts, M.D. (2011) Microclimate and habitat heterogeneity through the oil palm lifecycle. *Basic and Applied Ecology* 12, 540–551.

McGuire, K.L., D'Angelo H., Brearley F.Q., Gedallovich S.M., Babar N., Yang N. *et al.* (2015) Responses of soil fungi to logging and oil palm agriculture in Southeast Asian tropical forests. *Microbial Ecology* 69, 733–747.

Manokaran, N. and LaFrankie, J.V. (1990) Stand structure of Pasoh Forest Reserve, a lowland rain forest in peninsular Malaysia. *Journal of Tropical Forest Science* 3, 14–24.

Manokaran, N., Seng, Q.E., Ashton, P.S., LaFrankie, J.V., Noor, N.S.M., Ahmad, W.M.S.W. et al. (2004) Pasoh forest dynamics plot, peninsular Malaysia. In: Losos, E.C. and Leigh, E.G. (eds) *Tropical Forest Diversity and Dynamism: Findings from a Large-Scale Plot Network.* The University of Chicago Press, Chicago, Illinois, pp. 585–598.

Meyfroidt, P. and Lambin, E.F. (2011) Global forest transition: Prospects for an end to deforestation. *Annual Review of Environment and Resources* 36, 343–371.

Munyanziza, E., Kehri, H.K. and Bagyaraj, D.J. (1997) Agricultural intensification, soil biodiversity and agro-ecosystem function in the tropics: The role of mycorrhiza in crops and trees. *Applied Soil Ecology* 6, 77–85.

Nilsson, L.O., Wallander, H. and Gundersen, P. (2011) Changes in microbial activities and biomasses over a forest floor gradient in C-to-N ratio. *Plant and Soil* 355, 75–86.

Okuda, T., Suzuki, M., Adachi, N., Quah, E.S., Hussein, N.A. and Manokaran, N. (2003) Effect of selective logging on canopy and stand structure and tree species composition in a lowland dipterocarp forest in peninsular Malaysia. *Forest Ecology and Management* 175, 297–320.

Pan, Y., Birdsey, R.A., Fang, J., Houghton, R., Kauppi, P.E., Kurz, W.A. *et al.* (2011) A large and persistent carbon sink in the world's forests. *Science* 333, 988–993.

Peh, K.S.-H., Jong, J.D. Sodhi, N.S., Lim, S.L.H. and Yap, C.A.M. (2005) Lowland rainforest avifauna and human disturbance: Persistence of primary forest birds in selectively logged forests and mixed-rural habitats of southern Peninsular Malaysia. *Biological Conservation* 123, 489–505.

Peh, K.S.-H., Sodhi, N.S., de Jong, J., Sekercioglu, C.H., Yap, C.A.M. and Lim, S.L.H. (2006) Conservation value of degraded habitats for forest birds in southern Peninsular Malaysia. *Diversity and Distributions* 12, 572–581.

Pfeiffer, M., Cheng Tuck, H. and Chong Lay, T. (2008) Exploring arboreal ant community composition and co-occurrence patterns in plantations of oil palm *Elaeis guineensis* in Borneo and Peninsular Malaysia. *Ecography* 31, 21–32.

Phosri, C., Rodriguez, A., Sanders, I.R. and Jeffries, P. (2010) The role of mycorrhizas in more sustainable oil palm cultivation. *Agriculture Ecosystems and Environment* 135, 187–193.

Putz, F.E. and Redford, K.H. (2009) Dangers of carbon-based conservation. *Global Environmental Change-Human and Policy Dimensions* 19, 400–401.

Rousk, J., Brookes, P.C. and Bååth, E. (2011) Fungal and bacterial growth responses to N fertilization and pH in the 150-year 'Park Grass' UK grassland experiment. *FEMS Microbiology Ecology* 76, 89–99.

Sahani, U. and Behera, N. (1999) Soil characteristics under different plantations. *Sambalpur University Journal of Science Technology* 2(A), 23–27.

Sala, O.E., Chapin, F.S., Armesto, J.J., Berlow, E., Bloomfield, J., Dirzo, R. *et al.* (2000) Biodiversity – global biodiversity scenarios for the year 2100. *Science* 287, 1770–1774.

Schimel, J., Balser, T.C. and Wallenstein, M. (2007) Microbial stress-response physiology and its implications for ecosystem function. *Ecology* 88, 1386–1394.

Smith, J.E. and Read, D. (2008) *Mycorrhizal Symbiosis* (3rd edition). Academic Press, San Diego, California.

Sodhi, N.S., Koh, L.P., Brook, B.W. and Ng, P. (2004) Southeast Asian biodiversity: An impending disaster. *Trends in Ecology and Evolution* 19, 654–660.

Treseder, K.K. (2004) A meta-analysis of mycorrhizal responses to nitrogen, phosphorus, and atmospheric CO_2 in field studies. *New Phytologist* 164, 347–355.

Treseder, K.K. (2008) Nitrogen additions and microbial biomass: A meta-analysis of ecosystem studies. *Ecology Letters* 11, 1111–1120.

Waldrop, M.P., Balser, T.C. and Firestone, M.K. (2000) Linking microbial community composition to function in a tropical soil. *Soil Biology and Biochemistry* 32, 1837–1846.

Wardle, D.A. and Parkinson, D. (1990) Interactions between microclimatic variables and the soil microbial biomass. *Biology and Fertility of Soils* 9, 273–280.

Whalen, J.K. and Sampedro, L. (2010) *Soil Ecology and Management.* CAB International, Cambridge, Massachusetts.

White, D.C., Stair, J.O. and Ringelberg, D.B. (1996) Quantitative comparisons of in situ microbial biodiversity by signature biomarker analysis. *Journal of Industrial Microbiology and Biotechnology* 17, 185–196.

Wicke, B., Sikkema, R., Dornburg, V. and Faaij, A. (2011) Exploring land use changes and the role of palm oil production in Indonesia and Malaysia. *Land Use Policy* 28, 193–206.

Wilcove, D.S. and Koh, L.P. (2010) Addressing the threats to biodiversity from oil-palm agriculture. *Biodiversity and Conservation* 19, 999–1007.

Wood, T.E., Cavaleri, M.A. and Reed, S.C. (2012) Tropical forest carbon balance in a warmer world: A critical review spanning microbial- to ecosystem-scale processes. *Biological Reviews* 87, 912–927.

Woodcock, P., Edwards, D.P., Fayle, T.M., Newton, R.J., Khen, C.V., Bottrell, S.H. and Hamer, K.C. (2011) The conservation value of South East Asia's highly degraded forests: Evidence from leaf-litter ants. *Philosophical Transactions of the Royal Society B: Biological Sciences* 366, 3256–3264.

Zelles, L., Bai, Q.Y., Beck, T. and Beese, F. (1992) Signature fatty acids in phospholipids and lipopolysaccharides as indicators of microbial biomass and community structure in agricultural soils. *Soil Biology and Biochemistry* 24, 317–323.

4 Microbial Functioning in Response to a Simulated Drought in Malaysian Rain Forest and Oil Palm Soils

Francis Q. Brearley*
School of Science and the Environment, Manchester Metropolitan University, Manchester, UK

4.1 Introduction

Land-use change is known to affect the diversity and composition of soil microbial communities in a range of tropical ecosystems (Bossio *et al.*, 2005; Verchot, 2010; Dinesh *et al.*, Chapter 2, this volume; Mendes *et al.*, Chapter 5, this volume). In addition, land-use change also affects soil nutrient status through a reduction in carbon (C) input leading to knock-on changes mediated through soil microbial communities. While the focus of most of the research into the effects of land-use change on soils has been on their chemical properties and on the diversity of soil organisms/microbes, much less work has been conducted on the impacts on ecosystem functioning, such as functional stability.

One major land-use transition currently occurring in South-east Asia is the conversion of forested lands to oil palm plantations, which are a highly profitable agricultural crop (Koh *et al.*, 2011; Carlson *et al.*, 2012). Much of the research on the conversion to oil palm has focused on the changes in biodiversity of insects and larger animals (Fitzherbert *et al.*, 2008; Turner *et al.*, 2011) and there has been little attention paid to the organisms that live in soils. Furthermore, droughts, especially those related to El Niño–Southern Oscillation events, are predicted to increase in this region (Timmerman *et al.*, 1999; Kumagai *et al.*, 2013) leading to uncertain impacts on the soil environment and soil microbial communities. In this chapter, the comparative responses of soil microbial communities from undisturbed primary rain forest and an adjacent oil palm plantation to a simulated drought are determined in order to evaluate their resistance and resilience to drought stress.

Soil moisture plays an essential role in microbial activity and, hence, biogeochemical processes (Billings and Phillips, 2011). In the vast majority of studies, microbial activity is reduced in dry soils (Manzoni *et al.*, 2012; Moyano *et al.*, 2013). Upon rewetting, there is regularly a rapid release of CO_2 due to a soil microbial respiration pulse (often to levels higher than that found in moist soils) and increased nitrogen (N) mineralization rates (Borken and Matzner, 2009; Kim *et al.*, 2012). This is due to the 'microbial stress' response (*sensu* Xiang *et al.*, 2008) whereby microbial cells are ruptured following osmotic shock and there is a concomitant release of cell solutes either by cell lysis or actively in response to changing osmotic conditions; these compounds are then rapidly taken up by growing microbial cells. It is also

* Corresponding author: f.q.brearley@mmu.ac.uk

due to the 'substrate supply' mechanism in which changes in soil structure and disaggregation of soil organic matter during soil rewetting lead to more C being made labile in the soil that is, again, taken up by growing microbial cells (Fierer and Schimel, 2003). Many studies have examined the response of individual soils to such drying–rewetting stresses but there is little understanding of how land-use change affects microbial responses to these stresses (but see Gordon *et al.*, 2008; Huygens *et al.*, 2011; de Vries *et al.*, 2012).

In this study, soils were collected from a primary forest and an adjacent oil palm plantation and subjected to a brief simulated drought and subsequent rewetting to assess the short-term resistance and resilience of the microbial community from the different land-use types. Fungi are generally considered more tolerant of soil drying (Harris, 1980; Manzoni *et al.*, 2012) and the key hypothesis under examination here is that the primary forest soil, having a more diverse fungal community (McGuire *et al.*, 2015), will show a greater resistance and resilience to drought than the oil palm plantation soil community. While laboratory-based experiments, such as this one, allow more precise control of experimental conditions, they lack some of the ecological reality of those conducted under field conditions and, as such, the results should be considered with this in mind.

4.2 Methods

Soils (Ultisols) were collected from Pasoh Forest Reserve in south-central Malaysia (2°58′N; 102°18′E; Okuda *et al.*, 2003) and an adjacent oil palm plantation (about 8 years since planting; Fig. 4.1) in February 2011. This area is one of the driest in Malaysia and receives around 1800–2000 mm rainfall a year with extended dry periods (around 3 weeks) occurring typically in late January or late July (Manokaran *et al.*, 2004).

Ten soil samples were collected (to 8 cm depth), equally spaced along two 40-m transects in each land-use type. They were sieved to approximately 8 mm to remove any stones and roots, pooled by land-use type and returned to the laboratory in Malaysia where they were maintained in a fresh state in a refrigerator for 2 weeks before being returned to the UK. About 180 g of pooled soil (with four replicates per forest type) were placed in plastic containers (9.5 cm tall; 7.5 cm upper diameter, 5.5 cm lower diameter with a 7-mm diameter hole in the lid for aeration) in an

Fig. 4.1. Oil palm plantation adjacent to Pasoh Forest Reserve in south-central Malaysia.

incubator at 25°C in the dark. They were maintained at the water holding capacity at the time of collection for 1 week before being adjusted to 70% water holding capacity for 1 further week before the start of the experiment. At the start of the experiment, fluorescein diacetate hydrolysis, fungal and bacterial colony forming units, urease activity, phosphatase activity, ammonium and nitrate concentrations, soil pH, total C and N, extractable phosphorus (P) and exchangeable potassium (K), calcium (Ca) and magnesium (Mg) were all determined (see below for methods). The soils were then dried in the incubator for 2 weeks to simulate a drought by removing the lids of the plastic containers and the same measurements retaken at the end of this period. The soils were then rewet to 70% water holding capacity to simulate the end of the drought and left for 2 weeks more before the final set of measurements were completed.

Measurement of fluorescein diacetate hydrolysis (as a broad measure of total microbial activity) followed Margesin (2005): 1 g of fresh soil was added to 10 ml of NaH_2PO_4/Na_2HPO_4 buffer (30 mM of each) at pH7 and 100 µl of fluorescein diacetate substrate in acetone (2 mg ml^{-1}) was added before the tubes were vortexed and placed in an incubator at 25°C. After 2 h, they were removed, 10 ml of acetone was added to stop the reaction, they were centrifuged and absorbance of the solution measured at 490 nm (Cecil CE1020 spectrophotometer) with appropriate fluorescein standards and blanks.

The culturable fungal and bacterial numbers in soil were assessed by the number of colony forming units (CFUs) determined by a serial dilution viable count technique (Madigan and Martinko, 2006). About 1 g of soil was added to 10 ml of sterile water and then serially diluted (ten-fold) with 100 µl spread plated out on petri dishes containing potato dextrose agar (with 0.05% (w/v) chloramphenicol) for fungi and tryptic soy agar (with 0.05% (w/v) cyclohexamide) for bacteria. The petri dishes were incubated for 72 h at 25°C before enumeration of CFUs.

The activity of two soil enzymes was assessed to determine microbial capacity for nutrient mineralization. Urease activity was measured following the protocol of Schaller (1993). Two grams of fresh soil was added to 0.5 ml toluene and pre-incubated at 34°C for 15 mins. One ml of 55 mM urea and 3 ml of deionized water was added, the soils were then vortexed and incubated at 34°C for 5 h. At the end of the incubation, 2 ml of 0.5 M K_2SO_4 was added, the soils were vortexed again, centrifuged and filtered through filter papers followed by a 0.2-µm filter and analysed for ammonium on a Dionex ICS-2000 ion chromatography system. Blanks had deionized water added instead of urea, and urease activity was calculated as the difference in ammonium concentrations between the two. Phosphatase activity was measured using a protocol based on that of Johnson *et al.* (1998). A total of 1 g of soil was added to 10 ml of deionized water to create a soil slurry. Next, 1 ml of this was added to 4 ml of deionized water and 1 ml of 0.01 M *p*-nitrophenyl phosphate and incubated at 37°C for 1 h while shaking at 200 rpm. Then 4 ml of 2 M NaOH was added to stop the reaction and the tubes were centrifuged before absorbance of the solution was measured at 410 nm (Cecil CE1020 spectrophotometer) with appropriate standards of *p*-nitrophenol.

Assessment of the different land uses and the drying and rewetting process was also made on soil chemical properties. Ammonium and nitrate were extracted from 5 g of fresh soil with 50 ml of 1 M KCl and shaken at 180 rpm for 1 h. Extracts were then filtered (0.2 µm) and analysed on a Dionex ICS-2000 ion chromatography system. Soil pH was measured by adding 5 g of air-dried soil to 12.5 ml of distilled water that was stirred and left to equilibrate for 1 h before pH was measured on a Sartorius PB-11 pH meter. Total C and N were analysed on a LECO TruSpec CN elemental analyser using 0.2 g air-dried soils. P was extracted from 1 g of air-dried soil with 10 ml of Mehlich 1 solution (0.05 M HCl plus 0.025 M H_2SO_4) by shaking for 5 mins and subsequently analysed on a Thermo iCAP 6300 Duo inductively coupled plasma optical emission spectrometer. Exchangeable Ca, K and Mg were determined by leaching 2.5 g air-dried samples with five successive additions of 10 ml of 1 M ammonium acetate; lanthanum chloride was added to the samples for Ca and Mg determination to create a 0.1% LaCl solution before analysis on a Thermo iCE 3300 atomic absorption spectrometer. All results were converted to a soil oven-dry basis by drying soils at 105°C for 24 h.

Data were analysed using a general linear model with a repeated measures design using Minitab 16.1; Tukey's tests were conducted following all analyses. Data were log-transformed if required and 1 was added to any zero CFU readings to avoid bimodality.

4.3 Results

Microbial activity was significantly greater in the primary forest soil than the oil palm soil ($P < 0.001$; Fig. 4.2). Drought reduced fluorescein diacetate hydrolysis to a similar magnitude in both soil types to around 25–30% of initial (pre-drought) conditions ($P < 0.001$). Microbial activity recovered to around two-thirds of pre-drought activity after rewetting. Primary forest soil microbial communities showed a better recovery after rewetting than the oil palm plantation communities (73% of initial value compared to 58% of initial value) as they were significantly different from the activity during the drought, whereas the oil palm plantation communities were not (Fig. 4.2).

Initial fungal CFU numbers were significantly higher in the primary forest soil ($P = 0.01$; Fig. 4.3a) and were reduced following the drought ($P < 0.001$), significantly so in the primary forest (leading to a forest × drought interaction, $P < 0.05$). Recovery showed a similar pattern in both forest types although fungal CFUs did not reach initial pre-drought numbers at the end of the experiment (Fig. 4.3a). Overall, the number of bacterial CFUs was not significantly different between the two forest types (Fig. 4.3b), but their response to the drought contrasted markedly with that of fungi as, although bacteria from the primary forest showed a greater reduction in numbers in response to the drought ($P < 0.001$), they recovered to higher numbers than previously, leading to a significant forest × drought interaction ($P < 0.01$). Bacterial CFUs in the oil palm soil showed much less of a change following the drought or during the rewetting period (Fig. 4.3b).

Both ammonium and nitrate concentrations were significantly greater in the primary forest soils than in the oil palm plantation soils (ammonium $P < 0.001$; nitrate $P = 0.01$; Fig. 4.4). There was an increase in ammonium concentrations under the drought conditions that was further increased after rewetting in both forest types ($P < 0.001$; Fig. 4.4a). This effect was significantly greater in the primary forest than in the oil palm plantation ($P < 0.001$). Nitrate concentrations increased (although not significantly) when droughted and were then reduced (again, not significantly) when rewetted, so overall there were no significant differences in nitrate concentrations between the beginning and end of the experiment ($P < 0.05$; Fig. 4.4b).

Soil pH was more acidic in the primary forest soils and both soil types showed a very large increase in acidity (*c.* 2 pH units) when soils were droughted and a subsequent decrease to a value intermediate between the initial pre-drought and

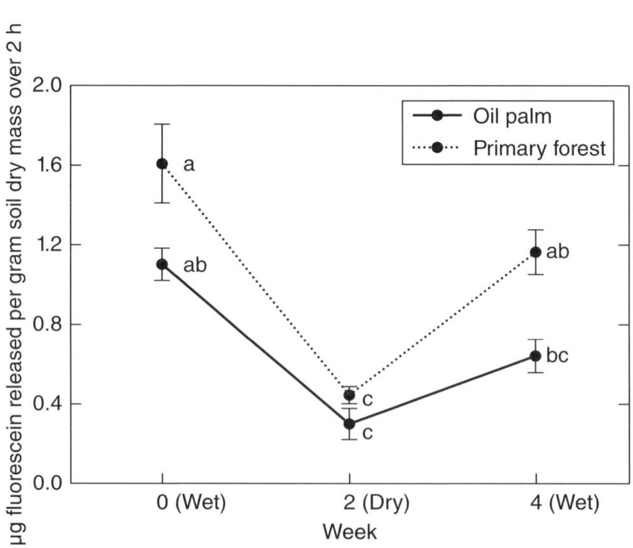

Fig. 4.2. Effects of drying and rewetting on fluorescein diacetate hydrolysis by soil microbial communities from microcosms of Malaysian primary forest and adjacent oil palm plantations (letters indicate significant differences at $P < 0.05$.

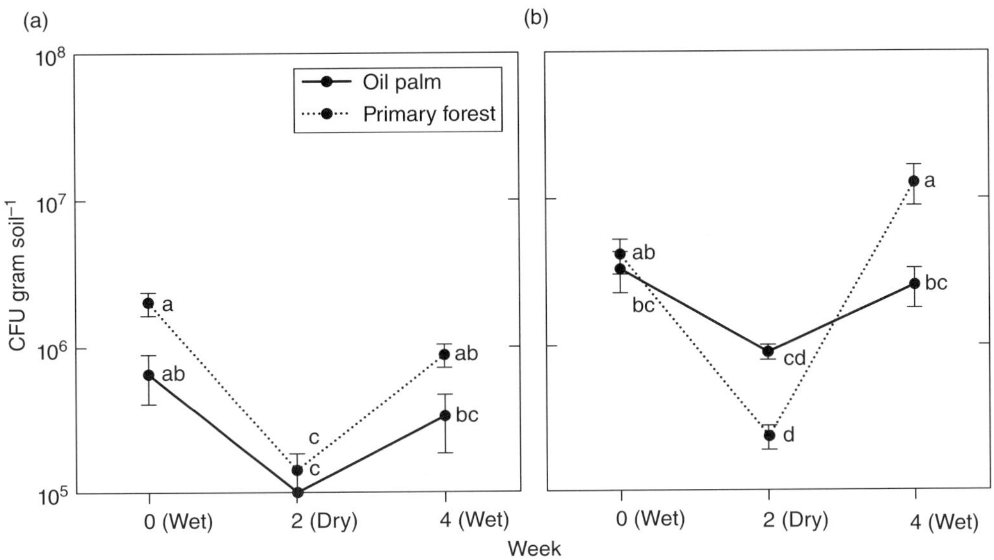

Fig. 4.3. Effects of drying and rewetting on soil (a) fungal and (b) bacterial populations in microcosms of Malaysian primary forest and adjacent oil palm plantations (letters indicate significant differences at $P < 0.05$).

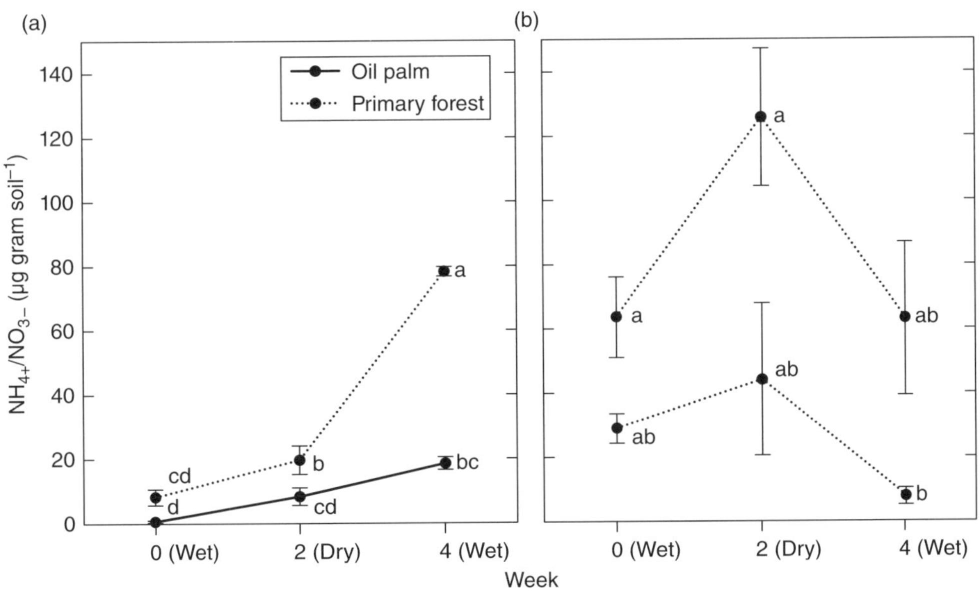

Fig. 4.4. Effects of drying and rewetting on soil (a) ammonium and (b) nitrate concentrations in microcosms of Malaysian primary forest and adjacent oil palm plantations (letters indicate significant differences at $P < 0.05$).

drought periods when rewetted (Table 4.1). Soil C and N were both higher in the primary forest soils; C did not change in response to the drying/rewetting treatment, whereas N was reduced when droughted and increased again following rewetting, with this pattern being more notable in the primary forest soils (Table 4.1). Extractable P was 16 times greater in the oil palm soils indicating fertilization with this element. It was reduced following drought to about 40% of initial

Table 4.1. Effects of drying and rewetting on soil chemical and enzymatic properties in microcosms of Malaysian primary forest (PF) and adjacent oil palm (OP) plantations.

	Land use	Week 0 Wet	Week 2 Dry	Week 4 Wet	Forest (*P*)	Time (*P*)	Forest × time (*P*)
pH	PF	5.66 ± 0.04[a]	3.21 ± 0.05[e]	4.13 ± 0.04[c]	<0.001	<0.001	<0.001
	OP	5.55 ± 0.08[a]	3.66 ± 0.01[d]	4.54 ± 0.03[b]			
C (%)	PF	2.03 ± 0.11[a]	1.80 ± 0.11[a]	1.82 ± 0.05[a]	<0.001	n.s.	n.s.
	OP	1.32 ± 0.03[b]	1.24 ± 0.01[b]	1.27 ± 0.04[b]			
N (%)	PF	0.21 ± 0.009[a]	0.14 ± 0.015[bc]	0.17 ± 0.005[ab]	<0.001	<0.001	n.s.
	OP	0.13 ± 0.003[cd]	0.10 ± 0.002[d]	0.11 ± 0.004[cd]			
P extractable	PF	3.15 ± 0.27[c]	3.12 ± 0.21[c]	2.03 ± 0.44[c]	<0.001	<0.001	<0.05
(mg kg^{-1})	OP	80.1 ± 7.3[a]	32.5 ± 3.7[b]	29.1 ± 4.7[b]			
K exchangeable	PF	0.12 ± 0.008[bc]	0.12 ± 0.006[c]	0.11 ± 0.009[c]	<0.05	n.s.	n.s.
($cmol_c$ kg^{-1})	OP	0.14 ± 0.004[ab]	0.13 ± 0.006[abc]	0.14 ± 0.004[a]			
Ca exchangeable	PF	0.10 ± 0.01[a]	0.12 ± 0.01[a]	0.11 ± 0.003[a]	<0.001	n.s.	n.s.
($cmol_c$ kg^{-1})	OP	0.46 ± 0.04[b]	0.51 ± 0.01[b]	0.50 ± 0.01[b]			
Mg exchangeable	PF	0.09 ± 0.05[e]	0.11 ± 0.003[cd]	0.10 ± 0.004[de]	<0.001	<0.001	n.s.
($cmol_c$ kg^{-1})	OP	0.12 ± 0.01[bc]	0.16 ± 0.004[a]	0.14 ± 0.004[ab]			
Phosphatase	PF	2.79 ± 0.43[ab]	2.88 ± 0.42[ab]	4.82 ± 1.53[a]	<0.01	<0.001	<0.01
(mM *p*-NPP produced g soil: 3 h incubation)	OP	0.50 ± 0.05[c]	1.68 ± 0.25[b]	2.37 ± 0.33[ab]			
Urease (μg NH_{4+}	PF	14.1 ± 0.75[a]	12.6 ± 2.29[a]	7.06 ± 1.47[ab]	<0.01	<0.05	n.s.
produced g soil: 5 h incubation)	OP	6.27 ± 1.38[b]	10.2 ± 0.76[ab]	5.48 ± 0.27[b]			

Note: letters indicate significant differences at $P < 0.05$.

values and remained at this lower level following rewetting in the oil palm soil; no changes were seen in the primary forest soil. Soil exchangeable K, Ca and Mg were all higher in the oil palm soils (Table 4.1); only Mg concentrations increased in response to drought and returned to pre-drought levels at the end of the experiment (Table 4.1).

Both phosphatase and urease were lower in the oil palm plantations but their response to the drying and rewetting was different (Table 4.1). Phosphatase activity in oil palm soils was lowest prior to the drought but increased during and following the drying, whereas there was no change in the primary forest soils (Table 4.1). Urease activity was idiosyncratic and patterns were not consistent or significantly different over time (Table 4.1).

4.4 Discussion

In common with many other studies, soil C and N were lower in the oil palm plantations when compared with the forest, but concentration of other nutrients did not differ. Lower soil C stocks are commonly found in soils under plantation crops (Verchot, 2010), including previous studies under oil palm plantations (Adachi *et al.*, 2006), due to increased rates of decomposition and reduced litter input. It cannot be confirmed whether the palms were fertilized or not, but the high values for extractable soil P make this highly likely; the increase in other cations may also be due to fertilization. There was a reduction in microbial biomass and microbial activity in oil palm soils compared to the primary forest soil (D'Angelo *et al.*, Chapter 3, this volume) as may be expected following the reduction in soil C during the process of land-use change (Dinesh *et al.*, Chapter 2, this volume).

In general, microbial activity was reduced in dry soils as expected (Manzoni *et al.*, 2012; Moyano *et al.*, 2013). However, when the patterns are unpicked a little further we can see some interesting and subtle patterns. When using fluorescein diacetate hydrolysis as a measure of microbial activity, there were indications that the primary forest soil microbial community functioning was more resilient to drought (i.e. recovered closer to pre-drought activity than the oil palm community). Similarly, Chaer *et al.* (2009) found microbial populations from primary forest soil to be more stable following a heat stress than those from adjacent agricultural land, with different microbial parameters either showing greater resistance or resilience in the primary forest soil. In contrast to the relative resilience of fungal populations, bacteria from the oil palm plantation soil appeared more resistant to the drought (i.e. populations did not change as much as in the primary forest community). This could be because the oil palm bacterial communities have adapted to an environment where drying and rewetting are more frequent due to a lower, more open canopy leading to a less humid environment (Luskin and Potts, 2011) and reduced soil moisture buffering capacity as there is less of an insulating litter layer (personal observation). In contrast to the bacteria from the oil palm plantation that exhibited a higher resistance to the drought by changing less in number, the bacteria in the primary forest soil exhibited a high resilience by increasing in number rapidly following rewetting after the drought. Data from fluorescein diacetate hydrolysis suggest that microbial communities from the primary forest are similarly resistant to drought to those of the oil palm plantation (i.e. show similar reductions in activity reduction) or perhaps less resistant (in the case of bacteria) but are more resilient (i.e. show closer recovery to initial activity). These overall results are somewhat in contrast to de Vries *et al.* (2012) who found more extensively managed grassland soils (i.e. fungal-dominated ecosystems) had a more resistant microbial community than wheat field soils (bacterial-dominated) when subjected to drought conditions.

Mineral N concentrations were greater in the primary forest soil than the oil palm soil. Ammonium and nitrate concentrations increased during the drought period. Ammonium increased further following the drought while nitrate concentrations then declined again (although these results were not significant for nitrate). This ammonification pulse was greater in primary forest soil than the oil palm soil. It is known that denitrification is high in moist tropical soils (Bai *et al.*, 2012) and is reduced by dry soil conditions. This may explain the accumulation of nitrate in the soils during the drying period that was then denitrified upon rewetting (Groffman and Teijde, 1988). Nitrifying bacteria appear to be particularly sensitive to drought conditions (Stark and Firestone, 1995) and this may explain the increase in ammonium concentrations during the drought conditions as the reduction in nitrifier activity during drying meant that additional ammonium released was not converted to nitrate.

These results compare favourably with those of Muhr *et al.* (2010) who showed a large increase in ammonium concentrations after drying and rewetting (*c.* ten-fold) as did Miller *et al.* (2005). In contrast, Fierer and Schimel (2002) did not see any effects of drying–rewetting on soil ammonium concentration. For nitrate, these results also compare with Fierer and Schimel (2002), who showed that nitrate concentrations were reduced by about 10–20% in their droughted treatments relative to their controls. Miller *et al.* (2005) also showed small reductions in nitrate concentrations, whereas Muhr *et al.* (2010) found an initial increase in nitrate after rewetting but that it was similar to pre-drying/rewetting values after about 20 days. Few studies have actually examined changes in mineral N concentrations throughout the wetting–drying cycle, but Muhr *et al.* (2010) showed an increase in ammonium and nitrate during the drying process, although this was not as pronounced as following the rewetting. However, there are differences between all of the above studies in the magnitude and timings of the differences in N concentrations, and the reasons for these differences are worthy of additional study but are likely to include soil mineralogy, length of drought and the ratio between ammonium and nitrate in addition to changes in land use.

The changes in soil pH with drying and rewetting were very large (greater than one pH unit). Reductions in soil pH with drying has been shown previously (Erich and Hoskins, 2011) although, in that case, the change in pH was considerably smaller (<0.5 pH units). Surprisingly, and in contrast, to other studies (e.g. Turner and Haygarth, 2000, 2003), the oil palm plantation soils showed a reduction in extractable P with drying. It is not clear why this is, but the addition of P as fertilizer may have led to increased sorption of P on soil surfaces with drying as reported by Haynes and Swift (1985).

Extracellular enzyme activity (of both phosphatase and urease) was lower in the oil palm plantation soils than the primary forest soils. Phosphatase activity was negatively correlated with the availability of P in the oil palm plantation soils; it was initially low when P concentrations were high and then decreased as P availability declined with the drying and rewetting. This result is in line with numerous other studies that have linked phosphatase activity to the availability of P in soil (Nannipieri *et al.*, 2011). The effects of the treatments on urease activity were somewhat idiosyncratic. This is maybe because urease does not appear to be sensitive to drought in tropical soils (Stark and Brearley, unpublished data) or it may reflect a process that is not yet understood. However, previous authors have found reductions in urease and phosphatase activities under drought conditions (Henry *et al.*, 2005; Sardans and Penuelas, 2005) although phosphatase is not always affected (Yavitt *et al.*, 2004). Furthermore, Zhao *et al.* (2010) found a reduction in urease activity with an increase in the number of drying–rewetting cycles. However, other authors have rarely considered the effects of drying and rewetting on microbial processes such as enzyme production and activity as the main areas of study are C (as CO_2 efflux) and mineral N fluxes.

4.5 Conclusions

This study has shown that conversion of Southeast Asian primary forest to oil palm plantation leads to changes in soil chemical and biological properties with general reductions in soil C and microbial activity in common with many other studies of tropical land-use change. It has also shown that disturbance to the soil system (in this case drying and rewetting) led to differential responses from the microbial communities in soils under the two land-use types. However, different populations and processes can show contrasting responses (e.g. fungi show similar responses in the two soil types, whereas bacteria showed quite different responses) and this is likely due to the soil properties and the environmental conditions as well as the structure of the microbial communities. These results therefore have implications for C, N and nutrient cycling and losses from the ecosystem. More broadly, changes in tropical land use and soil management have implications for biogeochemical cycling that will depend upon an improved understanding of both how microbial populations and their functioning are impacted by land-use changes and how this is differentially modulated by additional stresses and disturbances.

Acknowledgements

Thanks to David McKendry for analytical support and colleagues in Malaysia (especially Lee Su See and members of her laboratory) for assistance.

References

Adachi, M., Bekku, Y.S., Kadir, W.R., Okuda, T. and Koizumi, H. (2006) Differences in soil respiration between different tropical ecosystems. *Applied Soil Ecology* 34, 258–265.

Bai, E. Houlton, B.Z. and Wang, Y.P. (2012) Isotopic identification of nitrogen hotspots across natural terrestrial ecosystems. *Biogeosciences* 9, 3287–3304.

Billings, S.A. and Phillips, N. (2011) Forest biogeochemistry and drought. In: Levia, D.F., Carlyle-Moses, D.E. and Tanaka, T. (eds) *Forest Hydrology and Biogeochemistry: Synthesis of Past Research and Future Directions*, Ecological Studies 216. Springer-Verlag, Berlin, pp. 581–597.

Borken, W. and Matzner, E. (2009) Reappraisal of drying and wetting effects on C and N mineralization and fluxes in soils. *Global Change Biology* 15, 808–824.

Bossio, D.A., Girvan, M.S., Verchot, L., Bullimore, J., Borelli, T., Albrecht, A. *et al.* (2005) Soil microbial community response to land use change in a agricultural landscape of western Kenya. *Microbial Ecology* 49, 50–62.

Carlson, K.M., Curran, L.M., Asner, G.P., McDonald Pittman, A., Trigg, S.N. and Adeney, J.M. (2012) Carbon emissions from forest conversion by Kalimantan oil palm plantations. *Nature Climate Change* 3, 283–287.

Chaer, G., Fernandes, M., Myrold D. and Bottomley, P. (2009) Comparative resistance and resilience of soil microbial communities and enzyme activities in adjacent native forest and agricultural soils. *Microbial Ecology* 58, 414–424.

de Vries, F.T., Liiri, M.E., Bjørnlund, L., Bowker, M.A., Christensen, S., Setälä, H. and Bardgett, R.D. (2012) Land use alters the resistance and resilience of soil food webs to drought. *Nature Climate Change* 2, 276–280.

Erich, M.S. and Hoskins, B.R. (2011) Effects of soil drying on soil pH and nutrient extractability. *Communications in Soil Science and Plant Analysis* 42, 1167–1176.

Fierer, N. and Schimel, J.P. (2002) Effects of drying–rewetting frequency on soil carbon and nitrogen transformations. *Soil Biology and Biochemistry* 34, 777–787.

Fierer, N. and Schimel, J.P. (2003) A proposed mechanism for the pulse in CO_2 production commonly observed following the rapid rewetting of a dry soil. *Soil Science Society of America Journal* 67, 798–805.

Fitzherbert, E.B., Struebig, M.J., Morel, A., Danielsen, F., Brühl, C.A., Donald, P.F. and Phalan, B. (2008). How will oil palm expansion affect biodiversity? *Trends in Ecology and Evolution* 23, 538–545.

Gordon, H., Haygarth, P.M. and Bardgett, R.D. (2008) Drying and rewetting effects on soil microbial community composition and nutrient leaching. *Soil Biology and Biochemistry* 40, 302–311.

Groffman, P.M. and Teidje, J.M. (1988) Denitrification hysteresis during wetting and drying cycles in soil. *Soil Science Society of America Journal* 52, 1626–1629.

Harris, R.F. (1980) Effect of water potential on microbial growth and activity. In: Parr, F., Gardner, W.R. and Elliott, L.F. (eds) *Water Potential Relations in Soil Microbiology*. Soil Science Society of America, Madison, Wisconsin, pp. 23–96.

Haynes, R.J. and Swift, R.S. (1985) Effects of air-drying on the adsorption and desorption of phosphate and levels of extractable phosphate in a group of New Zealand soils. *Geoderma* 35, 145–157.

Henry, H.A.L., Juarez, J.D., Field, C.B. and Vitousek, P.M. (2005) Interactive effects of elevated CO_2, N deposition and climate change on extracellular enzyme activity and soil density fractionation in a California annual grassland. *Global Change Biology* 11, 1808–1815.

Huygens, D., Schouppe, J., Roobroeck, D., Alvarez, M., Balocchi, O., Valenzuela, E., *et al.* (2011) Drying–rewetting effects on N cycling in grassland soils of varying microbial community composition and management intensity in south central Chile. *Applied Soil Ecology* 48, 270–279.

Johnson, D., Leake, J.R., Lee, J.A. and Campbell, C.D. (1998) Changes in soil microbial biomass and microbial activities in response to seven years' simulated pollutant nitrogen deposition on a heathland and two grasslands. *Environmental Pollution* 103, 239–250.

Kim, D.-G., Vargas, R., Bond-Lamberty, B. and Turetsky, M.R. (2012) Effects of soil rewetting and thawing on soil gas fluxes: A review of current literature and suggestions for future research. *Biogeosciences* 9, 2459–2483.

Koh, L.P., Miettinen, J., Liew, S.C. and Ghazoul, J. (2011) Remotely sensed evidence of tropical peatland conversion to oil palm. *Proceedings of the National Academy of Science of the USA* 108, 5127–5132.

Kumagai, T., Kanamori, H. and Yasunari, T. (2013) Deforestation-induced reduction in rainfall. *Hydrological Processes* 27, 3811–3814.

Luskin, M.S. and Potts, M.D. (2011) Microclimate and habitat heterogeneity through the oil palm lifecycle. *Basic and Applied Ecology* 12, 540–551.

McGuire, K.L., D'Angelo H., Brearley F.Q., Gedallovich S.M., Babar N., Yang N. *et al.* (2015) Responses of soil fungi to logging and oil palm agriculture in Southeast Asian tropical forests. *Microbial Ecology* 69, 733–747.

Manzoni, S., Schimel, J.P. and Porporato, A. (2012) Responses of soil microbial communities to water stress: Results from a meta-analysis. *Ecology* 93, 930–938.

Madigan, M.T. and Martinko J.M. (2006) *Brock Biology of Microorganisms* (11th edition). Pearson Prentice Hall, New Jersey.

Manokaran, N. Quah, E.S., Ashton, P.S., LaFrankie, J.V., Noor, N.S.M., Ahmad, W.M.S.W. and Okuda, T. (2004) Pasoh Forest Dynamics Plot, peninsular Malaysia. In: Losos, E.C. and Leigh Jr, E.G. (eds) *Tropical Forest Diversity and Dynamism*. University of Chicago Press, Chicago, Illinois, pp. 585–598.

Margesin, R. (2005) Determination of enzyme activities in contaminated soils. In: Margesin, R. and Schinner, F. (eds) *Manual of Soil Analysis: Monitoring and Assessing Soil Bioremediation*. Springer-Verlag, Berlin, pp. 309–320.

Miller, A.E., Schimel, J.P., Meixner, T., Sickman, J.O. and Melack, J.M. (2005) Episodic rewetting enhances carbon and nitrogen release from chaparral soils. *Soil Biology and Biochemistry* 37, 2195–2204.

Moyano, F.E., Manzoni, S. and Chenu, C. (2013) Responses of soil heterotrophic respiration to moisture availability: An exploration of processes and models. *Soil Biology and Biochemistry* 59, 72–85.

Muhr, J., Franke, J. and Borken, W. (2010) Drying-rewetting events reduce C and N losses from a Norway spruce forest floor. *Soil Biology and Biochemistry* 42, 1303–1312.

Nannipieri, P., Giagnoni, L., Landi, L. and Renella, G. (2011) Role of phosphatase enzymes in soil. In: Bünemann, E.K., Oberson, A. and Frossard, E. (eds) *Phosphorus in Action: Biological Processes in Soil Phosphorus Cycling*. Springer-Verlag, Berlin, pp. 215–243.

Okuda, T., Manokaran, N., Matsumoto, Y., Niiyama, K., Thomas, S.C. and Ashton, P.S. (2003) *Pasoh: Ecology of a Lowland Rain Forest in Southeast Asia*. Springer-Verlag, Tokyo.

Sardans, J. and Peñuelas, J. (2005) Drought decreases soil enzyme activity in a Mediterranean *Quercus ilex* L. forest. *Soil Biology and Biochemistry* 37, 455–461.

Schaller, K. (1993) *Praktikum zur Bodenkunde und Pflanzenernährung*. Geisenheimer Berichte, Geisenheim, Germany.

Stark, J.M. and Firestone, M.K. (1995) Mechanisms for soil moisture effects on activity of nitrifying bacteria. *Applied and Environmental Microbiology* 61, 218–221.

Timmermann, A., Obernuber, J., Bacher, A., Esch, M., Latif, M. and Roeckner, E. (1999) Increased El Niño frequency in a climate model forced by future greenhouse warming. *Nature* 398, 694–697.

Turner, B.L. and Haygarth, P.M. (2000) Phosphorus solubilization in rewetted soils. *Nature* 411, 258.

Turner, B.L. and Haygarth, P.M. (2003) Changes in bicarbonate-extractable inorganic and organic phosphorus by drying pasture soils. *Soil Science Society of America Journal* 67, 344–350.

Turner, E.C., Snaddon, J.L., Ewers, R.M., Fayle, T.M. and Foster, W.A. (2011) The impact of oil palm expansion on environmental change: putting conservation research in context. In: Dos Santos Bernardes M.A. (ed.) *Environmental Impact of Biofuels*. InTech, Rijeka, Croatia, pp. 19–40. Available at: www.intechopen.com/books/environmental-impact-of-biofuels/the-impact-of-oil-palmexpansion-on-environmental-change-putting-conservation-research-in-context.pdf (accessed 6 March 2015).

Verchot, L.V. (2010) Impacts of forest conversion to agriculture on microbial communities and microbial function. In: Dion, P. (ed.) *Soil Biology and Agriculture in the Tropics*. Springer-Verlag, Berlin, pp. 45–63.

Yavitt, J.B., Wright, S.J. and Wieder, R.K. (2004) Seasonal drought and dry-season irrigation influence leaf-litter nutrients and soil enzymes in a moist, lowland forest in Panama. *Austral Ecology* 29, 177–188.

Xiang, S.-R., Doyle, A., Holden, P.A. and Schimel, J.P. (2008) Drying and rewetting effects on C and N mineralization and microbial activity in surface and subsurface California grassland soils. *Soil Biology and Biochemistry* 40, 2281–2289.

Zhao, B., Chen, J., Zhang, J. and Qin, S. (2010) Soil microbial biomass and activity response to repeated drying-rewetting cycles along a soil fertility gradient modified by long-term fertilization management practices. *Geoderma* 160, 218–224.

5 Impact of Land-use Changes in the Amazon on Bacterial Diversity, Composition and Distribution

Lucas W. Mendes,[1,2] Acácio A. Navarrete,[1,2] Clóvis D. Borges,[1] Eiko E. Kuramae[2] and Siu Mui Tsai[1]*

[1]Cell and Molecular Biology Laboratory, Center for Nuclear Energy in Agriculture CENA, University of São Paulo, São Paulo, Brazil; [2]Microbial Ecology Department, Netherlands Institute of Ecology NIOO-KNAW, Wageningen, the Netherlands.

5.1 Introduction

Soil-living microorganisms represent the largest biodiversity pool on Earth, with more than 10^{30} microbial cells and estimates of 10^4 to 10^6 species per gram of soil (Whitman *et al.*, 1998; Torsvik *et al.*, 2002; Roesch *et al.*, 2007). With their enormous numbers, large biomass and involvement in numerous key biogeochemical functions, soil microbial communities hold a central place in terrestrial ecosystems. Soil microbial communities carry out essential ecosystem functions (Bardgett *et al.*, 2008), including nutrient cycling, facilitating plant nutrition, disease suppression, water purification and biological attenuation of pollutants. Nowhere are soil microbial communities likely to be more complex than under tropical rain forests, which house the majority of plant diversity on Earth (Dirzo and Raven, 2003; Kreft and Jetz, 2007).

Biodiversity loss is happening rapidly in the tropics, much of which is due to expanding agricultural activities (Phalan *et al.*, 2013; Wilcove *et al.*, 2013). The process of land conversion and agricultural intensification is one of the most important causes of biodiversity loss, with negative effects both on the environment and on the sustainability of agricultural production. Deforestation and agricultural intensification are common land-use changes in tropical regions such as South-east Asia and the Amazon, which are hotspots of biodiversity (Erwin, 1983; Myers *et al.*, 2000). Due to anthropogenic activities and dramatic changes in land use, these regions have the highest deforestation rates (Sodhi *et al.*, 2004; Fearnside, 2006; Hansen *et al.*, 2008), which has impacted upon their rich and unique biodiversity, in particular aboveground diversity (Brook *et al.*, 2003; Sodhi *et al.*, 2004; Hoffmann *et al.*, 2010), for example, 5843 km^2 of Amazonia was deforested in 2013 (INPE, 2013). Therefore, increased attention has recently been paid to microbial communities resident in soils converted from forest into agricultural fields, which is needed to increase our knowledge about the effects of land-use changes on the microbial community structure, composition and functioning. Despite the increased appreciation of belowground microbial diversity in the tropics, little is known about bacterial taxa responses and much less about the functional responses to alterations in soil chemical properties and fertility following deforestation and agricultural management of tropical forest soils.

* Corresponding author: tsai@cena.usp.br

Deforestation and land degradation in the tropics has resulted in increased greenhouse gas emissions to the atmosphere, loss of biodiversity, a decrease in carbon (C) sequestration capacity and changes in many soil properties (Fearnside, 2005). Previous studies have shown that conversion of Amazonian forest to pasture causes an increase in soil pH, ammonium concentrations and bulk density, as well as a decrease in nitrate concentrations and soil porosity (Piccolo *et al.*, 1994; Reiners *et al.*, 1994). The consequent reduction in the diversity of the soil microbial community may cause a loss of ecosystem function, reducing the ability of ecosystems to withstand periods of stress and leading to undesirable environmental effects. The factors that drive the structure and composition of soil microbial communities are not well understood (Fierer and Lennon, 2011), and there is evidence that well-established patterns of plant diversity are not necessarily predictive of those for microorganisms, despite their obvious trophic linkages (Fierer *et al.*, 2011).

Because of the increasing role of large-scale agriculture in the Amazon and expansion of cropland into areas previously covered by forest, in this chapter we discuss the effects that land-use changes have on soil microbial diversity, composition and distribution in Amazonian tropical soils. We examine the effects of land-use change on aspects of the soil ecosystem from both a chemical and biological perspective.

5.2 Assessment of Soil Quality Based on Microbial Community Structure and Function in the Tropics

Classical microbiological methods combined with new molecular tools have proven to be a robust approach to measure soil parameters. Soil microbial biomass carbon (SMB-C), metabolic quotient (qCO_2), and total soil organic carbon (SOC) are widely accepted as indicators of soil quality, particularly when used to compare soils under different agricultural managements (Kaschuk *et al.*, 2010). Over three decades studying microbial biomass in soil, the results have shown that this is a sensitive parameter for detection of the effects of agricultural management and alterations in soil organic matter (SOM) quality and content (Carter, 2001; Roscoe *et al.*, 2006; Dinesh *et al.*, Chapter 2, this volume; D'Angelo *et al.*, Chapter 3, this volume) because it responds rapidly to environmental changes, often much earlier than soil physical and chemical parameters, including SOC and even crop productivity (Roscoe *et al.*, 2006; Franchini *et al.*, 2007; Hungria *et al.*, 2009).

Soil microbial biomass (SMB) has been defined by the portion of SOM constituted by archaea, bacteria and eukaryotes, excluding roots and animals smaller than 5×10^3 µm^3 (Jenkinson and Ladd, 1981), and it has been quantified through two different methods: fumigation–incubation (Jenkinson and Powlson, 1976), and fumigation–extraction (Vance *et al.*, 1987). The eco physiological indices (metabolic quotients) can be generated by basing physiological performance (respiration, C uptake and growth/death) on the total microbial biomass per unit of time (Anderson, 2003). In this sense, for example, an 'efficient' SMB would have a lower respiration rate compared to an 'ineffective' SMB.

Studies encompassing the Amazon biome have not determined indices such as qCO_2 for agricultural soils. Thus, we show results from agricultural soils in the Amazon and also in Cerrado, Atlantic, Caatinga and Pampa 'biomes' (Table 5.1). Low qCO_2 values indicate a high quality substrate used by microorganisms or a low microbial maintenance requirement.

In general, agricultural soils showed a decrease in SMB-C when compared with forests and in most, but not all, 'biomes', there was lower SOC in agricultural soils compared to forest soils (Table 5.1). The patterns for qCO_2 were not always consistent (Table 5.1) but we would expect the lowest qCO_2 values in forest soils because forest systems show lower disturbance to the soil microbial community. However, there is a lack of consistent information, especially from the tropics, about agricultural management effects on microbial community functioning. A major effect of increased SMB-C is an increase in C immobilization that is slowly released in an organic form according to plants' needs (Roscoe *et al.*, 2006). Thus, microbial properties allied to SOC content can be used to evaluate the sustainability of agricultural production. These properties are described as biological indicators capable of detecting changes in soil quality and its biological properties (Roscoe *et al.*, 2006).

Table 5.1. Soil microbial biomass carbon (SMB-C), metabolic quotient (qCO_2) and soil organic carbon (SOC) under different agricultural management regimes in five Brazilian 'biomes'.

'Biome'	Management	SMB-C (mg C kg^{-1} soil)	qCO_2 (mg C-CO_2 g^{-1} SMB-C h^{-1})	SOC g kg^{-1}	Reference
Amazon	Forest	503.4	1.60–1.90	17.3	Matoso *et al.*, 2012
	Sugarcane	nd	nd	nd	–
	Livestock	385.1	nd	14.7	Matoso *et al.*, 2012
	Soybean	290.1	nd	12.7	Matoso *et al.*, 2012
Cerrado	Forest	486.0–1330	0.04–1.41	15.8–18.0	Cardoso *et al.*, 2009; Frazão *et al.*, 2010
	Sugarcane	450.0	1.50	nd	Portilho *et al.*, 2011
	Livestock	113.9–500.0	1.90	4.6–19.0	D'Andréa *et al.*, 2002; Frazão *et al.*, 2010
	Soybean	387.8	1.61	18.6	Lourente *et al.*, 2010
Atlantic	Forest	444.0–1122	0.16–2.70	20.1–22.6	Niemeyer *et al.*, 2012; Souza *et al.*, 2012
	Sugarcane	328.4–700.0	nd	12.4–14.0	Pupin and Nahas 2011; Souza *et al.*, 2012
	Livestock	476.0–550.0	1.40–2.20	26.9	Ndaw *et al.*, 2009; Souza *et al.*, 2010
	Soybean	257.4–303.0	0.72	25.4	Babujia *et al.*, 2010; Borges *et al.*, 2012
Caatinga	Forest	99.0–500.0	0.44–1.50	9.8–25.0	Nunes *et al.*, 2012; Xavier *et al.*, 2006
	Sugarcane	125.0	nd	nd	Silva *et al.*, 2012
	Livestock	141.0	0.40	10.1	Xavier *et al.*, 2006
	Soybean	126.7–270.0	0.84	13.9–18.0	Leite *et al.*, 2010; Pragana *et al.*, 2012
Pampas	Forest	533.5	0.11	28.8	Dieckow *et al.*, 2009; Lupatini *et al.*, 2012
	Sugarcane	nd	nd	nd	–
	Livestock	280.3	0.08	22.6	Dieckow *et al.*, 2009; Lupatini *et al.*, 2012
	Soybean	153.3–233.4	0.21	11.5	Ferreira *et al.*, 2010; Lupatini *et al.*, 2012

Note: The values of SMB-C were obtained by the fumigation-extraction method with k_C and k_{EC} standardized at 0.40. nd, not determined.

5.3 Studies on Microbial Community Structure and Function in Amazonian Soils

In the past two decades, molecular tools have been allied to the classical microbiological methods to provide new insights into the microbial ecology of soils. Powerful molecular tools are becoming available to allow the examination of microbial communities through analysis of microbial DNA and RNA. For example, molecular approaches based on 16S ribosomal RNA (rRNA) sequence analysis allow direct investigation of the community structure, diversity and phylogeny of microorganisms in any environment, while quantification of the individual species of microorganisms or entire microbial communities may be addressed by nucleic acid hybridization techniques. Molecular sequence analysis of community DNA allows a new perception of the microbial diversity and function in a broad range of environments. The sequencing approach is a powerful tool for the study of microbial communities inhabiting soil and can be useful to predict changes in soil properties and quality. Next-generation sequencing technologies have allowed microbial ecologists to advance from analysing a few hundred sequences to millions of sequences per study. With the advance of such technologies, the length of the sequenced DNA fragment has increased, although they are shorter than the desirable length or the read length obtained from traditional Sanger sequencing (~1000 base pairs) (Luo *et al.*, 2012).

The assessment of microbial diversity will be advanced by the development of new technologies that answer some key questions about the 'who, what, when, where, why, and how' of microbial communities (Knight *et al.*, 2012). The rapid advance of sequence technologies allied to bioinformatics tools is increasing the possibility of massive studies on microbial ecology for a deeper comprehension of the composition and function that soil microorganisms play in a wide range of ecosystems. The new information available will be useful for a better understanding of soil quality and improve the sustainable use of ecosystems. Despite new technologies being used to better understand microbial communities in soil, there is a lack of data addressing this issue in tropical soils.

In this section, we present studies from Amazonian soils in order to illustrate how the assessment of soil microbial communities can contribute to our comprehension of soil management and its sustainable use in this tropical biome. The first description of the microbial diversity in Amazonian soils was made by Borneman and Triplett (1997) who used a sequencing approach to analyse sequences from forests and pastures. They were able to demonstrate the immense microbial diversity contained within the soils and differences in composition between forest and pasture. In a more recent study, Jesus *et al.* (2009) showed that changes in land use alter the structure of soil bacterial communities in the Amazon region. Phylogenetic analysis revealed that the groups *Actinobacteria*, *Bacteroidetes*, *Firmicutes* and *Proteobacteria* responded to the different land uses. The phylum *Bacteroidetes* were mostly found in crop and pasture soils, while *Firmicutes* were found in primary and secondary forests. In a finer taxonomic analysis, some sequences were indicators of environmental conditions, i.e. *Arthrobacter* and *Bradyrhizobium* were widespread among land-use systems, while *Acidobacteria* (subgroup I), *Sphingobacteriales* and *Sphingomonadaceae* were related to pasture sites, and *Acidobacteria* (subgroup I) and *Rhodoplanes* were related to forests. In a study in maize fields in the southern region of Brazil, Roesch *et al.* (2007) identified the classes α-*Proteobacteria* and β-*Proteobacteria* as being the most abundant in maize rhizosphere soils.

Navarrete *et al.* (2010) conducted a study on soil microbial ecology in the basin of Alto Solimões in the western Amazon. The study assessed the effects of land-use systems on microbial community structure in soils of different landscape types in the Brazilian Amazon that were previously covered with native tropical rain forest. The area studied encompassed agricultural systems of indigenous people primarily based on slash-and-burn, annual crops in shifting cultivation and a long fallow period. Soil samples were collected during 2 years in areas characterized by tropical rain forest, semi-permanent manioc cultivation, pasture and secondary forest at advanced stages of regeneration. The archaeal, bacterial and fungal communities were analysed by automated ribosomal intergenic spacer analysis (ARISA), and the results are represented graphically in Fig. 5.1. Based on ARISA data, soil archaeal, bacterial and fungal communities are heterogeneous and each capable of responding differently to environmental characteristics. ARISA data evidenced considerable difference in structure existing between microbial communities in forest and agricultural soils.

By observing the organization of the microbial community in relation to soil attributes, we can begin to generate and test hypotheses regarding the rules that govern the distribution of microorganisms in the different land-use systems. Navarrete *et al.* (2010) correlated the archaeal, bacterial and fungal community profiles obtained by the ARISA technique with soil chemical attributes, and their results show that pasture microbial communities were clearly separated from other land-use systems and were related to higher iron (Fe) concentrations detected in soil samples, for example.

Functional traits are also an important approach to defining microbiological parameters for monitoring soil quality in land-use systems. Navarrete *et al.* (2011) studied the composition of ammonia oxidizing archaea (*amo*A – ammonia monooxygenase gene) in tropical soils under primary and secondary forests, agriculture and pasture. Ammonia oxidation is the first step of

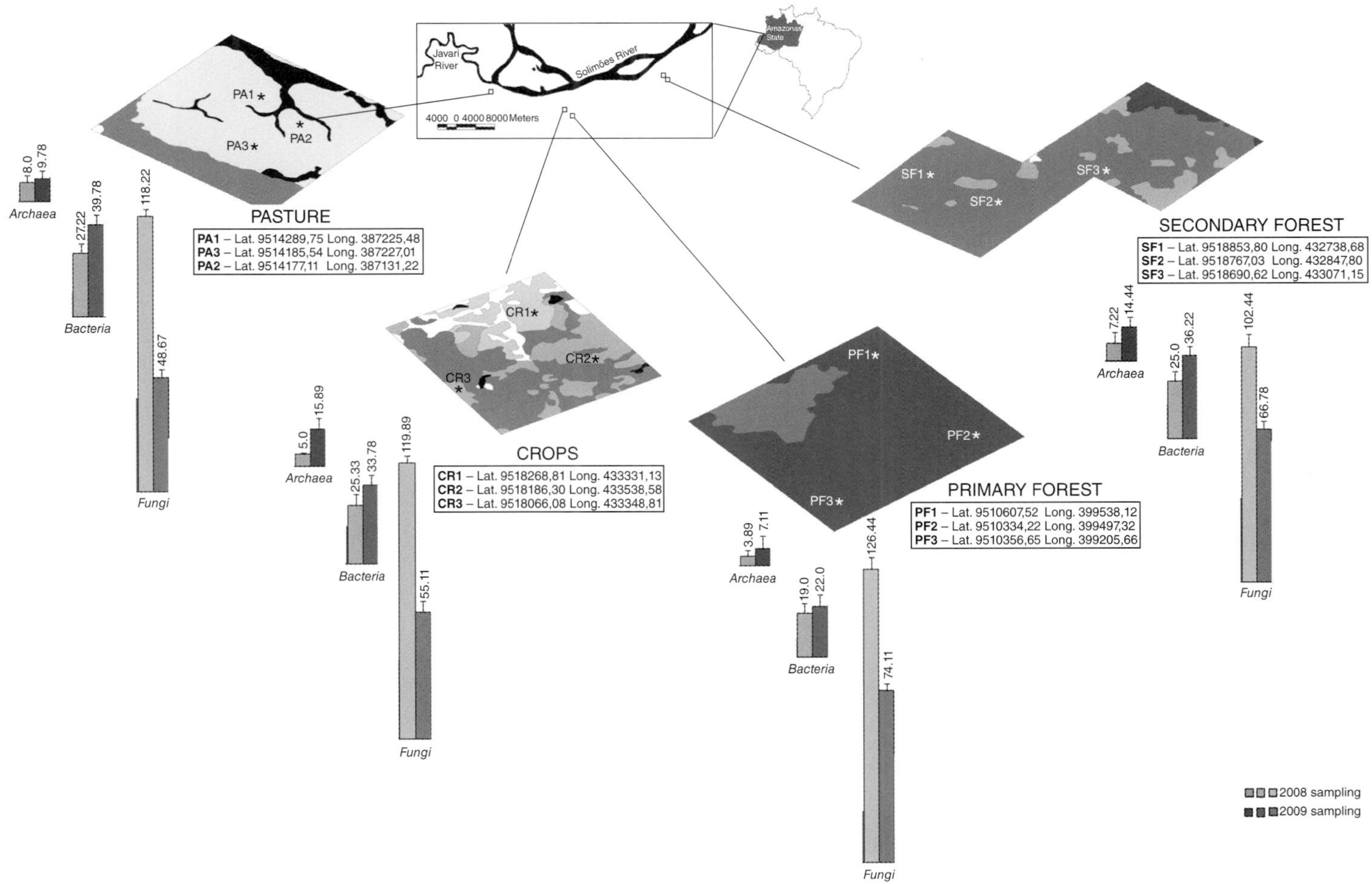

Fig. 5.1. Bacterial operational taxonomic unit richness based on intergenic spacer region length polymorphisms of the rRNA gene detected using automated ribosomal intergenic spacer analysis (ARISA) at sites located in an Amazonian landscape (Benjamin Constant municipality, Amazonas State, Brazil, along the Solimões River). Soil samples were collected at different points (*), at a depth of 0–20 cm from sites representing primary tropical rain forest, traditional crop systems of the indigenous people, pasture and secondary forest over a 2-year sampling period (March 2008 and January 2009). The richness values were calculated based on triplicate PCR products from the soil samples. (Adapted from Navarrete *et al.*, 2010.)

nitrification, an essential step in the global nitrogen (N) cycle. The study showed an increase in ammonia-oxidizing archaea richness in pasture soils, which could be related to N inputs through urine deposition by grazing animals. In another study, Taketani and Tsai (2010) explored the Amazonian Dark Earth (ADE) soils and found that soils subjected to agriculture displayed a higher number of *amo*A genes. In soils, nitrification can result in N loss from an ecosystem, thus knowledge about the microbial communities involved in nutrient cycles in soils can be useful for soil quality monitoring.

The analysis of functional diversity can also provide information on how microorganisms may influence the fertility of soils. Biodegradation due to bacterial activity is one of the most important processes occurring in soils and involves genes that may develop key roles in nutrient cycles. Germano *et al.* (2012) studied the diversity of the *ARHD* gene (aromatic ring-hydroxylating dioxygenase), which is responsible for the initial degradation of aromatic compounds. They found that secondary forest contained more unique operational protein clusters than agricultural sites, suggesting an effect of the management on the bacterial groups related to the C cycle, which can lead to a loss of important species that participate in this cycle in agriculture. In another study in Amazonian soils, Brossi *et al.* (2014) evaluated the diversity of the bacterial catabolic *BPH* gene (biphenyl dioxigenase), which is involved in aromatic hydrocarbon degradation. They found that soil chemical properties, such as high SOC, cation exchange capacity and more alkaline pH, were significantly correlated with the structure of BPH communities.

The first soil microbial ecology study using the 454 pyrosequencing approach was made by Roesch and collaborators (2007). They assessed the bacterial diversity in four soils across a large transect of the western hemisphere, collecting samples in Brazil, Florida, Illinois and Canada. The bacterial diversity of the forest soils was phylum-rich compared to agricultural soils, which were species-rich but phylum-poor. In a more recent study using pyrosequencing (Navarrete *et al.*, 2010), the researchers studied the bacterial diversity in the Brazilian ADE. Amazonian Dark Earth or *Terra Preta de Índio* of prehistoric origin is differentiated from surrounding soils by its darker colour, higher organic matter content, less acidic pH, greater total phosphorus (P) concentration, greater exchangeable calcium (Ca) and magnesium (Mg) and increased minor element concentrations (more information in Glaser *et al.*, 2001). The high fertility of these soils has been attributed to the high concentration of black carbon, which retains nutrients and water and increases the pH. The presence of black carbon in soils has also been shown to increase C storage by reducing C emissions (Marris, 2006).

Navarrete *et al.* (2010) studied these soils, comparing the microbial community of the bulk soil to the black carbon (BC). The pyrosequencing data indicate that, although the BC hosted lower richness of bacterial species than the ADE bulk soil, it still has a very high richness of unique (operational taxonomic units OTUs) (Fig. 5.2). On the other hand, from the total OTUs found in BC, 41% were not present in the ADE bulk soil, suggesting that this environment may be the substrate for specific microbial processes. These data bring new perspectives regarding the opportunity for reconstruction of tropical and degraded soils by contemporary human activities. *Proteobacteria* represented the dominant phylum in ADE soil, including representatives of the class α, β, γ and δ of *Proteobacteria* (Fig. 5.3). In contrast, *Acidobacteria* were the prominent phylum in BC with 21% of the bacterial sequences. The second most abundant phyla in ADE and BC were *Acidobacteria* (10%) and *Proteobacteria* (15%), respectively. Other prominent phyla in ADE were the *Actinobacteria*, *Planctomycetes* and *Verrucomicrobia*, which represented 8%, 7% and 4% of total sequences, respectively. In BC, the *Actinobacteria* were represented by more than 12% of the sequences;

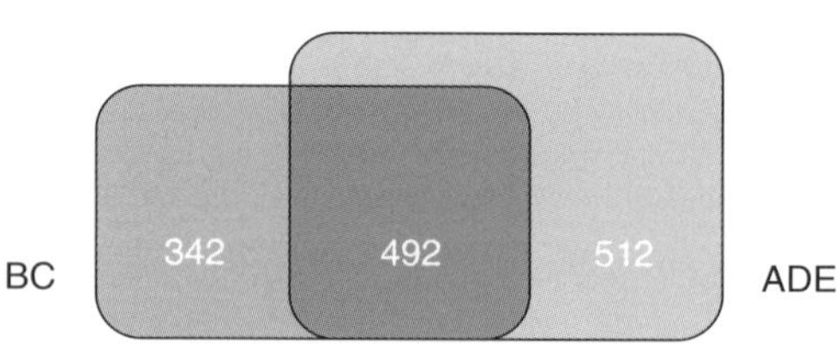

Fig. 5.2. Venn diagram of the bacterial operational taxonomic units (OTUs) based on the 16S rRNA gene analysed by pyrosequencing (454 Roche) found exclusively in Amazonian Dark Earths (ADE; 512), in black carbon (BC; 342) and the OTUs in held in common (492). (Adapted from Navarrete *et al.*, 2010.)

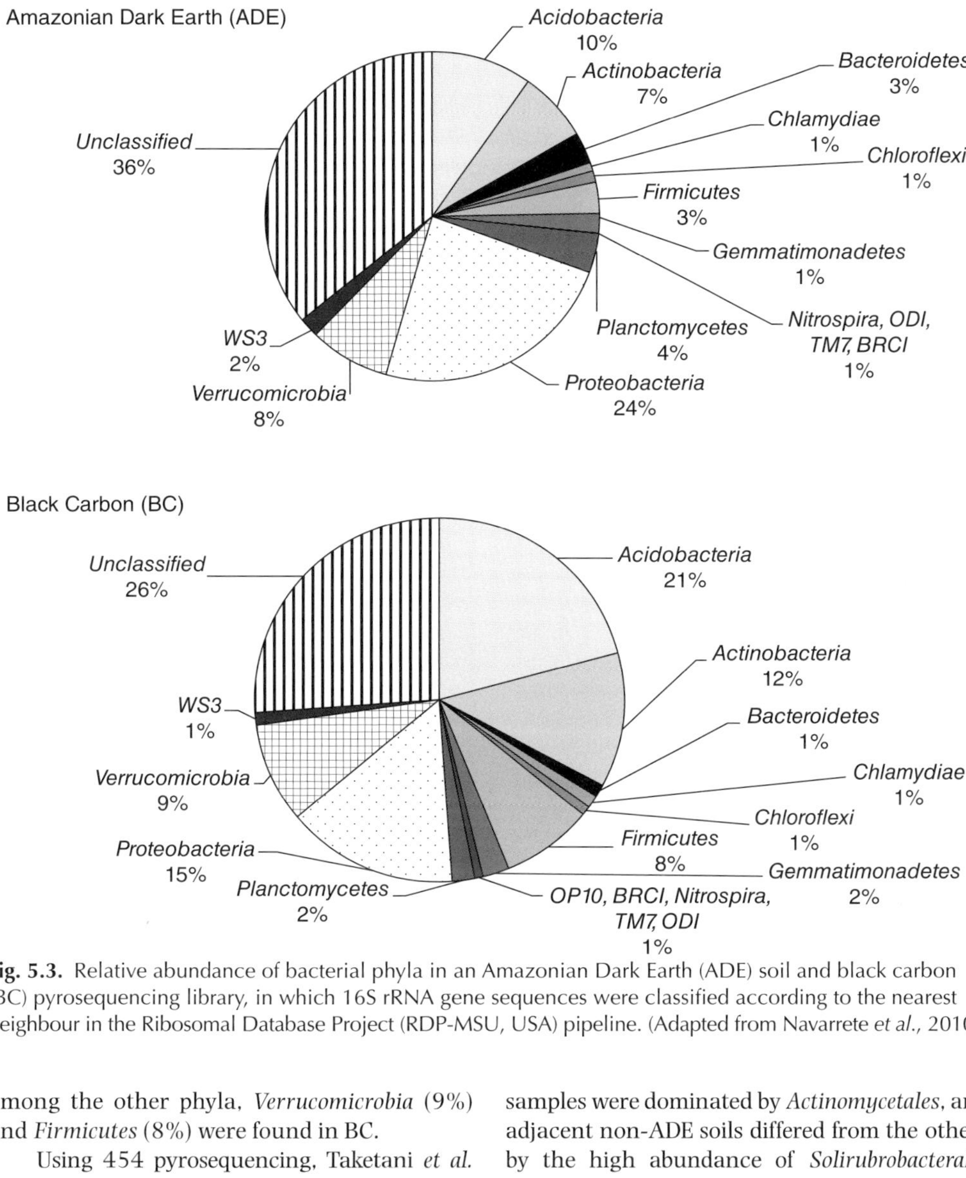

Fig. 5.3. Relative abundance of bacterial phyla in an Amazonian Dark Earth (ADE) soil and black carbon (BC) pyrosequencing library, in which 16S rRNA gene sequences were classified according to the nearest neighbour in the Ribosomal Database Project (RDP-MSU, USA) pipeline. (Adapted from Navarrete *et al.*, 2010.)

among the other phyla, *Verrucomicrobia* (9%) and *Firmicutes* (8%) were found in BC.

Using 454 pyrosequencing, Taketani *et al.* (2013) studied two ADE sites, one cultivated with manioc and another with secondary forest cover, with samples of BC, ADE and adjacent soils. The authors found that the most abundant bacterial phyla in all samples were *Acidobacteria*, *Actinobacteria*, *Proteobacteria* and *Verrucomicrobia*. Class composition varied among samples, highlighting the differences between BC, ADE and adjacent (non-ADE) soil. The soil properties had a stronger selective effect on the community composition than vegetation cover. Focusing on the two most dominant phyla, within the *Actinobacteria* all samples were dominated by *Actinomycetales*, and adjacent non-ADE soils differed from the others by the high abundance of *Solirubrobacterales* (>30%). Within the *Acidobacteria* phylum the adjacent soils were dominated by the Gp1, Gp2 and Gp3 classes, while BC and ADE were dominated by Gp5 and Gp6. Although the ADE soils are found in patches within the adjacent Ultisols, the contrasting characteristics found between them led to different microbial communities inhabiting each environment. In brief, the results suggest that the modifications made to these soils by ancient indigenous populations have affected soil physical, chemical and microbiological properties, which have acted together to create this fertile Amazonian soil. Amazonian Dark Earth has been conceived as a major example of sustainable agriculture because, since their initial description, these soils have been recognized as one of the most fertile soils known, even after extensive crop use. Their potential surpasses the use for the sustainable growing of food and fuel, and also has implications for C sequestration. Therefore, understanding how the microbial community is structured in such environments is key to a better management of this resource.

Because of the increasing role of large-scale agriculture in the Amazon and expansion of cropland (mainly soy) into areas previously covered by forest, Mendes *et al.* (2014) conducted a study in order to obtain better insight into the ecological process of bacterial community selection and assembly in the soybean rhizosphere in Amazon soils recently converted into cultivation (1 and 5 years after conversion). The studied area was located in the south-east Amazon, a region considered the largest agricultural frontier in the world (Macedo *et al.*, 2012), where native forests are converted into soybean crop fields, highlighting the social and economic importance of this system to Brazil. Therefore, using 454 pyrosequencing, they investigated the phylogenetic and metabolic diversity of microbial communities colonizing the rhizosphere and the bulk soil associated with the soybean crop field. From 24 soil samples from greenhouse mesocosm experiments, more than three million metagenomic sequences from soil of the first and fifth years of cultivation were obtained. The composition of the microbial community based on the taxonomic profile was distinct between the rhizosphere and bulk soil, however the metabolic structure did not show a clear separation, which may suggest that the taxonomy did not reflect the metabolic profile of the community. Rhizosphere samples showed an over-representation of bacterial phyla *Chlamydiae*, *Chloroflexi*, *Cyanobacteria*, *Firmicutes*, *Fusobacteria*, *Spirochaetes* and *Tenericutes*. In addition, the rhizosphere presented an over-representation of functional genes related to metabolism of N, P, Fe and potassium (K), with some specific bacterial groups linked to these functional categories found only in rhizosphere samples (Fig. 5.4). Still, the network correlations involving bacterial groups and metabolism were less complex in the rhizosphere than bulk soil, which suggests the specialization of some specific metabolic pathways. These results indicate a rhizosphere effect on the soil community, not only taxonomically, but also at a functional level, along with a selection of specific taxonomic groups related to some metabolic pathways, which could be related to plant nutrition and development (Fig. 5.4).

Recent studies have demonstrated that changes in soil microbial communities across space are often correlated with differences in soil chemistry, in particular soil pH (Fierer and Jackson, 2006; Jenkins *et al.*, 2009; Lauber *et al.*, 2009) and other soil factors such as Ca:Mg ratio, and P and aluminum (Al) concentrations (Lauber *et al.*, 2008; Kuramae *et al.*, 2012). This pattern holds both for overall bacterial community composition (Fierer and Jackson, 2006; Kuramae *et al.*, 2010; Lauber *et al.*, 200) and for the composition of individual bacterial groups (Jenkins *et al.*, 2009; Jones *et al.*, 2009). It also seems to hold across a variety of spatial scales (Fierer and Jackson, 2006; Lauber *et al.*, 2009; Philippot *et al.*, 2009) and across land-use types at a given location (Jenkins *et al.*, 2009; Jesus *et al.*, 2009; Lauber *et al.*, 2008). Although most of the previous studies have indicated the effect of soil pH on the diversity of bacterial communities, the dominant factors controlling soil bacterial community variation within the tropics are poorly known. Tripathi *et al.* (2012) compared the microbial communities of primary (unlogged), logged forests and crop and pasture lands in South-east Asia (Malaysia), and found soil pH was the best predictor of bacterial community composition and diversity across the various land-use types, with the highest diversity at close to neutral pH values. In addition, α-*Proteobacteria*, β-*Proteobacteria*, γ-*Proteobacteria*, *Acidobacteria* and *Actinobacteria* were significantly correlated with soil pH. The authors suggested that, unlike the general diversity pattern found for larger organisms, primary tropical forest is no richer in operational taxonomic units of soil bacteria than logged forest, and agricultural land (crop and pasture) is actually richer than primary forest, partly due to agricultural soils being more fertile and the effects of liming raising soil pH.

The *Acidobacteria* phylum has been shown to be highly abundant in Amazon soils (Kim *et al.* 2007; Jesus *et al.*, 2009; Navarrete *et al.* 2010). Previous studies have indicated that soil pH is the main driver of *Acidobacteria* abundance in different soil types (Jones *et al.*, 2009; Rousk *et al.*,

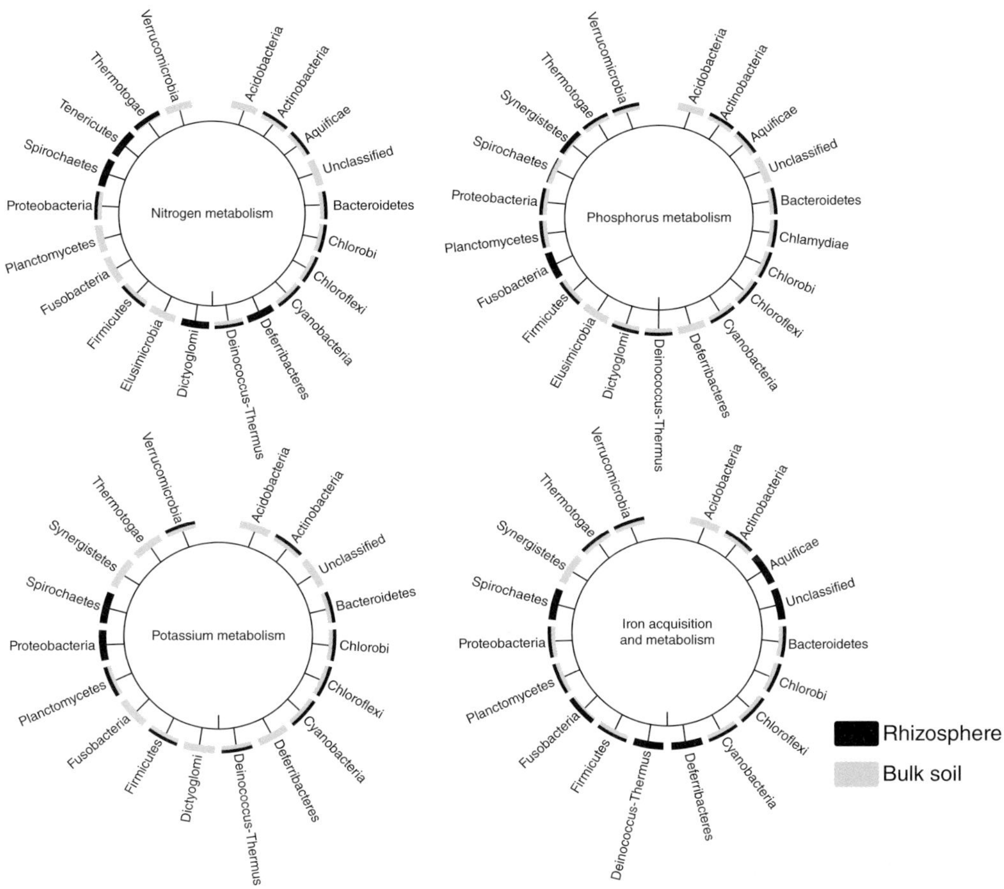

Fig. 5.4. Bacterial groups linked to specific metabolic pathways in soybean rhizosphere and bulk soil samples from the Brazilian Amazon, assessed by shotgun metagenomics. (Adapted from Mendes *et al.*, 2014.)

2010) and across spatial scales (Lauber *et al.*, 2009). However, soil pH is often correlated with many other factors, including nutrient availability, and causality among factors is not easily determined. Navarrete *et al.* (2013) showed differential responses of *Acidobacteria* subgroups to Amazon soil characteristics altered by the liming practice required in the agricultural management, such as pH, Al, Ca, Mg, K, boron and micronutrients. In the soil environment, pH is one of the most important variables and it is related to changes in other soil factors, such as Al concentration and nutrient availability. Aluminum toxicity has long been known to affect microbes as well as plants in tropical soils (Joner *et al.*, 2005). Tropical soils are usually rich in Al; hence, it is not surprising that Al and pH covary and correlate with acidobacterial communities in these soils. In addition, bacterial communities dominated by the *Acidobacteria* phylum changed significantly along gradients of base saturation, Al, and pH in western Amazonian soils (Jesus *et al.*, 2009).

Navarrete *et al.* (2013) identified sequences belonging to the *Acidobacteria* phylum subgroups 1–8 (Hugenholtz *et al.*, 1998); subgroups 9–11 (Zimmermann *et al.*, 2005); and subgroups 12–26 (Barns *et al.*, 2007). This study showed the responsiveness of *Acidobacteria* subgroups to abiotic soil factors in an Amazon region characterized by the expansion of cropland (mainly soybean) into areas previously covered by native forest (Fig. 5.5). In this geographical area, liming is practised based on the calculation of the necessity of Ca and Mg for the plant, tolerance to Al

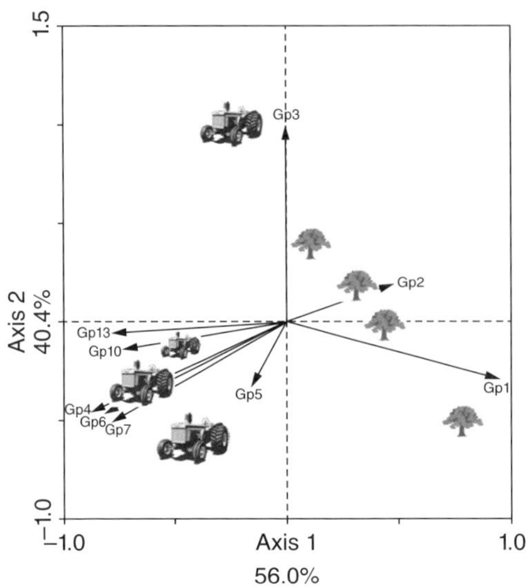

Fig. 5.5. Constrained ordination diagram for sample plots (soils from forest (denoted by trees) and soybean cropland (denoted by tractors) in the Brazilian Amazon) in the first two redundancy analysis (RDA) axes based on the soil chemical characteristics and their relationship with the relative abundance of *Acidobacteria* subgroups (1–7, 10 and 13). (Adapted from Navarrete *et al.*, 2013.)

and soil clay content. In tropical acid soils, changes in soil chemical properties due to the lime requirement for agricultural practices include a decrease in hydrogen ion activity, a decrease in Al, Al saturation, manganese toxicities, an increase in Ca and Mg availability, and benefits associated with Ca as a complementary ion in the cation exchange complex (Abruña *et al.*, 1964; Amedee and Peech, 1976; Oliveira and Pavan, 1996), which concomitantly decreases heavy metal toxicity (Ingerslev, 1997; Kreutzer, 1995). The liming appeared to have a direct effect on *Acidobacteria* subgroups 4, 6 and 7 since the abundance of these subgroups was negatively correlated with Al and Al saturation (Navarrete *et al.*, 2013). However, for *Acidobacteria* subgroup 6, Ca and base saturation also explained their abundance, while for *Acidobacteria* subgroup 7, Mg and sum of bases (Ca, Mg, K) were additional factors explaining their high abundance in cropland soils. Because of their recalcitrant culturability, the functions of the *Acidobacteria* (including subgroups 4, 6 and 7) in the ecosystems are not yet completely understood. Nevertheless, some studies (Ward *et al.*, 2009) revealed that *Acidobacteria* are metabolically active in soils and suggested that members of this phylum might contribute greatly to biogeochemical processes. Sequenced genomes of three cultivated *Acidobacteria* revealed sequences encoding proteins capable of degrading structural C, such as cellulose and hemicellulose (Ward *et al.*, 2009) and a recent acidobacterial isolate from peatland soils was capable of degrading cellulose, albeit at a slow rate (Pankratov *et al.*, 2011).

5.4 Conclusions

Soil habitats probably contain the greatest microbial diversity of all environments on Earth, and information about specific phylogenetic and functional groups and land-use systems and management options is fundamental for understanding soil quality and fertility. Therefore, the possibility for using indicator organisms or model groups as a proxy for biological soil status is tractable for monitoring soil ecosystems (Wessén and Hallin, 2011). The use of new technologies has opened a wide range of possibilities in exploring microbial ecology. Metagenomics allows deeper studies spanning both phylogenetic and functional diversity, however one of the major challenges for soil metagenomics is to develop methods to capture the heterogeneity and dynamics of microbial communities temporally and spatially (Daniel, 2005). However, in recent years, metagenomics has proven promising for discovering novel organisms and functional traits, revealing potential biomarkers for monitoring environmental changes.

The characterization and identification of microbial communities inhabiting soil in a wide range of environments has demonstrated important insights into ecosystem function and stability. The knowledge about which groups are associated with which management or crop system is the first step in defining microbial bio-indicators. However, accurate assessment of the diversity and phylogenetic composition of the soil microbial community is essential to characterize spatial and temporal patterns of diversity in response to land-use changes (Bent and Forney, 2008). When these responses are related to specific soil management factors, they may be used to develop early-warning bio-indicators for soil impacts. However, there are still few studies focusing on the responses of certain groups or subgroups of bacteria to environmental factors in the tropics. Ultimately, a better understanding of how agricultural management affects soil microbial ecology will support the development of more productive and sustainable systems. Therefore, it is important to understand the factors that influence the biodiversity of soil bacterial communities, to understand how these communities are structured and also to predict ecosystem responses to a changing environment.

Acknowledgements

Our studies are supported by Fundação de Amparo à Pesquisa do Estado de São Paulo (FAPESP-2008/58114-3, 2011/50914-3, 2011/51749-6), Coordenação de Aperfeiçoamento de Pessoal de Nível Superior (CAPES-Wageningen – 010/09, 2238/10-1, 1240/11-0) and Conselho Nacional de Pesquisa e Desenvolvimento (CNPq 485969/2012-2, 485801/2011-6, 304713/2010-6). Publication 5672 of NIOO-KNAW.

References

Abruña, F., Vicente-Chandler, J. and Pearson, R.W. (1964) Effects of liming on yields and composition of heavily fertilized grasses and on soil properties under humid tropical conditions. *Soil Science Society of America Journal* 28, 657–661.

Amedee, G. and Peech, M. (1976) Liming of highly weathered soils of the humid tropics. *Soil Science* 121, 259–266.

Anderson, T.H. (2003) Microbial eco-physiological indicators to asses soil quality. *Agriculture, Ecosystems and Environment* 98, 285–293.

Babujia, L.C., Hungria, M., Franchini, J.C. and Brookes, P.C. (2010) Microbial biomass and activity at various soil depths in a Brazilian oxisol after two decades of no-tillage and conventional tillage. *Soil Biology and Biochemistry* 42, 2174–2181.

Bardgett, R.D., Freeman, C. and Ostle, N.J. (2008) Microbial contributions to climate change through carbon cycle feedbacks. *The ISME Journal* 2, 805–814.

Barns, S.M., Cain, E.C., Sommerville, L. and Kuske, C.R. (2007) *Acidobacteria* phylum sequences in uranium-contaminated subsurface sediments greatly expand the known diversity within the phylum. *Applied Environmental Microbiology* 73, 3113–3116.

Bent, S.J. and Forney, L.J. (2008) The tragedy of uncommon: Understanding limitations in the analysis of microbial diversity. *The ISME Journal* 2, 689–695.

Borges, C.D., Corá, J.E., Barbosa, J.C. and Nahas, N. (2012) Soil microbiological attributes under summer/winter crops rotation in a no-tillage system. *Archives of Agronomy and Soil Science* 59, 1471–1485.

Borneman, J. and Triplett, E.W. (1997) Molecular microbial diversity in soils from Eastern Amazonia: Evidence for unusual microorganisms and microbial population shifts associated with deforestation. *Applied Environmental Microbiology* 63, 2647–2653.

Brook, B.W., Sodhi, N.S. and Ng, P.K.L. (2003) Catastrophic extinctions follow deforestation in Singapore. *Nature* 424, 420–423.

Brossi, M.J.L., Mendes, L.W., Germano, M.G., Lima, A.B. and Tsai, S.M. (2014) Assessment of bacterial *bph* gene in Amazonian Dark Earth and their adjacent soil. *PLoS One* 9, e99597.

Cardoso, E.L., Silva, M.L.N., Moreira, F.M.S. and Curi, N. (2009) Atributos biológicos indicadores da qualidade do solo em pastagem cultivada e nativa no Pantanal. *Pesquisa Agropecuária Brasileira* 44, 631–637.

Carter, M.R. (2001) Organic matter and sustainability. In: Rees, R.M., Ball, B.C., Campbell, C.D. and Watson, C.A. (eds) *Sustainable Management of Soil Organic Matter*. CAB International, New York, pp. 9–22.

D'Andrea, A.F., Silva, M.L.N., Curi, N., Siqueira, J.O., Carneiro, M.A.C. (2002) Atributos biológicos indicadores da qualidade do solo em sistemas de manejo na região do Cerrado no sul do estado de Goiás. *Revista Brasileira Ciência do Solo* 26, 913–923.

Daniel, R. (2005) The metagenomics of soil. *Nature Reviews* 3, 470–478.

Dieckow, J., Bayer, C., Conceição, P.C., Zanatta, J.A., Martin-Neto, L., Milori, D.B.M. *et al.* (2009) Land use, tillage, texture and organic matter stock and composition in tropical and subtropical Brazilian soils. *European Journal of Soil Science* 60, 240–249.

Dirzo, R. and Raven, P.H. (2003) Global state of biodiversity and loss. *Annual Review of Environment and Resources* 28, 137–167.

Erwin, T.L. (1983) Beetles and other arthropods of the tropical forest canopies at Manaus, Brasil, sampled with insecticidal fogging techniques. In: Sutton, S.L., Whitmore, T.C. and Chadwick, A.C. (eds) *Tropical Rain Forests: Ecology and Management*. Blackwell Scientific Publications, Oxford, UK, pp. 59–75.

Fearnside, P.M. (2005) Deforestation in Brazilian Amazonia: History, rates, and consequences. *Conservation Biology* 19, 680–688.

Fearnside, P.M. (2006) Uso da terra na Amazônia e as mudanças climáticas globais. *Brazilian Journal of Ecology* 7, 83–100.

Ferreira, E.P.B., Santos, H.P., Costa, J.R., De-Polli, H. and Rumjanek, N.G. (2010) Microbial soil quality indicators under different crop rotations and tillage management. *Revista Ciência Agronômica* 41, 177–183.

Fierer, N. and Jackson, R.B. (2006) The diversity and biogeography of soil bacterial communities. *Proceedings of the National Academy of Sciences of the USA* 103, 626–631.

Fierer, N. and Lennon, J. (2011) The generation and maintenance of diversity in microbial communities. *American Journal of Botany* 98, 439–448.

Fierer, N., McCain, C.M., Meir, P., Zimmermann, M., Rapp, J.M., Silman, M.R. and Knight, R. (2011) Microbial elevational diversity does not follow the biogeographical trends of plants and animals. *Ecology* 92, 797–804.

Franchini, J.C., Crispino, C.C., Souza, R.A., Torres, E. and Hungria, M. (2007) Microbiological parameters as indicators soil quality under various soil management and crop rotation systems in southern Brazil. *Soil Tillage and Research* 92, 18–29.

Frazão, L.A., Piccolo, M.C., Feigl, B.J., Cerri, C.C. and Cerri, C.E.P. (2010) Inorganic nitrogen, microbial biomass and microbial activity of a Sandy Brazilian Cerrado soil under different land uses. *Agriculture, Ecosystems and Environment* 135, 161–167.

Germano, M.G., Cannavan, F.S., Mendes, L.W., Lima, A.B., Teixeira, W.G., Pellizari, V.H. and Tsai, S.M. (2012) Functional diversity of bacterial genes associated with aromatic hydrocarbon degradation in anthropogenic dark earth of Amazonia. *Pesquisa Agropecuária Brasileira* 47, 654–664.

Glaser, B., Guggenberger, G., Haumaier, L. and Zech, W. (2001) Persistence of soil organic matter in archaeological soils (terra preta) of the Brazilian Amazon region. In: Rees, R.M., Ball, B.C., Campbell, C.D. and Watson, C.A. (eds) *Sustainable Management of Soil Organic Matter*. CAB International, Wallingford, UK, pp. 190–194.

Hansen, M.C., Stehman, S.V., Potapov, P.V., Loveland, T.R., Townshend, J.R.G., DeFries, R.S. *et al.* (2008) Humid tropical forest clearing from 2000 to 2005 quantified by using multitemporal and multiresolution remotely sensed data. *Proceedings of the National Academy of Sciences of the USA* 105, 9439–9444.

Hoffmann, M., Hilton-Taylor C. Angulo, A., Böhm, M, Brooks, T.M., Butchart, S.H.M. *et al.* (2010) The impact of conservation on the status of the world's vertebrates. *Science* 330, 1503–1509.

Hugenholtz, P., Goebel, B.M. and Pace, N.R. (1998) Impact of culture-independent studies on the emerging phylogenetic view of bacterial diversity. *Journal of Bacteriology* 180, 4765–4774.

Hungria, M., Franchini, J.C., Brandão-Junior, O., Kaschuk, G. and Souza, R.A. (2009) Soil microbial activity and crop sustainability in a long-term experiment with tree soil-tillage and two crop-rotation systems. *Applied Soil Ecology* 42, 288–296.

Ingerslev, M. (1997) Effects of liming and fertilization on growth, soil chemistry and soil water chemistry in a Norway spruce plantation on a nutrient-poor soil in Denmark. *Forest Ecology and Management* 92, 55–66.

Instituto Nacional de Pesquisas Espaciais (INPE) (2013) Available at www.inpe.br/noticias/noticia.php?Cod_Noticia=3443. Accessed on 11 March 2015.

Jenkins, S.N., Waite, I.S., Blackburn, A., Husband, R., Rushton, S.P., Manning, D.C. *et al.* (2009) Actinobacterial community dynamics in long term managed grasslands. *Antonie van Leeuwenhoek* 95, 319–334.

Jenkinson, D.S. and Ladd, J.N. (1981) Microbial biomass in soil: Measurement and turnover. In: Paul, E.A. and Ladd, J.N. (eds) *Soil Biochemistry*, vol 5. Marcel Dekker, New York, Basel, pp. 415–471.

Jenkinson, D.S. and Powlson, D.S. (1976) The effects of biocidal treatment on metabolism in soil. I. Fumigation with chloroform. *Soil Biology and Biochemistry* 8, 167–177.

Jesus, E.C., Marsh, T.L., Tiedje, J.M. and Moreira, F.M.S. (2009) Changes in land use alter the structure of bacterial communities in Western Amazon soils. *The ISME Journal* 3, 1004–1011.

Joner, E.J., Eldhuset, T.D., Lange, H. and Frostegard, A. (2005) Changes in the microbial community in a forest soil amended with aluminum *in situ*. *Plant Soil* 275, 295–304.

Jones, R.T., Robeson, M.S., Lauber, C.L., Hamady, M., Knight, R. and Fierer, N. (2009) A comprehensive survey of soil acidobacterial diversity using pyrosequencing and clone library analyses. *The ISME Journal* 3, 442–453.

Kaschuk, G., Alberton, O. and Hungria, M. (2010) Three decades of soil microbial biomass studies in Brazilian ecosystems: Lessons learned about soil quality and indications for improving sustainability. *Soil Biology and Biochemistry* 42, 1–13.

Kim, J.S., Sparovek, G., Longo, R.M., De Melo, W.J. and Crowley, D. (2007) Bacteria diversity of terra preta and pristine forest soil from the western Amazon. *Soil Biology and Biochemistry* 39, 684–690.

Knight, R., Jansson, J., Field, D., Fierer, N., Desai, N., Fuhrman, J.A., Hugenholtz, P. *et al.* (2012) Unlocking the potential of metagenomics through replicated experimental design. *Nature Biotechnology* 30, 513–520.

Kreft, H. and Jetz, W. (2007) Global patterns and determinants of vascular plant diversity. *Proceedings of the National Academy of Science of the USA* 104, 5925–5930.

Kreutzer, K. (1995) Effects of forest liming on soil processes. *Plant and Soil* 168, 447–470.

Kuramae, E.E., Gamper, H.A., Regeau, E., Piceno, Y.M., Brodie, E.L., DeSantis, T.Z. *et al.* (2010) Microbial secondary succession in a chronosequence of chalk grasslands. *The ISME Journal* 4, 711–715.

Kuramae, E.E., Yergeau, E., Wong, L., Pijl, A., van Veen, J.A. and Kowalchuk, G.A. (2012) Soil characteristics more strongly influence soil microbial communities than land management. *FEMS Microbiology Ecology* 79, 12–24.

Lauber, C.L., Strickland, M.S., Bradford, M.A. and Fierer, N. (2008) The influence of soil properties on the structure of bacterial and fungal communities across land-use types. *Soil Biology and Biochemistry* 40, 2407–2415.

Lauber, C.L., Hamady, M., Knight, R. and Fierer, N. (2009) Pyrosequencing-based assessment of soil pH as a predictor of soil bacterial community composition at the continental scale. *Applied Environmental Microbiology* 75, 5111–5120.

Leite, L.F.C., Galvão, S.R.S., Neto, M.R.H., Araújo, F.S. and Iwata, B.F. (2010) Atributos químicos e estoques de carbono em Latossolo sob plantio direto no cerrado do Piauí. *Revista Brasileira de Engenharia Agrícola e Ambiental* 14, 1273–1280.

Lourente, E.R.P., Mercante, F.M., Marchetti, M.E., Souza, L.C.F., Souza, C.M.A., Gonçalves, M.C. and Silva, M.A.G. (2010) Rotação de culturas e relações com atributos químicos e microbiológicos do solo e produtividade do milho. *Semina: Ciências Agrárias* 31, 829–842.

Luo, C., Tsementzi, D., Kyrpides, N., Read, T. and Konstantinidis, K.T. (2012) Direct comparison of Illumina vs. Roche 454 sequencing technologies on the same microbial community DNA sample. *PLoS One* 7, e30087.

Lupatini, M., Jacques, R.J.S., Antoniolli, Z.I., Suleiman, A.K.A., Fulthorpe, R.R. and Roesch, L.F.W. (2012) Land-use change and soil type are drivers of fungal and archaeal communities in the Pampa biome. *World Journal of Microbiology and Biotechnology* 29, 223–233.

Macedo, M.N., DeFries, R.S., Morto, D.C., Stickler, C.M., Galford, G.L. and Shimabukuro, Y.E. (2012) Decoupling of deforestation and soy production in the southern Amazon during the late 2000s. *Proceedings of the National Academy of Sciences of the USA* 109, 1341–1346.

Marris, E. (2006) Putting the carbon back: Black is the new green. *Nature* 442, 624–626.

Matoso, S.C.G., Silva, A.N., Fiorelli-Pereira, E.C., Colleta, Q.P. and Maia, E. (2012) Frações de carbono e nitrogênio de um Latossolo Vermelho-Amarelo distrófico sob diferentes usos na Amazônia brasileira. *Acta Amazonica* 42, 231–240.

Mendes, L.W., Kuramae, E.E., Navarrete, A.A., van Veen, J.A. and Tsai, S.M. (2014) Taxonomical and functional microbial community selection in soybean rhizosphere. *The ISME Journal* 8, 1577–1587.

Myers, N., Mittermeier, R.A., Mittermeier, C.G., da Fonseca, G.A.B. and Kent, J. (2000) Biodiversity hotspots for conservation priorities. *Nature* 403, 853–858.

Navarrete, A.A., Cannavan, F.S., Taketani, R.G. and Tsai, S.M. (2010) A molecular survey of the diversity of microbial communities in different Amazonian agricultural model systems. *Diversity* 2, 787–809.

Navarrete, A.A., Taketani, R.G., Mendes, L.W., Cannavan, F.S., Moreira, F.M.S. and Tsai, S.M. (2011) Land-use systems affect archaeal community structure and functional diversity in western Amazon soils. *Revista Brasileira de Ciência do Solo* 35, 1527–1540.

Navarrete, A.A., Kuramae, E.E., de Hollander, M., Pijl, A.S., van Veen, J.A. and Tsai, S.M. (2013) Acidobacterial community responses to agricultural management of soybean in Amazon forest soils. *FEMS Microbiology Ecology* 83, 607–621.

Ndaw, S.M., Gama-Rodrigues, A.C., Gama-Rodrigues, E.F., Sales, K.R.N. and Rosado, A.S. (2009) Relationships between bacterial diversity microbial biomass and litter quality in soils under different plant covers in northern Rio de Janeiro State Brazil. *Canadian Journal of Microbiology* 55, 1089–1095.

Niemeyer, J.C., Lolata, J.B., Carvalho, G.M., Silva, E.M., Sousa, J.P., Nogueira, M.A. (2012) Microbial indicators of soil health as tools for ecological risk assessment of a metal contaminated site in Brazil. *Applied Soil Ecology* 59, 96–105.

Nunes, J.S., Araujo, A.S.F., Nunes, L.A.P.L., Lima, L.M., Carneiro, R.F.V., Salviano, A.A.C. and Tsai, S.M. (2012) Impact of land degradation on soil microbial biomass and activity in Northeast Brazil. *Pedosphere* 22, 88–95.

Oliveira, E.L. and Pavan, M.A. (1996) Control of soil acidity in no-tillage system for soybean production. *Soil Tillage and Research* 38, 47–57.

Pankratov, T.A., Ivanova, A.O., Dedysh, S.N. and Liesack, W (2011) Bacterial populations and environmental factors controlling cellulose degradation in an acidic Sphagnum peat. *Environmental Microbiology* 13, 1800–1814.

Phalan, B., Bertzky, M., Butchart, S.H.M., Donald, P.F., Scharlemann, J.P.W. *et al.* (2013) Crop expansion and conservation priorities in tropical countries. *PLoS One* 8, e51759.

Philippot, L., Čuhel, J., Saby, N.P.A., Cheneby, A., Chroňáková, A. Bru, D. *et al.* (2009) Mapping field-scale spatial distribution patterns of size and activity of the denitrifier community. *Environmental Microbiology* 11, 1518–1526.

Piccolo, M.C., Neill, C. and Cerri, C.C. (1994) Natural-abundance of n-15 in soils along forest-to-pasture chronosequences in the western Brazilian Amazon basin. *Oecologia* 99, 112–117.

Portilho, I.I.R., Borges, C.D., Costa, A.R., Salton, J.C. and Mercante, F.M. (2011) Resíduos da cultura da cana-de-açúcar e seus efeitos sobre a fauna invertebrada epigeica. *Semina: Ciências Agrárias* 32, 959–970.

Pragana, R.B., Nóbrega, R.S.A., Ribeiro, M.R. and Lustosa Filho, J.F. (2012) Atributos biológicos e dinâmica da matéria organica em latossolos amarelos na região do Cerrado Piauiense sob sistema plantio diretor. *Revista Brasileira de Ciência do Solo* 36, 851–858.

Pupin, B. and Nahas, E. (2011) Impact of successive sugarcane harvests and trash management practices on soil microbiological properties. *Soil Research* 49, 183–189.

Reiners, W.A., Bouwman, A.F., Parsons, W.F.J. and Keller, M. (1994) tropical rain-forest conversion to pasture – Changes in vegetation and soil properties. *Ecological Applications* 4, 363–377.

Roesch, L.F.W., Camargo, F.A.O., Bento, F.M. and Triplett, E.W. (2007) Biodiversity of diazotrophic bacteria within the soil, root and stem of field-grown maize. *Plant Soil* 302, 91–104.

Roscoe, R., Mercante, F.M., Mendes, I.C., Reis-Júnior, F.B., Santos, J.C.F. and Hungria, M. (2006) Biomassa microbiana do solo: Fração mais ativa da matéria orgânica. In: Roscoe, R., Mercante, F.M. and Salton, J.C. (eds) *Dinâmica da matéria orgânica do solo em sistemas conservacionistas – modelagem matemática e métodos auxiliares.* Embrapa Agropecuária Oeste, Dourados, Brazil, pp. 163–200.

Rousk, J., Bååth, E., Brookes, P.C., Lauber, C.L., Lozupone, C., Caporaso, *et al.* (2010) Soil bacterial and fungal communities across a pH gradient in an arable soil. *The ISME Journal* 4, 1340–1351.

Silva, D.K.A., Freitas, N.O., Souza, R.G., Silva, F.S.B., Araujo, A.S.F. and Maia, L.C. (2012) Soil microbial biomass and activity under natural and regenerated forest and conventional sugarcane plantations in Brazil. *Geoderma* 189–190, 257–261.

Sodhi, N.S., Koh, L.P., Brook, B.W. and Ng, P.K.L. (2004) Southeast Asian biodiversity: An impending disaster. *Trends in Ecology and Evolution* 19, 654–660.

Souza, D.E., Costa, A.S.E.V.G., Anghinoni, I., Lima, C.V.S., Carvalho, P.C.F. and Martins, A.P. (2010) Biomassa microbiana do solo em sistema de integração lavoura-pecuária em plantio direto, submetido a intensidades de pastejo. *Revista Brasileira de Ciência do Solo* 34, 79–88.

Souza, R.A., Telles, T.S., Machado, W., Hungria, M., Tavares-Filho, J. and Guimarães, M.F. (2012) Effects of sugarcane harvesting with burning on the chemical and microbiological properties of the soil. *Agriculture, Ecosystems and Environment* 155, 1–6.

Taketani, R.G. and Tsai, S.M. (2010) The influence of different land uses on the structure of Archaeal communities in Amazonian Anthrosols based on 16S rRNA and *amo*A genes. *Microbial Ecology* 59, 734–743.

Taketani, R.G., Lima, A.B., Jesus, E.C., Teixeira, W.G., Tiedje, J.M. and Tsai, S.M. (2013) Bacterial community composition of anthropogenic biochar and Amazonian anthrosols assessed by 16S rRNA gene 454 pyrosequencing. *Antonie van Leeuwenhoek* 104, 233–242.

Torsvik, V., Ovreas, L. and Thingstad, T.F. (2002) Prokaryotic diversity – Magnitude, dynamics, and controlling factors. *Science* 296, 1064–1066.

Tripathi, B.M., Kim, M., Singh, D., Lee-Cruz, L., Lai-Hoe, A., Ainuddin, A.N. *et al.* (2012) Tropical soil bacterial communities in Malaysia: pH dominates in the equatorial tropics too. *Microbial Ecology* 64, 474–484.

Vance, E.D., Brookes, P.C. and Jenkinson, D.S. (1987) An extraction method for measuring soil microbial biomass. *Soil Biology and Biochemistry* 19, 703–707.

Ward, N.L., Challacombe, J.F., Janssen, P.H. *et al.* (2009) Three genomes from phylum *Acidobacteria* provide insight into the lifestyles of these microorganisms in soils. *Applied Environmental Microbiology* 75, 2046–2056.

Wessén, E. and Hallin, S. (2011) Abundance of archaeal and bacterial ammonia oxidizers – possible bioindicator for soil monitoring. *Ecological Indicators* 11, 1696–1698.

Whitman, W.B., Coleman, D.C. and Wiebe, W.J. (1998) Prokaryotes: The unseen majority. *Proceedings of the National Academy of Science of the USA* 95, 6578–6583.

Wilcove, D.S., Giam, X., Edwards, D.P., Fisher, B. and Koh, L.P. (2013) Navjot's nightmare revisited: Logging, agriculture and biodiversity in Southeast Asia. *Trends in Ecology and Evolution* 28, 531–540.

Xavier, F.A.S., Maia, S.M.F., Oliveira, T.S. and Mendonça, E.S. (2006) Biomassa microbiana e matéria orgânica leve em solos sob sistemas agrícolas orgânico e convencional na Chapada da Ibiapaba-CE. *Revista Brasileira de Ciência do Solo* 30, 247–258.

Zimmermann, J., Gonzalez, J.M., Saiz-Jimenez, C. and Ludwig, W. (2005) Detection and phylogenetic relationships of highly diverse uncultured acidobacterial communities in Altamira cave using 23S rRNA sequence analyses. *Geomicrobiology Journal* 22, 379–388.

6 Acidification of Tropical Soils under Forest and Continuous Cropping in Thailand and Indonesia

Kazumichi Fujii,[1,2]* Chie Hayakawa,[2,3] Shinya Funakawa[2] and Takashi Kosaki[4]

[1]Forestry and Forest Products Research Institute, Tsukuba, Japan; [2]Graduate School of Agriculture, Kyoto University, Kyoto, Japan; [3]National Institute for Agro-Environmental Science, Ibaraki, Japan; [4]Department of Tourism Science, Tokyo Metropolitan University, Tokyo, Japan

6.1 Introduction

In the humid tropics, shifting cultivation is an extensive farming system on typically highly weathered and leached soils (Nye and Greenland, 1960; Kyuma and Pairintra, 1983; Mertz *et al.*, 2009). Owing to rapid population growth, traditional shifting cultivation with an adequately long fallow period has been replaced with more intensive cropping systems with shorter fallow periods or continuous cropping (Kyuma and Pairintra, 1983; Mertz *et al.*, 2009). Since restoration of soil fertility is dependent upon a sufficiently long fallow period, continuous cropping risks widespread soil degradation and reductions in plant productivity in Asian countries.

Soil degradation associated with conversion of forest fallow to cropland typically includes the depletion of soil organic matter (SOM) and acceleration of soil acidification and erosion (Funakawa *et al.*, 2006). Soil acidification is generally a natural process under a humid climate where precipitation exceeds evapotranspiration, but it can be accelerated by agricultural practices (Helyar and Porter, 1989; Juo *et al.*, 1996). In croplands, a number of factors can accelerate soil acidification such as the enhanced net mineralization of soil organic nitrogen (N) (Tanaka *et al.*, 1997; Funakawa *et al.*, 2006), limited plant uptake of mineralized nitrate at the beginning of the cropping season (Poss *et al.*, 1995) and N fertilization (Bouman *et al.*, 1995). This contrasts with forests, where excess cation accumulation in organic matter (biomass and SOM) contributes to soil acidification (Hallbäcken and Tamm, 1986).

On the other hand, the loss of SOM has the potential to increase acid neutralization in soils by mineralization of organic anions. Since most proton-generating and -consuming processes are associated with organic matter cycles under low-input agriculture, soil acidification is strongly influenced by the cropping-induced changes in SOM cycles. In tropical regions, SOM cycles, especially SOM decomposition, can change dramatically within relatively short periods in response to cropping following deforestation (Bruun *et al.*, 2009).

Soil acidification is also influenced by proton sources other than nitrification, i.e. acidic deposition, dissociation of organic and carbonic acids and excess uptake of cations over anions by plants. Therefore, all proton generation and

* Corresponding author: fjkazumichi@gmail.com

consumption mechanisms need to be quantified in order to evaluate cropping-induced changes in soil acidification. The utilization of a proton budget approach in combination with measurement of SOM decomposition rates can allow us to estimate proton consumption associated with the loss of SOM (Poss *et al.*, 1995).

In the present study, the effects of continuous cropping on soil acidification were assessed by quantifying proton budgets in a soil–vegetation system including solute leaching, vegetation uptake and SOM decomposition in croplands and adjacent forests in Thailand and Indonesia. We tested whether SOM loss can neutralize acidity generated from nitrification under continuous cropping.

6.2 Materials and Methods

6.2.1 Study sites

Experimental plots consisted of one forest and one cropland plot in both Thailand and Indonesia (Fig. 6.1). The forest and cropland plots in Thailand (RPf and RPc, respectively) were located in Ban Rakpaendin, Chiang Rai Province (Fig. 6.1; 19° 50′ N, 100° 20′ E; 697 m a.s.l.), where the mean annual precipitation and temperature are 2084 mm and 25.0°C, respectively. There are distinct dry and wet seasons: the dry season is from November to March and the wet season is from April to October. The vegetation was dominated by *Lithocarpus* sp. (Fagaceae) and *Eugenia* sp. (Myrtaceae) in RPf, while maize (*Zea mays*; Poaceae) had been cultivated during the wet season without fertilization in RPc for 3 years since the conversion of forest to cropland. Soils were derived from sedimentary rocks (RPf) and sedimentary rocks associated with granite intrusion (RPc), and classified as Typic Haplustults (Ultisols).

The forest and cropland plots in Indonesia (BSf and BSc, respectively) were located in the Experimental Forest of the Tropical Rain forest Research Center, Mulawarman University, Bukit Soeharto, East Kalimantan Province (Fig. 6.1; 0° 51′ S, 117° 06′ E; 99 m a.s.l.), where the mean annual precipitation and temperature are 1977 mm and 26.8°C, respectively. The vegetation is dominated by *Shorea laevis* and *Dipterocarpus cornutus* (both Dipterocarpaceae) in BSf, while chili (*Capsicum* sp.; Solanaceae) had been cultivated for 2 years after deforestation in BSc. Soils were derived from sedimentary rocks and classified as Typic Paleudults (Ultisols). Poultry manure (0.71 Mg ha^{-1} $year^{-1}$ (dry weight)) was applied at the beginning of the cropping season (October 2004 to October 2005).

6.2.2 Sampling and analytical methods for soil and plant materials

Five composite soil samples were air-dried and ground to pass through a 2-mm sieve. Soil pH was measured using a soil to solution (water) ratio of 1:5 (w/v) after shaking for 1 h. Carbon (C) and N concentrations were determined using a CN analyser (NC-800-13N; Sumika Chemical Analysis Service). Particle size distribution was determined by the pipette method. Exchangeable base cation, i.e. calcium (Ca), magnesium (Mg), sodium (Na) and potassium (K), concentrations and cation exchange capacity (CEC) were determined using the ammonium acetate (1 M at pH7.0) method; Na and K concentrations were determined by flame photometry, and Mg and Ca concentrations were determined by atomic absorption spectroscopy.

In the forest plots, the aboveground biomasses in RPf and BSf were estimated by applying the diameters of stems at breast height (1.3 m) to the regression equations obtained by Tsutsumi *et al.* (1983) and Yamakura *et al.* (1986) for the Thailand and Indonesia sites, respectively. Litterfall was collected by circular litter traps of 60-cm diameter in five replicates over a 1-year period. Wood increment was estimated by tree ring analysis, after obtaining core samples using an increment borer, and using the regression equation for estimating tree biomass according to Johnson and Risser (1974). In the cropland plots, plant biomass (leaves, stems and fruits) was collected at the end of the wet season. Fine roots in the O horizon were collected in 30 cm × 30 cm quadrats in five replicates, while those in mineral soil horizons (0–30 cm) were collected at 5-cm intervals by taking 0.1-l cores in five replicates. Roots were rinsed in distilled water to remove soil material. Plant samples were oven-dried at 70°C for 48 h, weighed and milled. The C and N in plant materials were determined using a CN analyser. For concentrations of other elements

Fig. 6.1. Landscapes and soils of the experimental sites in Thailand (RP) and Indonesia (BS). (a) Use of fire in conversion of forest fallow to cropland, (b) cultivation of maize or upland rice, (c) clayey Ultisol at RP and (d) tropical lowland forests, and (e) acidic Ultisol at BS.

in plant samples, we conducted nitric-sulfuric acid (1:1) wet digestion. The Na and K concentrations were determined by flame photometry, Mg and Ca by atomic absorption spectroscopy, iron (Fe) and aluminium (Al) by inductively coupled plasma atomic emission spectrometry (ICP-AES, SPS1500; Seiko Instruments Inc.), and phosphorus (P) using the colorimetric method (molybdate blue method, UV-VIS spectrophotometer UV-1200; Shimadzu) The chlorine (Cl) and sulfur (S) concentrations were determined by high performance liquid chromatography (HPLC) after combustion according to Busman *et al.* (1983).

6.2.3 Measurements of decomposition rates and stocks of organic matter

Soil respiration consists of gases produced during SOM decomposition by microorganisms and by root respiration (Kuzyakov, 2006). Organic matter decomposition rates can therefore be estimated by excluding root respiration using the trenching method (Shinjo *et al.*, 2006). The rates of SOM decomposition were estimated from soil CO_2 efflux after trenching. Carbon dioxide efflux was measured monthly in five replicates using a closed-chamber method in the trenched plots, for which collars with a diameter of 15 cm and a height of 40 cm were installed in the soil to a depth of 20 cm. Gases in the headspace of the soil collars were sampled 0, 10, 20 and 30 min after the tops of the collars were covered with plastic sheets. Sampled gases were analysed with an infrared CO_2 controller (ZFP9; Fuji Electric Instruments Co., Ltd). The CO_2 efflux was calculated using linear regression from the initial CO_2 concentration increase and air temperature measured within the chamber before measurement. At the same time, soil temperature at a depth of 5 cm was measured with a thermister probe (107 Temperature Probe; Campbell Scientific, Inc.), while volumetric water contents in soils at depths of 5, 15 and 45 cm (5, 15 and 30 cm in BSf and BSc) were measured with time domain reflectometer probes (CS615 Water Content Reflectometer; Campbell Scientific Inc.). Data were recorded using dataloggers at 30-min intervals (CR-10X; Campbell Scientific, Inc.) during the experiments.

Soil C stocks in the organic and mineral horizons (0–20 cm) were quantified on an area basis in five replicates. The C stocks in the organic horizons (forest sites) were collected from 20 cm × 20 cm quadrats in five replicates. The C stocks in the mineral horizons were calculated using bulk density and soil C concentrations at the depths of 0–5, 5–10, 10–15 and 15–20 cm. The core samples (0.1 l) were collected in five replicates.

6.2.4 Sampling and analytical methods of precipitation, throughfall and soil solution

Soil solutions were collected using a tension-free lysimeter, draining a surface area of 200 cm^2 in five replicates beneath the A, BA and Bt horizons (7, 20 and 45 cm depths) in RPf and RPc; the O, A and B1 horizons (0, 5 and 30 cm depths) in BSf; and the Ap1, Ap2 and B1 horizons (5, 20 and 30 cm depths) in BSc. Precipitation and throughfall were collected separately using a precipitation collector in five replicates. Sample solutions were filtered through 0.45-μm filters before analysis. The concentrations of hydrogen ions (H^+) in solution were determined with a glass electrode. The concentrations of ammonium, nitrate, Na^+, K^+, Mg^{2+}, Ca^{2+}, Cl^- and SO_4^{2-} in solution were determined by HPLC. The concentrations of Fe^{3+} and Al^{n+} in solution were determined by ICP-AES. The total charge equivalent of Al ions (Al^{n+}) was calculated as the equivalent sum of Al^{3+}, $AlOH^{2+}$ and $Al(OH)_2^+$. The concentrations of dissolved organic C (DOC) and inorganic C were determined using a total organic carbon analyser (TOC-V_{CSH}; Shimadzu). The concentrations of HCO_3^- in solution were determined from the solution pH and inorganic C concentration, based on pKa = 6.3. The anion deficit, if any, was assigned to the negative charge of organic acids (Org^{n-}).

The fluxes of ions from each soil horizon were calculated by multiplying the water fluxes by the concentrations of ions in precipitation, throughfall and soil solution. The water fluxes of precipitation and throughfall were measured using a precipitation collector, while the fluxes of soil water percolating from the surface (5–15 cm depth) and subsurface soil horizons (15–45 and 15–30 cm depths in RP and BS, respectively) were estimated by applying Darcy's law to the unsaturated hydraulic conductivity and the gradient of the hydraulic heads. A detailed description for the calculation of water fluxes was presented in a previous paper (Fujii *et al.*, 2008).

6.2.5 Calculation of proton budgets

Proton budgets associated with solutes leaching from the system and those associated with vegetation uptake can be successfully quantified in forests (Van Breemen *et al.*, 1983, 1984). In croplands, where continuous cropping results in the loss of SOM, proton consumption owing to net mineralization of organic anions should also be quantified (Helyar and Porter, 1989; Poss *et al.*, 1995).

Net proton generation (NPG) resulting from excess cation uptake by vegetation (NPG_{Bio}), transformation of N (NPG_{Ntr}), dissociation of organic acids (NPG_{Org}), dissociation of carbonic acid (NPG_{Car}) and net proton influx from the overlying horizon ($\{(H^+)_{in}-(H^+)_{out}\}$) can be calculated based on the input–output budget of ions in the soil (Bredemeier *et al.*, 1990). Representative processes and equations of NPG are presented in Table 6.1 and Fig. 6.2. NPG_{Ntr}, NPG_{Org} and NPG_{Car} can be calculated from the fluxes of ions entering and leaving the soil horizon compartment, e.g. an increase in fluxes of organic acids in the O horizon. NPG_{Bio}, which consists of wood increment (forest), crop removal (cropland) and litterfall, is calculated from ion fluxes ($kmol_c$ ha^{-1} $year^{-1}$) caused by vegetation uptake of cations (Na^+, K^+, Mg^{2+}, Ca^{2+}, Fe^{3+} and Al^{n+}) and anions (Cl^-, $H_2PO_4^-$ and SO_4^{2-}).

Protons are also consumed by net mineralization of organic anions in croplands (Table 6.1). Net proton consumption owing to mineralization of organic anions (NPC_{Min}) is caused by decomposition of litterfall, manure and SOM. NPC_{Min} from litter and manure is calculated from the difference between cations and anions of litterfall and manure entering into soil, while NPC_{Min} associated with the loss of SOM is calculated using

Table 6.1. Proton-producing and proton-consuming processes in soils.

H^+ budget	Representative reaction	Proton budget calculation
Proton-generating processes		
(1) H^+ input (e.g. acid rain)		H^+ input = $(H^+)_{in} - (H^+)_{out}$
(2) Cation excess uptake by plants	$Ca^{2+} + 2R\text{-}OH \rightarrow (R\text{-}O)_2Ca$ (org) + $2H^+$	$NPG_{Bio} = (Cat)_{bio} - (Ani)_{bio}$
(3) N transformation (e.g. nitrification)	$NH_4^+ + H_2O \rightarrow NO_3^- + 2H^+ + H_2O$	$NPG_{Ntr} = (NH_4^+)_{in} - (NH_4^+)_{out} + (NO_3^-)_{out} - (NO_3^-)_{in}$
(4) Dissociation of carbonic acid	$H_2CO_3 \rightarrow HCO_3^- + H^+$	$NPG_{Car} = (HCO_3^-)_{out} - (HCO_3^-)_{in}$
(5) Dissociation of organic acid	$2CH_2O + 3/2O_2 \rightarrow HC_2O_4^- + H^+ + H_2O$	$NPG_{Org} = (Org^{n-})_{out} - (Org^{n-})_{in}$
Proton-consuming processes		
(6) Mineralization of organic matter	$(R\text{-}O)_2Ca$ (org) $+ 2H^+ \rightarrow Ca^{2+} + 2R\text{-}OH$	$NPC_{Min} = \Delta SOC \times 1.8 \times CEC_{SOM}$
(7) Weathering and cation exchange reaction	$n/2M_{2/n}O_{(s)} + nH^+ \rightarrow M^{n+} + n/2H_2O$	

Note: The suffix 'bio' represents ion fluxes caused by vegetation uptake, assuming that vegetation uptake was equal to the sum of wood increment and litterfall. The suffixes 'in' and 'out' represent ion fluxes entering the soil horizon (e.g. throughfall for the O horizon) and leaving the horizon, respectively. Cat and Ani represent cations (Na^+, K^+, Mg^{2+}, Ca^{2+}, Fe^{3+}, Al^{n+}) and anions (Cl^-, SO_4^{2-}, $H_2PO_4^-$), respectively. ΔSOC represents the rates of SOC loss (Mg C ha^{-1} $year^{-1}$), whereas CEC_{SOM} represents the cation exchange capacity of SOM ($kmol_c$ kg^{-1} SOM). The SOM to SOC ratio of 1.8 was used. CEC_{SOM} was calculated using the soil pH-CEC_{SOM} equation of Helyar and Porter (1989); $CEC_{SOM} = 32 \times$ (soil pH–1.5).

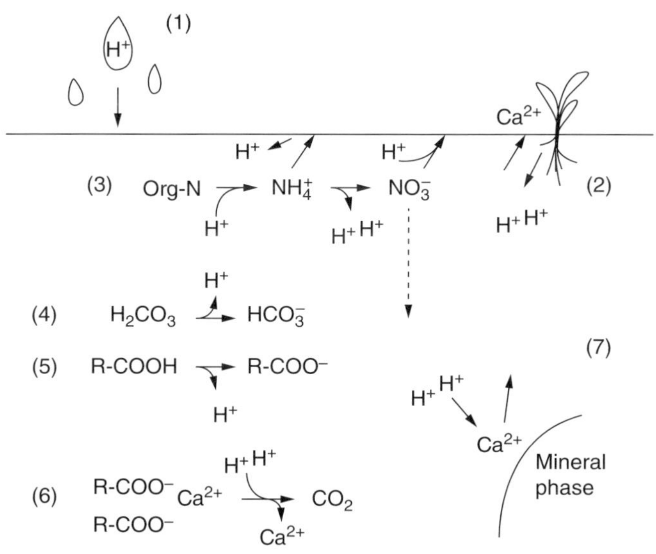

Fig. 6.2. Conceptual diagram of proton-generating and consuming processes in soils. The numbers in parentheses correspond to those in Table 6.1.

the rates of soil organic carbon (SOC) loss (Mg C ha^{-1} $year^{-1}$), the SOM to SOC ratio of 1.8 and the cation exchange capacity of SOM (CEC_{SOM}) ($kmol_c$ kg^{-1} SOM) (Helyar and Porter, 1989; Poss *et al.*, 1995). The principles and calculation methods are described in detail in Fujii *et al.* (2008, 2009a, 2012).

6.2.6 Calculations and statistical analyses

All results were expressed on an oven-dried soil (24 h, 105°C) weight basis and as means ± standard error (SE) of five determinations for SOM decomposition rates, C inputs and C stocks. The significance of differences in mean values between sites was tested using analysis of variance (ANOVA) at the $P < 0.05$ significance level for ion concentrations. Spearman's linear correlations were used to assess the relationships between SOM decomposition rates and volumetric water content in soil. The statistical analyses were performed using Systat 11.0 (SPSS Inc., 2008).

6.3 Results

6.3.1 Physicochemical properties of soils

Soil pH was more acidic throughout the profiles in BSf and BSc than in RPf and RPc (Table 6.2). The RPf and RPc soils were rich in clay (>40%). The SOC concentrations in the A horizons of the cropland plots were significantly lower than in the adjacent forest plots ($P < 0.05$; Tables 6.2 and 6.3). The SOC stocks in the cropland soil (0–20 cm) were greater than in the respective adjacent forest soil ($P < 0.05$; Table 6.3).

6.3.2 Decomposition rates of soil organic matter

Rates of SOM decomposition were positively correlated with volumetric water content in RP (RPf: $r = 0.91$, $P < 0.01$, RPc: $r = 0.63$, $P < 0.10$; Fig. 6.3), while they were independent of volumetric water content in BS (Fujii *et al.*, 2009a). Therefore, the annual rates of SOM decomposition in RPf and RPc were calculated as totals of the C fluxes simulated using the regression equations and soil moisture data (Fig. 6.3) according to Funakawa *et al.* (2006) (Table 6.3). In the forest plots, the annual rate of SOM decomposition was high (5.4–5.5 Mg C ha^{-1} year^{-1}) compared to the reported values in tropical seasonal forest (4.1 Mg C ha^{-1} year^{-1}; Funakawa *et al.*, 2006) and tropical rain forest (4.8–8.9 Mg C ha^{-1} year^{-1}; Bond-Lamberty *et al.*, 2004), respectively. Assuming that annual rates of root litter input were 20% of the fine root biomass (Nakane, 1980), the annual rates of SOM decomposition were almost balanced with C inputs as the sum of litterfall and root litter input in the forest plots (Table 6.3).

In the cropland plots, the annual rates of SOM decomposition were significantly higher than C inputs ($P < 0.05$; Table 6.3). This suggests a loss of SOM in the cropland plots (Table 6.3). This is consistent with Funakawa *et al.* (1997, 2006), in which continuous cropping after deforestation resulted in a loss of SOM in Thailand (5.0 Mg C ha^{-1} year^{-1}). The SOM loss caused by continuous cropping can account for the lower SOM stock in the cropland plots (Table 6.3).

6.3.3 Composition of soil solution

In the BS site, the soil solution was highly acidic in the forest plot (BSf), with the less acidic pH in the cropland soil (BSc) being caused by poultry manure application (Table 6.4). In RP, the soil solution pH was less acidic irrespective of land use, consistent with higher base saturation than BS (Tables 6.2 and 6.4). According to the annual mean concentrations of ions in soil solution (Table 6.4), bicarbonate was present at all plots except for BSf, where it was negligible owing to the low pH of the soil solution. The concentrations and fluxes of organic acids in the surface soil solution were significantly higher in BSf and BSc, compared to RPf and RPc ($P < 0.05$; Table 6.4). The higher DOC concentrations of BSf and BSc, as compared to RPf and RPc, contributed to proton generation through dissociation of one acidic functional group for 11.5 and 10.0 C atoms of DOC, respectively.

In the cropland plots, the concentrations of nitrate in soil solution were significantly higher in both RPc and BSc, when compared to the respective adjacent forest plots ($P < 0.05$; Table 6.4; Fig. 6.4). The concentrations of nitrate in the Ap horizons were highest, especially at the beginning of the cropping season in RPc (April 2004–June 2004) and BSc (October 2004–January 2005), owing to the enhanced decomposition of SOM and the small amount of biomass (Fig. 6.4).

6.3.4 Net proton generation and consumption

Proton budgets were calculated using the ion fluxes associated with solute leaching and plant uptake (Table 6.1). Cation contents exceeded anion contents in litter and wood materials in all

Table 6.2. Physicochemical properties of forest and cropland soils in Thailand and Indonesia.

	Horizon	Depth (cm)	pH	Total C[a] (g kg^{-1})	Total N[a] (g kg^{-1})	Exchangeable cations[a] Ca (cmol$_c$ kg^{-1})	Mg	K	Na	CEC[a] (cmol$_c$ kg^{-1})	Particle size distribution[b] Clay (%)	Silt	Sand
Thailand													
RPf	O	+1–0											
forest	A	0–7	5.0	62.6	3.8	2.98	2.23	0.54	0.02	27.6	70	25	5
	BA	7–20	4.9	19.8	1.5	0.97	0.72	0.22	0.04	19.9	73	23	4
	Bt	20–45+	4.6	8.9	1.0	0.66	0.47	0.15	0.09	20.1	75	19	6
RPc	Ap	0–7	5.4	26.8	2.0	5.66	2.03	0.23	0.03	11.5	40	18	42
cropland	BA	7–20	5.5	15.0	1.3	4.32	1.40	0.08	0.11	14.4	55	13	32
	Bt	20–45+	5.5	10.8	1.0	3.26	0.97	0.05	0.00	14.4	51	10	39
Indonesia													
BSf	O	+2–0											
Forest	A	0–5	4.0	22.9	1.4	0.60	0.11	0.41	1.11	8.5	23	25	52
	BA(E)	5–25	3.8	4.2	0.6	0.63	0.13	0.04	0.00	6.2	24	27	49
	B1	25–40	4.0	3.5	0.5	0.59	0.19	0.04	0.03	5.0	27	30	43
	Bt	40+	4.3	2.5	0.5	0.60	0.14	0.09	0.14	5.0	31	35	34
BSc	Ap1	0–5	4.3	14.3	1.6	1.54	0.12	0.15	0.29	6.8	19	16	64
Cropland	Ap2	5–20	4.2	4.8	0.5	1.26	0.41	0.05	0.04	8.7	28	19	54
	B1	20–40	4.2	3.7	0.5	0.85	0.22	0.06	0.10	10.3	33	21	46
	Bt	40–60+	4.3	3.2	0.4	0.68	0.16	0.06	0.00	11.3	26	19	54

Note: [a]Oven-dry basis. [b]Clay (<0.002 mm); silt (0.002–0.05 mm); sand (0.050–2 mm).

Table 6.3. Stocks and annual fluxes of C in forest and cropland ecosystems in Thailand and Indonesia.

		Thailand		Indonesia	
		Forest	Cropland	Forest	Cropland
		RPf	RPc	BSf	BSc
C stock (Mg C ha^{-1})					
Aboveground biomass		169.1 (–)	5.8 (0.8)	292.6 (–)	1.2 (0.2)
O horizon		2.6 (0.1)	–	3.5 (0.3)	–
Mineral soil horizon (0–20 cm)		55.9 (4.9)	43.5 (2.0)	21.2 (0.7)	18.9 (0.3)
Fine root biomass		6.2 (0.4)	0.2 (0.1)	4.7 (0.6)	0.3 (0.1)
O, A horizon[a]		2.4 (0.3)	0.1 (0.0)	1.3 (0.4)	0.2 (0.1)
BA horizon[b]		2.2 (0.2)	0.1 (0.0)	0.9 (0.3)	0.1 (0.0)
B1 horizon[c]		1.7 (0.1)	–	2.6 (0.4)	–
C flux (Mg C ha^{-1} year^{-1})					
Organic matter decomposition	(a)	5.5 (0.3)	8.2 (0.3)	5.4 (0.5)	3.6 (0.3)
Organic matter production					
C input[a]	(b)	5.2 (0.2)	4.3 (0.7)	5.0 (0.5)	1.4 (0.1)
Litterfall	(c)	4.0 (0.2)	4.1 (0.7)	4.1 (0.5)	1.1 (0.1)
Root litter[b]	(d)	1.2 (0.1)	0.2 (0.1)	0.9 (0.1)	0.3 (0.1)
Crop removal		–	1.7 (0.3)	–	0.2 (0.0)
Wood increment		5.8 (–)	–	10.7 (–)	–
C budget in soil[c]	(e)	–0.2 (0.4)	–3.9 (0.8)	–0.4 (0.8)	–2.2 (0.3)

Note: The figures in parenthesis represent standard errors ($n = 5$). [a]C input was calculated as the sum of litterfall and root litter (b = c + d). [b]The annual rates of root litter input were assumed to be 20% of the fine root biomass in the forests (Nakane, 1980). [c]budget in soil was calculated as the difference between organic matter decomposition and C input (e = b – a).

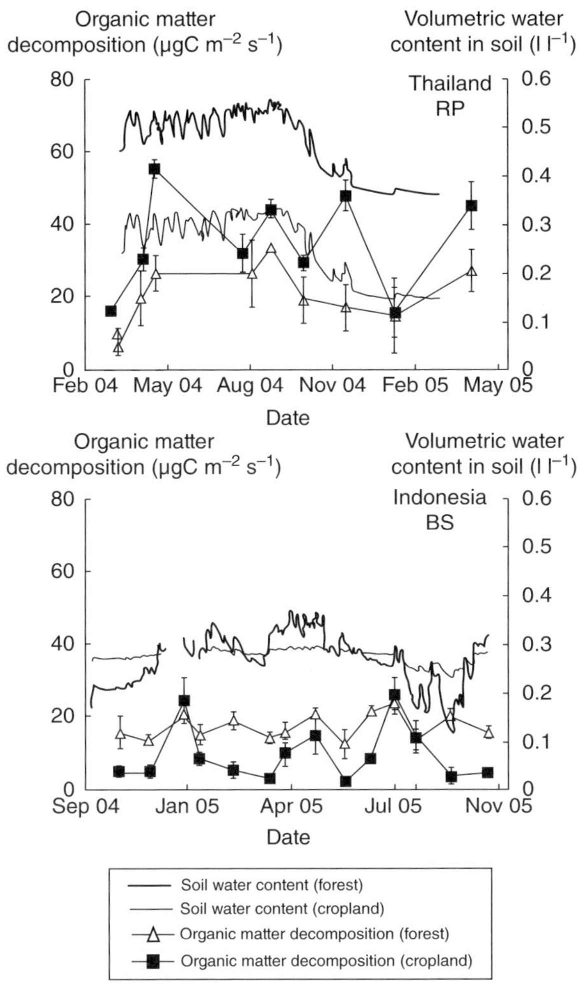

Fig. 6.3. Seasonal fluctuations of soil moisture and organic matter decomposition rates in the forest and cropland sites of Thailand (RP) and Indonesia (BS). Bars indicate standard errors ($n = 5$).

plots (Fig. 6.6). Excess cation charge was compensated for by the net proton load to the soil as NPG_{Bio} (= cation uptake by plants). Within plant uptake, since litterfall was the circulating fraction, NPG_{Bio} attributable to wood increment (forest) or crop removal (cropland) was counted in net proton generation. NPG_{Bio} in each of the soil horizons was calculated by allocating it based on the distribution of the fine root biomass (Table 6.3), according to Shibata *et al.* (1998). Using NPG_{Ntr} (net proton generation by nitrification), NPG_{Car} (dissociation of carbonic acid), NPG_{Org} (dissociation of organic acids), NPG_{Bio} (cation loss caused by wood increment or crop removal) and NPC_{Min} (proton consumption by SOM loss and manure) (Table 6.1; Figs 6.5 and 6.6), the proton budgets were calculated in each of the soil horizons (Fig. 6.7).

In RPf, the fluxes of organic acids and nitrate in soil solution were small owing to the adsorption of organic acids and rapid nitrate uptake by vegetation, respectively (Fig. 6.5), and thus, contribution of nitric and organic acids to acidification was negligible (Fig. 6.7). Although bicarbonate was one of the dominant anions in soil solution in RPf (Table 6.4; Fig. 6.5), the contribution of carbonic acid to soil acidification was minor. In RPf, net proton generation for wood increment was found in the mineral soil horizons and it was

Table 6.4. Annual volume-weighed mean concentrations of ions in precipitation, throughfall and soil solution in forest and cropland soils of Thailand and Indonesia.

Site	Horizon	Water flux (mm year^{-1})	pH	DOC (Fujii et al., 2013) (mg C l^{-1})	HCO_3^-	Cl^-	NO_3^-	SO_4^{2-}	Org^{n-} [b]	H^+	Na^+	NH_4^+	K^+	Mg^{2+}	Ca^{2+}	Fe^{2+}	Al^{n+} [c]
					(mmol$_c$ l^{-1})												
Thailand																	
RPf	TF[a]	2083	6.1	2.7	0.03	0.05	0.02	0.01	0.01	0.00	0.01	0.01	0.04	0.03	0.03	0.00	0.00
Forest	A	1602	6.2	3.9	0.03	0.06	0.02	0.02	0.02	0.00	0.02	0.01	0.05	0.03	0.03	0.01	0.01
	BA	1162	6.1	3.1	0.03	0.05	0.03	0.01	0.01	0.00	0.02	0.01	0.04	0.03	0.02	0.01	0.01
	Bt	825	6.1	3.1	0.03	0.06	0.02	0.01	0.01	0.00	0.02	0.01	0.04	0.03	0.02	0.01	0.00
RPc	P[a]	2223	6.2	2.9	0.03	0.04	0.01	0.02	0.01	0.00	0.01	0.01	0.01	0.02	0.05	0.00	0.00
Cropland	Ap	1414	5.8	4.5	0.03	0.06	0.37	0.06	0.02	0.00	0.04	0.01	0.03	0.19	0.26	0.00	0.00
	BA	1064	5.7	2.2	0.05	0.05	0.38	0.02	0.01	0.00	0.03	0.01	0.01	0.16	0.28	0.00	0.01
	B1	1064	5.8	1.9	0.05	0.07	0.34	0.03	0.01	0.00	0.04	0.01	0.01	0.15	0.24	0.00	0.01
Indonesia																	
BSf	TF[a]	2031	5.2	9.0	0.01	0.04	0.04	0.06	0.10	0.01	0.03	0.04	0.08	0.05	0.04	0.00	0.01
Forest	O	1619	4.4	34.7	0.00	0.06	0.04	0.07	0.26	0.04	0.04	0.09	0.11	0.07	0.04	0.02	0.04
	A	1196	4.2	17.2	0.00	0.05	0.04	0.05	0.20	0.06	0.04	0.04	0.07	0.05	0.03	0.01	0.04
	B1	545	4.4	9.9	0.00	0.06	0.03	0.06	0.12	0.04	0.04	0.02	0.06	0.04	0.04	0.00	0.03
BSc	P[a]	2187	6.1	4.2	0.02	0.01	0.01	0.01	0.02	0.00	0.01	0.01	0.00	0.01	0.02	0.00	0.00
Cropland	Ap1	1144	5.6	9.7	0.04	0.04	0.14	0.05	0.13	0.00	0.03	0.02	0.05	0.08	0.14	0.01	0.05
	Ap2	824	5.6	9.3	0.05	0.04	0.14	0.05	0.09	0.00	0.04	0.02	0.04	0.06	0.15	0.01	0.03
	B1	824	5.2	5.9	0.04	0.04	0.06	0.05	0.04	0.01	0.04	0.02	0.03	0.03	0.08	0.01	0.02

[a]P and TF represent precipitation and throughfall, respectively. O, A, Ap, Ap1, Ap2 and BA, B, B1 and Bt represent soil horizons. [b]Org^{n-} represents anion deficit, the negative charge of organic acids. [c]Al^+ represents total charge equivalent of Al ions (Al^{3+}, $AlOH^{2+}$ and $Al(OH)_2^+$).

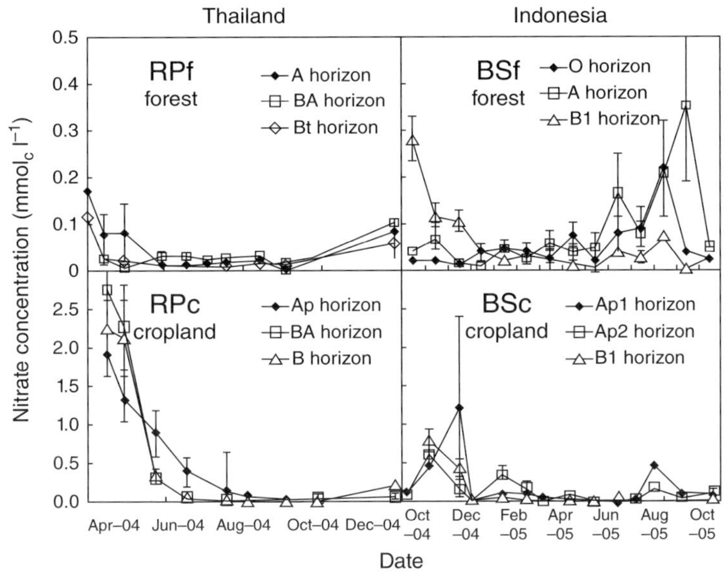

Fig. 6.4. Seasonal fluctuations of the concentrations of nitrate in soil solution of forest and cropland soils in Thailand (RP) and Indonesia (BS). Bars indicate standard errors (n = 5).

a dominant proton source in each soil horizon (Fig. 6.7). In BSf, protons were produced by the dissociation of organic acids in the O horizon, although they were consumed owing to the mineralization and adsorption of organic acids in the A and B horizons (Fig. 6.7). Proton generation by plant uptake (wood increment) was also found in the O horizon, as well as in the A and B horizons, and contributed to acidification of the O horizon (Fig. 6.7).

In RPc and BSc, protons were produced by nitrification in the Ap horizons, most of which were neutralized by basic cations in the Ap horizons and this resulted in leaching of basic cations (Figs 6.5 and 6.7). Owing to the fine root biomass concentrated in the Ap horizons (Table 6.3), NPG_{Bio} (crop removal) was concentrated in the Ap horizons in the cropland plots (Fig. 6.6). Although NPG_{Org} in RPc and BSc was lower than the respective forest plots, organic acids also contributed to proton generation in the Ap horizon of BSc. In the cropland plots, assuming the loss of SOM from the Ap horizons (Table 6.3) and CEC_{SOM} to be 124 and 90 cmol$_c$ kg^{-1} SOM in RPc and BSc, respectively, NPC_{Min} values associated with SOM loss were estimated to be 8.8 ± 1.8 and 3.6 ± 0.5 kmol$_c$ ha^{-1} year^{-1} in RPc and BSc, respectively. There was a minor contribution to the value of NPC_{Min} by poultry manure (0.2 kmol$_c$ ha^{-1} year^{-1}) in BSc (Fig. 6.6). NPC_{Min} was higher than the sum of NPGs and it resulted in soil alkalinization in the Ap horizons (Fig. 6.7).

6.4 Discussion

6.4.1 Natural acidification processes of tropical forest soils

The importance of organic acids to soil acidification was different between forest sites. Proton generation of organic acids in BSf is as high as those in Spodosol soils under temperate forests (1.2–3.7 kmol$_c$ ha^{-1} year^{-1} from Guggenberger and Kaiser (1998) and Fujii *et al.* (2008)). In BSf, the larger production of DOC in the O horizon resulted in acidification of surface soils by organic acids. Ammonium leaching from the O horizon and its uptake in the A horizon also contributed to the vertical variation of acid load in the soil profile of BSf, while the concurrence of mineralization and uptake by plants resulted in small fluxes of protons associated with nitrification in RPf.

Since nitric and organic acids are working only within the surface soil horizons, their contribution to acidification in the entire soil

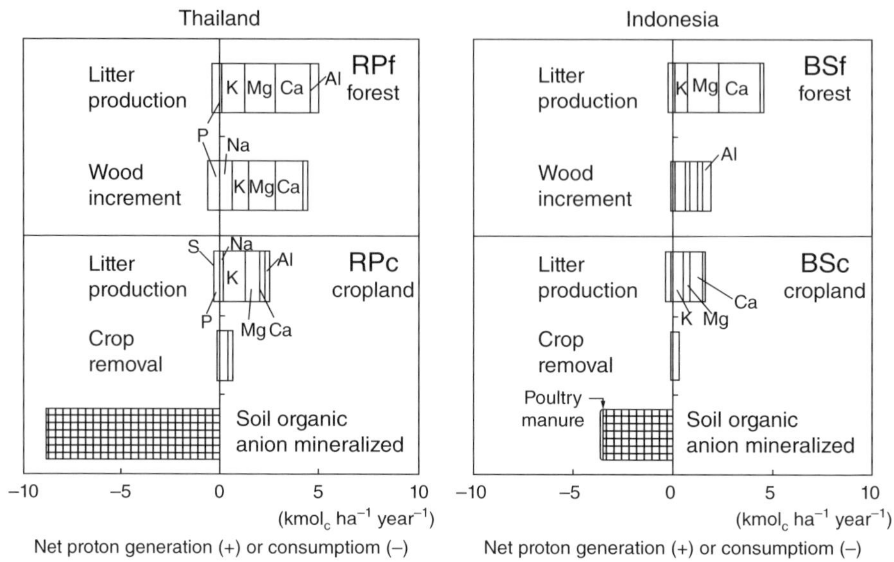

Fig. 6.5. Proton generation by vegetation uptake and proton consumption by mineralization of soil organic anions in forest and cropland soils in Thailand (RP) and Indonesia (BS).

profile is minor in the forests studied. Excess cation accumulation in wood is commonly a dominant proton source in the growth stage of the RP and BS forests. The small loss of cations from the soil profile and large fluxes of plant uptake appeared to be common to tropical forests (Fujii, 2014).

6.4.2 Soil acidification processes in cropland soils

In cropland soils, the higher temperature than found in forest soils could increase microbial activity of SOM decomposition (Hayakawa *et al.*, 2014). Loss of SOM is considered to contribute to nitrification predominating over vegetation uptake in the Ap horizons of the croplands. This could contribute to acidification of cropland soils. Despite a lack of fertilization, the magnitude of proton generation by nitrification in RPc is comparable to the higher reported values in fertilized croplands (1.4–11.5 kmol$_c$ ha^{-1} year^{-1} from Ridley *et al.* (1990), Bouman *et al.* (1995), Poss *et al.* (1995), Lesturgez *et al.* (2006) and Noble *et al.* (2008)). Since proton generation owing to nitrification is caused by the net mineralization and nitrification of soil organic N, the higher proton generation in the RPc soil is considered to be caused by the higher rate of SOM loss, compared to BSc. The lower NPG_{Ntr} in BSc is considered to be partly attributed to the lower soil pH (Kemmitt *et al.*, 2006), as nitrification is generally suppressed in acidic soils (Kemmitt *et al.*, 2006). Although acidification is increased by nitrification after continuous cropping, its extent varies depending on soil pH, as well as the rates of SOM loss.

Although continuous cropping results in a decrease in the DOC concentration in the A horizon owing to a loss of the O horizon, the DOC concentrations in the Ap horizon of BSc were still higher compared to RPc. The higher DOC concentration results in higher proton generation owing to organic acid dissociation in BSc compared to RPc. In the RPc soil, the contribution of organic acids to acidification was negligible owing to the greater adsorption of DOC on the clay-rich soil.

In the croplands, crop removal, as well as nitrification, contributed to soil acidification. This is consistent with the pattern of soil acidification in pastures (Bolan *et al.*, 1991). However, proton budget analysis showed that an intensive acid load by nitrification could be neutralized by proton consumption associated with a loss of SOM

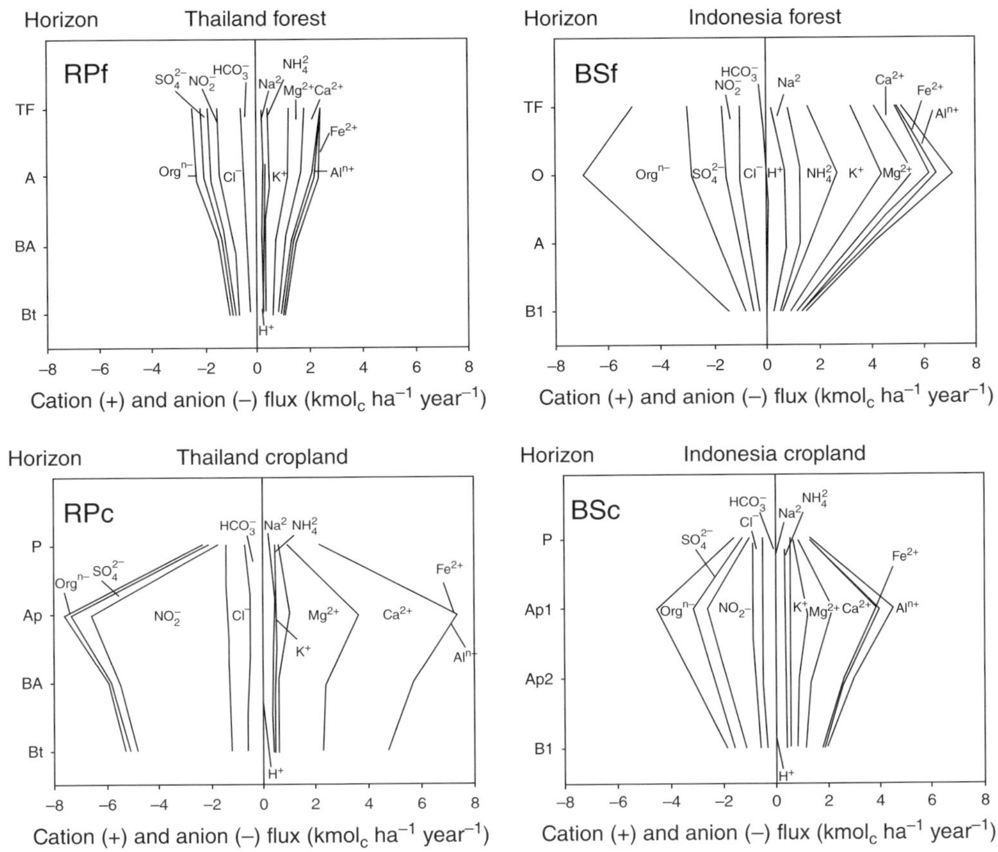

Fig. 6.6. Fluxes of solutes at each horizon of forest and cropland soils in Thailand (RP) and Indonesia (BS). P and TF represent precipitation and throughfall, respectively. O, A, Ap, Ap1, Ap2 and BA, B, B1 and Bt represent soil horizons.

in cropland soil in the initial stages of cropping following deforestation in the present study.

6.4.3 Effects of continuous cropping on organic matter cycles and soil acidification

In the forest soil profiles, the contribution of nitrification to acidification was minor. Judging from the almost even balance between SOM decomposition and C input, a complete cycle of N is considered to be balanced with no net proton fluxes in forest ecosystems (Binkley and Richter, 1987). On the other hand, in the croplands, where SOM decomposition was greater than C, inputs resulting in a loss of SOM, net mineralization of soil organic nitrogen and limited vegetation uptake at the beginning of the cropping season contributed to net proton generation.

Concurrently, the loss of SOM could contribute to proton consumption because of the mineralization of organic anions in croplands. This contrasts with the case of forests, where excess cation accumulation in organic matter (biomass and humus) contributes to soil acidification. In the present study, organic matter production contributed to proton generation at the rate of 0.004–0.020 mol$_c$ for the production of 1 mol organic C. On the other hand, in the cropland soils, a loss of SOM consumed protons at rates ranging from 0.019 to 0.026 mol$_c$ for the loss of 1 mol soil organic C (Fujii *et al.*, 2009a). Both proton generation and consumption were greater in RPc than in BSc owing to the higher nitrification activity and the higher CEC_{SOM} of the less acidic soils of RPc. Net proton consumption

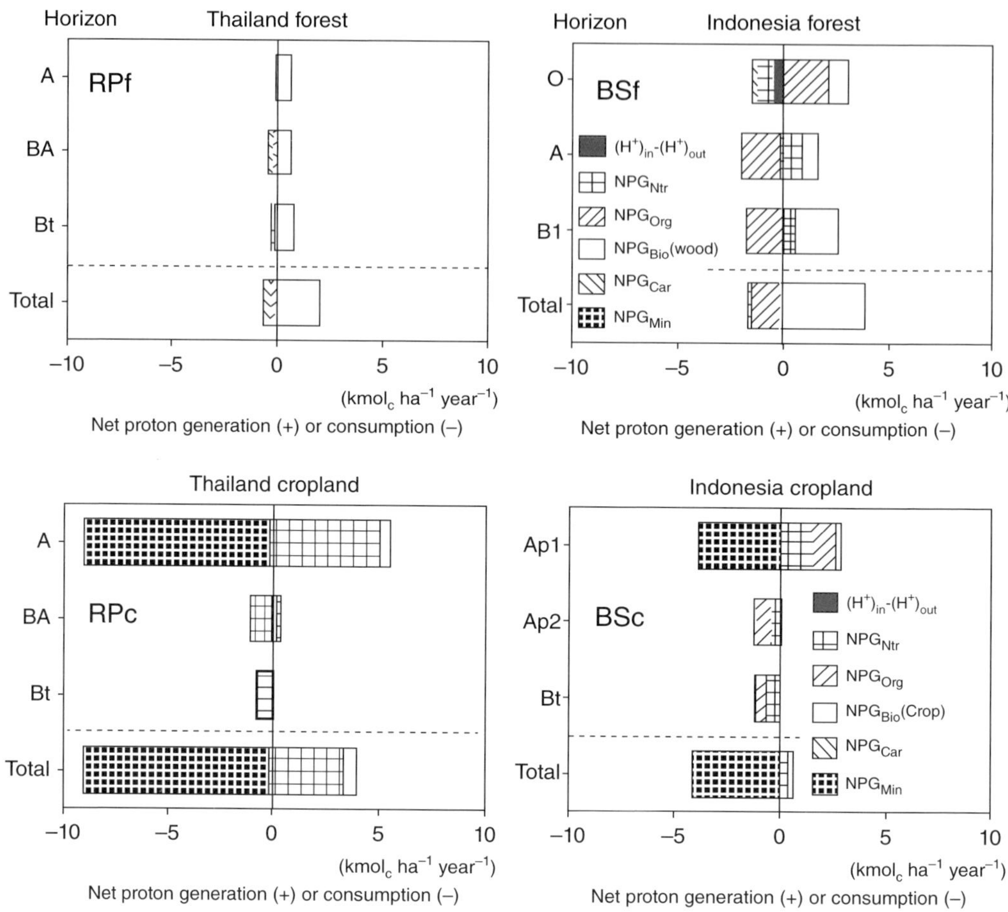

Fig. 6.7. Net proton generation and soil acidification in the soil profiles from Thailand (RP) and Indonesia (BS). O, A, Ap, Ap1, Ap2 and BA, B, B1 and Bt represent soil horizons. NPG_{Bio}: net proton generation (NPG) resulting from excess cation uptake by vegetation, NPG_{Ntr}: NPG owing to transformation of nitrogen, NPG_{Org}: NPG owing to dissociation of organic acids, NPG_{Car}: NPG owing to dissociation of carbonic acid and $\{(H^+)_{in}\text{-}(H^+)_{out}\}$: net proton influx from the overlying horizon.

owing to net mineralization of organic anions could increase with increasing pH and SOC content (Poss *et al.*, 1995). The effect of continuous cropping on soil acidification could vary from soil to soil, depending on the organic matter budgets and soil pH and texture. Under low-input tropical agriculture, SOM cycles have a strong influence on acidification processes. Soil acidification associated with intensive cropping after deforestation can be ameliorated by the build-up of SOM during forest fallow periods. Case-by-case land-use strategies need to be developed by adjusting the length of forest fallow periods between cropping for maximizing the harvest or benefits of farmers.

6.4.4 Soil acidification and implications for amelioration strategy in Asian countries

Increasing application of fertilizer N (especially ammonium sulfate), as well as net mineralization of SOM, can accelerate soil acidification (Bolan *et al.*, 1991; Bouman *et al.*, 1995). In Asia in particular, soil pH has decreased by 0.3 to 1.1 units within the last few decades (Ali *et al.*, 1997; Darmawan *et al.*, 2006; Guo *et al.*, 2010) (Fig. 6.8). In Bangladesh and Indonesia (Java), the use of high-yielding varieties of rice and chemical fertilizers following the green revolution has not only increased rice production but has also

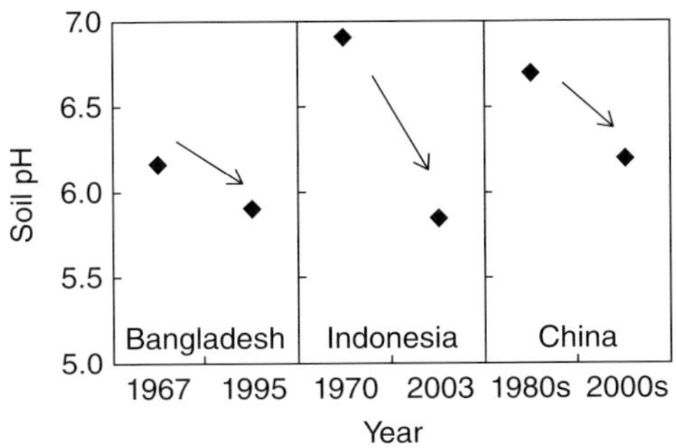

Fig. 6.8. Soil pH changes over recent decades in selected Asian countries. (Data sources for Bangladesh, Indonesia and China are Ali *et al.* (1997), Darmawan *et al.* (2006) and Guo *et al.* (2010), respectively.)

accelerated soil acidification (Ali *et al.*, 1997; Darmawan *et al.*, 2006). The increased input of N fertilizer without sufficient application of lime and organic matter results in a low to moderate rate of soil acidification (~2 $kmol_c\ ha^{-1}\ year^{-1}$) in these regions (Ali *et al.*, 1997; Darmawan *et al.*, 2006). In China, the conversion of crop production systems to cash crop production (e.g. vegetables) has resulted in rapid soil acidification through nitrification of excessive ammonium fertilizer N (500–4000 kg N $ha^{-1}\ year^{-1}$) and crop removal (Guo *et al.*, 2010). Proton generation due to N cycling (20–221 $kmol_c\ ha^{-1}\ year^{-1}$) in China is extremely high compared with the lower N fertilizer rates in other regions (1.4–11.5 $kmol_c\ ha^{-1}\ year^{-1}$).

The impact of agriculture on soil acidification may be ameliorated by improving N fertilization strategies and maintaining SOM levels. For example, strategies to improve N use efficiency and reduce excessive N fertilization have been proposed in China (Ju *et al.*, 2009). Proton transfer associated with SOM cycles in cropland soils suggests that the maintenance of SOM levels will neutralize protons produced by nitrification (Fujii *et al.*, 2009a). Proton budget analysis is a promising approach for identifying the dominant acidifying processes and for proposing site-specific and feasible countermeasures against soil acidification.

6.5 Conclusions

Under continuous cropping of the deforested land in northern Thailand, the enhanced SOM decomposition and crop removal result in a loss of SOM. The mineralization of soil organic nitrogen can enhance proton generation owing to nitrification. At the same time, SOM loss also enhanced proton consumption through mineralization of soil organic anions. Acidity contributed by nitrification can be neutralized by depletion of SOM at least during the initial stage of cropping after deforestation in tropical regions. However, it should be noted that the capacities of acid neutralization associated with SOM are reduced by continuous cropping in tropical regions, where the SOM stock is originally small compared to the temperate regions. In some regions of Asia, the conversion of traditional agricultural systems to intensive farming has accelerated soil acidification through nitrification of excessive fertilizer N within the past few decades. The impact of agriculture on soil acidification may be ameliorated by improving N fertilization strategies and maintaining SOM levels by assessing the effects of cropping on proton budgets. Under low-input tropical agriculture, SOM cycles have a strong influence on acidification processes. Soil acidification associated with intensive cropping after deforestation can be ameliorated by the build-up of SOM during forest fallow periods. Thus, case-by-case land-use strategies need to be developed by adjusting the length of forest fallow periods between cropping for maximizing the harvest or benefits to farmers through a fuller understanding of tropical soil chemical processes.

References

Ali, M.M., Saheed, S.M., Kubota, D., Matsunaga, T. and Wakatsuki, T. (1997) Soil degradation during the period 1967–1995 in Bangladesh. II. Selected chemical characters. *Soil Science and Plant Nutrition* 43, 879–890.

Binkley, D. and Richter, D. (1987) Nutrient cycles and H^+ budgets of forest ecosystems. *Advances in Ecological Research* 16, 1–51.

Bolan, N.S., Hedley, M.J. and White, R.E. (1991) Processes of soil acidification during nitrogen cycling with emphasis on legume based pastures. *Plant and Soil* 134, 53–63.

Bond-Lamberty, B., Wang, C. and Gower, S.T. (2004) A global relationship between the heterotrophic and autotrophic components of soil respiration? *Global Change Biology* 10, 1756–1766.

Bouman, O.T., Curtin, D., Campbell, C.A., Biederbeck, V.O. and Ukrainetz, H. (1995) Soil acidification from long-term use of anhydrous ammonia and urea. *Soil Science Society of America Journal* 59, 1488–1494.

Bredemeier, M., Matzner, E. and Ulrich, B. (1990) Internal and external proton load to forest soils in Northern Germany. *Journal of Environmental Quality* 19, 469–477.

Busman, L.M., Dick, B.P. and Tabatabai, M.A. (1983) Determination of total sulfur and chlorine in plant materials by ion chromatography. *Soil Science Society of America Journal* 47, 1167–1170.

Bruun, T.B., de Neergaard, A., Lawrence, D. and Ziegler, A. (2009) Environmental consequences of the demise in swidden agriculture in Southeast Asia: Carbon storage and soil quality. *Human Ecology* 37, 375–388.

Darmawan, K.K., Saleh, A., Masunaga, T. and Wakatsuki, T. (2006) Effect of green revolution technology during the period 1970–2003 on sawah soil properties in Java, Indonesia: II. Changes in the chemical properties of soils. *Soil Science and Plant Nutrition* 52, 645–653.

Fujii, K. (2014) Soil acidification and adaptations of plants and microorganisms in Bornean tropical forests. *Ecological Research* 29, 371–381.

Fujii, K., Funakawa, S., Hayakawa, C. and Kosaki, T. (2008) Contribution of different proton sources to pedogenetic soil acidification in forested ecosystems in Japan. *Geoderma* 144, 478–490.

Fujii, K., Funakawa, S., Hayakawa, C. and Sukartiningsih, K.T. (2009a) Quantification of proton budgets in soils of cropland and adjacent forest in Thailand and Indonesia. *Plant and Soil* 316, 241–255.

Fujii, K., Uemura, M., Funakawa, S., Hayakawa, C., Sukartiningsih, K.T. and Ohta, S. (2009b) Fluxes of dissolved organic carbon in two tropical forest ecosystems of East Kalimantan, Indonesia. *Geoderma* 152, 127–136.

Fujii, K., Funakawa, S. and Kosaki, T. (2012) Soil acidification: Natural process and human impacts. *Pedologist* 55, 415–425.

Fujii, K., Hayakawa, C., Funakawa, S. and Sukartiningsih, K.T. (2013) Fluxes of dissolved organic carbon and nitrogen in cropland and adjacent forest in Thailand and Indonesia. *Soil and Tillage Research* 126, 267–275.

Funakawa, S., Tanaka, S., Shinjo, H., Kaewkhongkha, T., Hattori, T. and Yonebayashi, K. (1997) Ecological study on the dynamics of soil organic matter and its related properties in shifting cultivation systems of northern Thailand. *Soil Science and Plant Nutrition* 43, 681–693.

Funakawa, S., Hayashi, Y., Tazaki, I., Sawada, K. and Kosaki, T. (2006) The main functions of the fallow phase in shifting cultivation by Karen people in northern Thailand – A quantitative analysis of soil organic matter dynamics. *Tropics* 15, 1–27.

Guggenberger, G. and Kaiser, K. (1998) Significance of DOM in the translocation of cations and acidity in acid forest soils. *Zeitschrift für Pflanzenernaehrung und Bodenkunde* 161, 95–99.

Guo, J.H., Liu, X.J., Zhang, Y., Shen, J.L., Han, W.X., Zhang, W.F., *et al.* (2010) Significant acidification in major Chinese croplands. *Science* 327, 1008–1010.

Hallbäcken, L. and Tamm, C.O. (1986) Changes in soil acidity from 1927 to 1982–1984 in a forest area of south-west Sweden. *Scandinavian Journal of Forest Research* 1, 219–232.

Hayakawa, C., Funakawa, S., Fujii, K., Kadono, A. and Kosaki, T. (2014) Effects of climatic and soil properties on cellulose decomposition rates in temperate and tropical forests. *Biology and Fertility of Soils* 50, 633–643.

Helyar, K.R. and Porter, W.M. (1989) Soil acidification, its measurement and the processes involved. In: Robson, A.D. (ed.) *Soil Acidity and Plant Growth*. Academic Press, London, pp. 61–101.

Johnson, F.L. and Risser, P.G. (1974) Biomass, annual net primary production, and dynamics of six mineral elements in a post oak–blackjack oak forest. *Ecology* 55, 1246–1258.

Ju, X.T., Xing, G.X., Chen, X.P., Zhang, S.L., Zhang, L.J., Liu, X.J. *et al.* (2009) Reducing environmental risk by improving N management in intensive Chinese agricultural systems. *Proceedings of the National Academy of Science of the USA* 106, 3041–3046.

Juo, A.S.R., Franzluebbers, K., Dabiri, A. and Ikhile, B. (1996) Soil properties and crop performance on a kaolinitic Alfisol after 15 years of fallow and continuous cultivation. *Plant and Soil* 180, 209–217.

Kemmitt, S.J., Wright, D., Goulding, K.W.T. and Jones, D.L. (2006) pH regulation of carbon and nitrogen dynamics in two agricultural soils. *Soil Biology and Biochemistry* 38, 898–911.

Kuzyakov, Y. (2006) Sources of CO_2 efflux from soil and review of partitioning methods. *Soil Biology and Biochemistry* 38, 425–448.

Kyuma, K. and Pairintra, C. (1983) Shifting Cultivation. *An Experiment at Nam Phrom, Northeast Thailand, and its Implications for Upland Farming in the Monsoon Tropics*. Kyoto University, Kyoto, Japan.

Lesturgez, G., Poss, R., Noble, A., Grunberger, O., Chintachao, W. and Tessier, D. (2006) Soil acidification without pH drop under intensive cropping systems in Northeast Thailand. *Agriculture, Ecosystems and Environment* 114, 239–248.

Mertz, O., Padoch, C., Fox, J., Cramb, R.A., Leisz, S.J., Lam, N.T. and Vien, T.D. (2009) Swidden change in southeast Asia: understanding causes and consequences. *Human Ecology* 37, 259–264.

Nakane, K. (1980) Comparative studies of cycling of soil organic carbon in three primeval moist forests. *Japanese Journal of Ecology* 30, 155–172.

Noble, A.D., Suzuki, S., Soda, W., Ruaysoongnern, S. and Berthelsen, S. (2008) Soil acidification and carbon storage in fertilized pastures of Northeast Thailand. *Geoderma* 144, 248–255.

Nye, P.H. and Greenland, D.J. (1960) *The Soil under Shifting Cultivation*. CBS Tech. Commun. No. 51, Harpenden, UK.

Poss, R., Smith, C.J., Dunin, F.X. and Angus, J.F. (1995) Rate of soil acidification under wheat in a semi-arid environment. *Plant and Soil* 177, 85–100.

Ridley, A.M., Slattery, W.J., Helyar, K.R. and Cowling, A. (1990) The importance of the carbon cycle to acidification of a grazed annual pasture. *Australian Journal of Experimental Agriculture* 30, 529–537.

Shibata, H., Kirikae, M., Tanaka, Y., Sakuma, T. and Hatano, R. (1998) Proton budgets of forest ecosystem on canogenous Regosols in Hokkaido. *Water, Air and Soil Pollution* 105, 63–72.

Shinjo, H., Kato, A., Fujii, K., Mori, K., Funakawa, S. and Kosaki, T. (2006) Carbon dioxide emission derived from soil organic matter decomposition and root respiration in Japanese forests under different ecological conditions. *Soil Science and Plant Nutrition* 52, 233–242.

Tanaka, S., Funakawa, S., Kaewkhongkha, T., Hattori, T. and Yonebayashi, K. (1997) Soil ecological study on dynamics of K, Mg, and Ca, and soil acidity in shifting cultivation in northern Thailand. *Soil Science and Plant Nutrition* 43, 695–708.

Tsutsumi, T., Yoda, K., Sahunalu, P., Dhanmanonda, P. and Prachaiyo, B. (1983) Forest: Felling, burning and regeneration. In: Kyuma, K. and Pairintra, C. (eds) *An Experiment at Nam Phrom, Northeast Thailand, and Its Implications for Upland Farming in the Monsoon Tropics*. Kyoto University, Kyoto, Japan.

Van Breemen, N., Mulder, J. and Driscoll, C.T. (1983) Acidification and alkalization of soils. *Plant and Soil* 75, 283–308.

Van Breemen, N., Driscoll, C.T. and Mulder, J. (1984) Acidic deposition and internal proton in acidification of soils and waters. *Nature* 307, 599–604.

Yamakura, T., Hagihara, A., Sukardjo, S. and Ogawa, H. (1986) Aboveground biomass of tropical rain forest stands in Indonesia Borneo. *Vegetatio* 68, 71–82.

7 The Importance of Soil Quality in the Safe Practice of Urban Agriculture in Zimbabwe, Kenya and South Africa

Lovemore Chipungu,[1] Hangwelani H. Magidimisha,[2] Michael Hardman[3]* and Luke Beesley[4]

[1]School of the Built Environment and Development Studies, University of KwaZulu-Natal, Durban, South Africa; [2]Human Science Research Council, Demography Governance and Service Delivery, Durban, South Africa; [3]School of Environment and Life Sciences, University of Salford, Salford, UK; [4]Environmental and Biochemical Sciences, The James Hutton Institute, Craigiebuckler, Aberdeen, UK

7.1 Introduction: Urban Soils as Vital Pseudo-natural Capital

Urban soils, waters and wastes are a valuable natural capital asset for the world's burgeoning urban population. The utilization of this capital is beginning to be recognized as fundamental to strategies for ensuring a safe and secure food supply in many countries. This idea of growing in the city, or 'urban agriculture', is a relatively new concept in certain parts of the world, although ample literature exists on the practice in the North American context (see, e.g. Mougeot, 1999; Viljoen, 2005; Wiskerke and Viljoen, 2012). This chapter explores urban agriculture in Africa, which to date has had relatively little exposure in the academic literature.

Within this chapter, we focus on three major cities across the continent: Harare (Zimbabwe), Nairobi (Kenya) and Johannesburg (South Africa). In each case, we provide an overview of urban agricultural activities, authority responses to the practice and an evaluation of the soil properties in which the agriculture occurs. This overview of urban agriculture in selected African cities, particularly the exploration of soil quality and contamination issues breaks new ground since it features one of the first reviews of urban agriculture and soils. Ultimately, this piece aims to provide an insight into practices on the African continent and pave the way for subsequent research.

7.1.1 Urban soils as distinct entities

Soils are the basis of life: providing vital nutrients, harbouring diverse ecological communities, storing carbon and maintaining water supplies. While the United States Department of Agriculture (USDA) soil taxonomy system does not recognize anthropogenic soils at higher orders of taxonomy, the World Reference Base classification system (IUSS Working Group WRB, 2006) does recognize soils strongly influenced by human intervention as distinct and existing

* Corresponding author: m.hardman@salford.ac.uk

in their own right and categorizes them as those: (i) formed or strongly modified by human activity (Anthrosols), and (ii) predominantly containing artefacts derived from human activities (Technosols) (Rossiter, 2007). Figure 7.1 displays a representation of a typical composition of various waste materials supplementing native soil-forming materials and added organic amendment (Beesley, 2012). Both Anthrosol and Technosol designations may apply to urban soils and have a profound influence on the productivity and safety of agricultural activities practised thereon, although Technosols are more likely to exist in urban areas, due to the locally derived artefact contribution.

7.1.2 Risks associated with cultivation of urban soils

In various urban topsoils, concentrations of heavy metals have been shown to vary considerably over relatively small geographical scales (Madrid *et al.*, 2002, 2007) creating sporadically elevated levels of 'risk elements', which are difficult to identify, map and avoid (Hursthouse *et al.*, 2004; Thums *et al.*, 2008). In themselves, elevated concentrations of 'risk elements' – those that induce toxicity symptoms if ingested – present minimal health concern without transfer from soil to people. Contemporary reviews, documenting the environmental context of urban agricultural activities in developed countries, highlight the enhanced possibilities for the passage of contaminants from soils to humans due to cultivating urban soils (Cook *et al.*, 2005). For instance, Wortman and Taylor-Lovell (2013) summarize various sources of contamination to a typical vacant urban space in agricultural use, including emissions from traffic, storm water (often used for irrigation) and artefacts, such as paint fragments containing lead (Pb), that are all sources of contamination to soils.

Fig. 7.1. Typical urban 'Technosol' profile composed of native soil-forming materials, waste artefacts and organic amendments.

Mok *et al.* (2014), reviewing urban agricultural practices in developed countries, highlight chemical contamination of soils as a key risk associated with food produced in urban areas, stating that best management practices to mitigate such risks include rigorous monitoring of sites, bioremediation of contaminated areas and removal of soils. If one considers 'formalized' or 'planned' urban agriculture, such as allotments, a burden of responsibility lies with the local administration to ensure that soils are fit for purpose, and a raft of local and national regulatory frameworks exist for this purpose, some of which are specifically related to urban soils with agricultural uses (see Hardman and Larkham, 2014). However, in situations where urban agriculture is practised 'informally' or where no regulations exist or are not upheld, there is a clear possibility that the quality of soils, and their management, could result in contaminant loads exceeding safe levels in foods. Where soil quality is already poor, climatic conditions unfavourable or demand for food exceeds supply, the pressure to cultivate marginal soils, whether that is those with low fertility or elevated 'risk element' loadings, is more likely.

7.1.3 Emerging pressures on urban natural capital

Reliance on urban agriculture as a predominant source of food differs with economic development (cf. Caputo, 2012). Gumbo (2000), quoting the

UNDP (1996), argues that 15% of total world food production comes from urban agriculture (mainly farming, horticulture, animal husbandry and fishing) and that over 1 billion people are actively engaged in urban agriculture with 200 million being full-time farmers producing food for market. Data collected from 15 developing and transition countries by the United Nations Food and Agriculture Organization (UN-FAO) (2008) show that the share of income from agriculture by urban poor households is increasing. Within the African context, urban agriculture among poor urban households has risen. This is against the backdrop of a population increase in Africa of around 60% between 2010 and 2050, with the urban population tripling to 1.23 billion during the same period (UN-Habitat, 2013). This will inevitably introduce a huge number of challenges (cf. Odendaal, 2012), many of which will stretch natural capital to its limits and beyond. This increase in population is associated, at least partially, with the 'urbanization of poverty' typified by the development of informal settlements (UN-Habitat, 2003). Yet despite these issues, and Africa being on the frontline of the global food crisis, there has been little exploration of urban agriculture within this context. Rapid urbanization is one of Africa's greatest challenges and, coupled with the need to achieve sustainable growth in all sectors of the urban economy (UN-Habitat, 2013; Zeeuw and Dubbling, 2009), urban agricultural activities are only likely to increase further.

Generally, Africans already have a lack of access to the most basic of services, from educational and social services, employment, safe and affordable water to sanitation, housing and more. Yet the recent financial crisis has resulted in more pressure on the continent's population; the rise in food and fuel prices is perhaps the most severe arising from the crisis. Food prices in Harare (Zimbabwe), for example, increased by 53% between 1991 and 1992 due to the removal of subsidies and price controls, spurring poor urban consumers to obtain access to food outside of market channels through home production or bartering (Tevera, 1996).

However, in as much as cities concentrate poverty, they are also the best hope of escaping it, with urban agriculture representing a huge opportunity, given considerate exploitation of natural capital (Choguill, 1995). In this instance, we argue that urban agriculture, if practised effectively, could be a tool for enabling certain elements of the African community to have greater access to fresh produce. Havana (Cuba), a city that faces similar issues to many of those experienced in Africa, demonstrates the effectiveness of urban agriculture: in this case, the authorities initially resisted the practice, but upon realizing its benefits and potential, began to encourage it across the capital (Angotti, 2013). Havana, alongside Detroit (USA), is now seen as an exemplar of how urban agriculture can be used as a mechanism to alleviate food poverty and enable populations to have greater access to fresh fruit and vegetables (Hardman and Larkham, 2014).

7.2 The Development of Urban Agriculture in Africa

7.2.1 Origins and progress

Urban agriculture is not a recent development in Africa, dating to the colonial era when urban farming was seen as a viable way of supplying food to the local population. For most Africans, urban agriculture, though small scale, provides continuity to their rural lifestyle, which has largely been historically agricultural in nature. For example, Yoshikuni (2007), in his studies on Zimbabwe, noted that when black workers were finally allowed to live in towns, most wanted settlements that resembled their rural homes where they could cultivate crops. However, for decades, the history of urban agriculture is a history punctuated by negligence and isolation mainly arising out of lack of recognition by urban authorities. Arku *et al.* (2012) observed that in most cities of the developing world, the practice of urban agriculture was either ignored or, at best, treated as marginal.

It should also be noted that the historical development of most African cities coincided with the modernist era whose town planning principles excluded farming as an urban activity while at the same time trying to protect property values in the interest of European settlers (Chipungu, 2011). This was aggravated by the perception that urban agriculture had a minimal impact on the urban economy; rather it was considered a public health nuisance, mainly due to odours and the spread of diseases. Yet despite these setbacks, urban agriculture has continued to grow, with its contribution to urban food security and the economy now viewed as of clear importance (Haysom, 2007; Lynch *et al.*, 2013). In fact, evidence gathered within a variety of African cities has demonstrated that urban agriculture is directly contributing towards the enhancement of the urban poor's welfare. Kisner (2008), for instance, noted that in Harare, 25% of the urban poor are dependent on urban agriculture, which in turn contributed 60% to their total food consumption. Similarly, Arku *et al.* (2012) observed that a large percentage of the urban population in Dar es Salaam (Tanzania) is increasingly becoming dependent on crops grown in urban spaces for food and income.

These studies suggest that the practice of urban agriculture not only improves food security among urban poor, but also contributes towards the diversity of the produce consumed. Similarly, with an increased dependency placed on urban soils, waters and wastes as natural assets to sustain yields and produce marketable goods, and with new generations of urban farmers dis-attached from the agricultural practices of previous generations, there is the potential for soil degradation due to agricultural intensification.

7.2.2 The economic value of urban agriculture in Africa

Urban agriculture has also emerged as an opportunity to ease local unemployment, with an estimated 40% of the urban population in Africa currently employed in urban agriculture and its related sectors (Arku *et al.*, 2012). While low-income households practise urban agriculture out of necessity, more affluent households with larger plots steer production towards marketable produce. A breakdown of earnings from selected cities (Table 7.1) demonstrates the importance of urban agriculture in contributing towards income security as well as offering employment opportunities to some urban households.

7.2.3 Planning, policy and opposition to urban agriculture

Most existing urban development plans and policies in Africa do not view urban agriculture as a priority; indeed from a purely land zoning perspective, its inclusion is seen as a waste of urban land that is already in short supply. Where it is grudgingly accommodated, security concerns have been raised, with produce liable to theft or damage. Concerns over its impact on sanitation and 'fragile urban ecologies' have also been raised by officials (Magidimisha *et al.*, 2012).

Such concerns have forced practitioners of urban agriculture to engage in the activity illegally, thereby infringing on existing laws and regulations and introducing problems pertaining to lack of infrastructure, capital, land and tenure. Zeeuw (2004) attributed these problems to the nature of the activity, which he argued is yet to establish its relevance beyond and above the household level and receive acknowledgement within the urban economy. This position is substantiated by Moustier and Danso (2006), who argue that urban agriculture in African cities is highly subsistence in nature, and mainly driven by families on home-based plots and in illegal spaces.

7.2.4 Urban agriculture and its impact on African soils

At the centre of urban agriculture in African cities is the need to understand how this practice impacts on soils, especially since the majority of those partaking in the activity are from poor households who often view the practice as a survival strategy; this is predominantly due to the challenges and limited opportunities within the urban economy. This, in turn, places urban soils under intense pressure, since they are supposed to yield adequate harvests in order to contribute towards food security for households as well as to sustain their livelihoods. Hence, in order to meet such diversified expectations, a variety of strategies are employed by households and firms that, in the long run, have both positive and negative implications for soils.

In terms of the negative impacts, perhaps the most fundamental issue is that urban agriculture is generally practised illegally and pursued through subsistence means. Lagerkvist (2014) argues that, due to this illegal approach, households are not prepared to invest their resources into the land, let alone into soil management, since they do not own the space. Further negative practices range from the use of contaminated

Table 7.1. Mean monthly income from urban agriculture in selected African cities. (Adapted from Arku *et al.*, 2012.)

City (country)	Percentage of households practising urban agriculture	Mean net income per month (US$) from urban agriculture
Accra (Ghana)	46	27
Bissau (Guinea-Bissau)	30	12
Dar es Salaam (Tanzania)	20	24
Kumasi (Ghana)	57	27
Nairobi (Kenya)	30	33
Oagadougou (Mali)	36	25
Yaoundé (Cameroon)	35	53

water, to the overuse of inorganic fertilizers and overexploitation of land. In turn, these have had consequences for soils, such as loss of soil structure, soil erosion and loss of soil fertility. Wopereis and Maatman (2002) add to this by estimating that some 17% of 3 million hectares of cultivated land, which is being seriously degraded, can be attributed to urban agricultural practices. This observation is substantiated by Sanchez and Swaminathan (2005) who note that over the past decades, Africa's small-scale farmers have removed large quantities of nutrients from their soils without returning them in sufficient quantities as manure or fertilizer.

However, the gradual relaxation of urban policies, in recognition of the value of urban agriculture by some African cities, is contributing towards good practices with positive repercussions on soils. Provision of vacant land and peri-urban land for urban agriculture, observable in some African cities such as Cape Town (South Africa) and Bulawayo (Zimbabwe), allows for the stabilization of this practice due to formal tenure arrangements that contribute towards better investment in land and soil management. In addition, the provision of training and education on urban agriculture, normally conducted by NGOs, is also contributing towards better soil management. Zeeuw *et al.* (2009) observe that some training programmes include identifying adaptable cultivars, water saving techniques and training in the production of organic bio-fertilizers. Such intervention measures, while benefiting urban farmers, also contribute immensely towards sustaining urban soils. However, these measures are only operative in selected countries – a factor that leaves most African soils under urban agriculture at the mercy of competing human interests.

7.3 Urban Agriculture in Africa: A Selection of Case Studies

The diversity in urban agricultural practices at the continental level is also mirrored at regional and national levels. Hence, there is no uniformity in the nature of urban agriculture practised within a country, largely because of the absence of national legislation to guide the practice coupled with *ad hoc* responses by local authorities. Therefore, case studies depicted in this chapter are based on selected major cities of respective countries that, to a large extent, experience varied developmental dynamics, with urban agriculture being one of them. The driving assumption is that such cities are 'seats' of development where both positive and negative practices converge and cascade to other urban centres within the country. Thus their choice, to some extent, mirrors practices common in a given country. The three case studies selected for this chapter are Zimbabwe, Kenya and South Africa: providing a 'flavour' of urban agricultural activity on the African continent.

In each case study, emerging urban agricultural practices evolved out of defying stringent town planning systems that were opposed to accommodating such a 'rural' practice within urban areas. It is this resilience of urban agriculture that finds meaning in these countries, which therefore serve as an indicator of the importance of this activity.

7.3.1 The practice of urban agriculture in Harare – Zimbabwe

Urban agriculture in Harare is an activity that can be traced back to the inception of the city in 1890. It survived under stringent urban colonial rules when cities were exclusive to European settlers as well as under the majority government that came into power in 1980 and upheld colonial land-use policies. The increase in the urban population, caused by an expectation of better opportunities in urban areas, coupled with challenges arising out of the poor performance of the economy, has seen urban agriculture becoming a mainstay of livelihoods, especially among the low-income households in high density areas (Kamete, 2007).

Cultivation in Harare is largely of crops for domestic consumption, such as maize (*Zea mays*; Poaceae), sweet potatoes (*Ipomoea batatas*; Convolvulaceae) and a variety of other vegetables. Toriro (2006) notes that those practising the activity include low-income families unable to meet their basic needs through cash income derived from formal employment. There was an average annual increase of 67% in land area cultivated by the urban poor between 2005 and 2007 (Kamete, 2007). Kamete (2007) continues to argue that this increase has been attributed to general population increase, rural–urban migration, a growing number of dependents per household, burgeoning unemployment and inflation resulting in a lowering of income in real terms.

Inevitably, this mix of economic hardship, demands on urban land and increasing population has resulted in the cultivation of soils that will produce the best yields. Urban agriculture in Harare is practised on approximately 87 km^2 of land with 57% being on *vlei* areas or fragile wetlands which, though susceptible to different environmental changes, have high soil fertility (Gumbo, 2000). Other areas where urban agriculture is practised include roadsides, railway sides and on household plots. Although these locations provide access to available soils, their close proximity to road and railway pollutants, as well as general household waste in domestic plots, may increase contaminant loadings such that the safety of produce grown in those locations is compromised (Cook *et al.*, 2005).

Legislation and urban agriculture in Harare

In terms of legislation, Bowyer and Tengbeth (1995) note that urban agriculture in Harare takes place in areas that are not specifically zoned for that purpose, thus off-plot cultivation is mostly illegal and often crops are subsequently destroyed by the local authority. The most actively implemented legislation for control of this cultivation is the Environmental Management Act of 2002, which forbids cultivation within 30 m of any watercourse. By doing so, it aims to implement a degree of catchment management to maintain vegetation cover that reduces erosion and deposition of soil into watercourses. This legislation also prohibits cultivation of wetlands that are viewed as ecologically sensitive and whose disturbance has adverse effects on both flora and fauna conservation. In addition to this, the legislation also forbids cultivation without written approval from the local authority and, where crops have already been grown, written approval is also required for their destruction (Bowyer and Tengbeth, 1995; Toriro, 2006).

Kamete (2007) observes that the enforcement of legislation, either through the destruction of crops or levying of fines, has been mollified to some extent by the opportunities for residents to form cooperatives and as such, to apply for the permission to use designated urban land for the cultivation of crops. Although the procedures for this are lengthy and cumbersome, some groups have obtained permission. Yet bureaucratic confusion has resulted in incidents where legal crops have been slashed and cooperatives have requested financial compensation running into thousands of Zimbabwean dollars to account for their losses (Kamete, 2007).

In summary, the authorities still suppress illegal cultivation in Harare and the legislative justification for such action remains. At the same time, the need to practise urban agriculture has become even more pressing as the impact of the economic structural adjustment programme and political uncertainty has percolated deeper into society (Kanyenze, 2004). As with other cities in Africa, there is also increasing evidence that middle-income families are becoming involved in urban agriculture, not always as direct producers or consumers, but as organizers of production and employers of individuals from poorer households as labourers (Toriro, 2006); creating an interesting trickle-up effect on the popularity of urban agriculture.

Soils and soil management in Harare

In Zimbabwe, large areas of the country's soils are derived from granitic parent material and as a result are sandy, nutrient-deficient and contain only small amounts of soil organic carbon (SOC).

Although greenstone soils (related to Oxisols) have a higher nutrient content, continuous cropping has depleted them; a factor that, to a large extent, is exacerbated by poor farming practices (Barber, 1991). Intensive cropping and prolonged periods of insufficient nutrient replenishment have also increased soil fertility problems (Nyamapfene, 1991). Thus, as mentioned earlier in this section, much of the urban agriculture in Zimbabwe is practised on the more SOC-rich and fertile *vlei* areas. This continual cultivation of *vlei* areas can lead to a deterioration of soil structure and alter its hydrological functioning (Gumbo, 2000). *Vlei* watercourses can also become choked with sediments eroded from the catchment area due to cultivation. Mutyavaviri (2006) also observed that *vlei* sites under cultivation have exposed (unvegetated) surfaces leading to increased soil temperatures, aeration and decomposition, and net reductions in SOC. A decrease in organic matter input to the soil due to crop stalk harvesting, a practice common amongst urban farmers on some *vlei* sites, may also be a factor reducing SOC stocks (Mutyavaviri, 2006).

Soil erosion is also a major problem in and around Harare (Mbiba, 2000). It leads to lowering of land surfaces; the direct and indirect loss of fertilizers though erosion and loss of cation exchange capacity; a reduction in SOC stocks; and the deposition of eroded sediments into watercourses. Above all, urban agriculture in Harare is highly dependent on intensive cultivation made possible by tillage, which adversely affects the structure of topsoils (Shumba *et al.*, 2010).

It has been observed that urban agricultural plots in Harare rely on inputs of organic fertilizers obtained from animal manures and composts, while 90% of off-plot farmers use chemical fertilizers to enhance soil fertility (Kisner, 2008). Other research has also noted that the use of animal manures and organic municipal wastes (sewage sludge) is also increasing (Nyamasoka *et al.*, 2010). Further ways of enhancing crop yield include the use of treated effluent, while the use of contaminated domestic and industrial effluent has been observed in gardens around the Mukuvisi River (Mapanda *et al.*, 2005). However, heavy application of fertilizers to increase crop yields results in traces of aluminium, cadmium (Cd), Pb and mercury in food; a factor that may pose a threat to the health of households (Saunyama, 2001).

Urban agriculture in Harare is spreading to middle- and high-income households; however, it could be argued that this is happening at the expense of soil degradation. Measures should be taken to deal with soil degradation arising from soil tillage, removal of crop residues and nutrient mining to ensure that demand for urban agricultural produce is met, but in a safe and environmentally friendly manner. However, it is yet to be seen how the local authority and the central government will respond in the face of such challenges.

7.3.2 The practice of urban agriculture in Nairobi – Kenya

An estimated 14% of land in Nairobi is used for urban agriculture, involving some 150,000 households (City of Nairobi, 2009). As in Harare, the exclusion of agriculture from the city's landscape is a legacy of colonial urban planning. According to Egziabher and Lee-Smith (2006), during pre-colonial times, land was in communal ownership in Kenya but this has changed since land is now either under public or private ownership. Public land is owned by either the municipal council or the central government and is either used for municipal or government purposes (MCN, 1999); such land does not include urban agriculture. According to municipal bylaws, farming practices are forbidden within the town's boundaries, but the practice is tolerated and bylaws permit it under stringent conditions, such as along road reserves, in the middle of roundabouts, between railway lines, in open spaces and parks, along rivers and river valleys, under power lines and within backyards of residential plots (Olima, 2006). This brings into focus the problem of access to land for urban agriculture in Kenya's urban areas, with a lack of formal support existing for the activity, and the use of polluted soils, which may exist along transportation corridors.

Legislation and urban agriculture in Nairobi

Although agriculture is not legally recognized as an urban land use, there is increasing evidence that farming is a widespread activity in Nairobi often practised by vulnerable groups, the majority of whom are women. Ayaga *et al.* (2005) noted that under the Local Government Act, local authorities in Kenya are given the power to lease, transfer or allocate land for temporary use. They also have the power to make bylaws necessary to prevent and suppress nuisance, maintain residents' health and safety as well as to regulate, control or prohibit any activity that is deemed contradictory to existing regulations. However, Nairobi City Council has used these powers to enact bylaws that prohibit cultivation on public streets and keeping livestock. Ayaga *et al.* (2005) also note that the Local Government Act prohibits cultivation by unauthorized persons on land that is not occupied or enclosed, or land belonging to private persons, government and local authorities. In addition to this, the Kenya Public Health Act empowers the prohibition of cultivation or irrigation within and around townships.

Due to the rapid increase of the practice of urban agriculture, the present day municipality allows crop cultivation, as long as the crop is less than 1 m high. Crops such as kale (*Brassica oleracea* var. *acephala*; Brassicaceae), cowpeas (*Vigna unguiculata*; Leguminosae) and spinach (*Spinacia oleracea*; Chenopodiaceae) are often cultivated for commercial purposes, since this type of produce is in high demand and a ready market is present. Urban agriculture in Nairobi is mainly supported by irrigation water from polluted rivers arising out of untreated or partially treated wastewater (Tibaijuki, 2007; Karanja *et al.*, 2010), thus attendant health risks can be implied from urban-grown produce.

Soil management in Nairobi

Nairobi's soil types include various clays, red silt, laterites, decomposed tuffs, alluvial and swamp soils (Onyancha *et al.*, 2011). They are the weathering products of volcanic parent material and most are classified as Andisols or Ultisols. Andisols are generally fertile, with high concentrates of available phosphorus (P), whereas Ultisols are acidic and highly weathered, containing low concentrations of plant available nutrients partly due to high concentrations of iron oxides and their ability to retain anions, such as P (Bationo *et al.*, 2006). It is because of the nature of the soils that Nairobi has a diversity of urban agriculture that includes households engaging in crop production, livestocking and the growing of flowers. However, lack of recognition of urban agriculture in land-use planning has rendered the activity to be largely informal and unregulated, a factor that contributes to environmental degradation arising from deforestation, poor agricultural and waste management practices, and water pollution (City of Nairobi, 2009).

Most areas where urban agriculture is practised in Nairobi depend on irrigation, with rivers and boreholes being the main sources of water. The majority of surface water, used for irrigation, is heavily polluted with domestic, industrial and solid waste that results in contamination when applied to soils. Results from a study by Karanja *et al.* (2010), on soils from irrigated plots in the Kibera area of Nairobi, showed a high degree of salinization and traces of Pb and Cd, which are toxic to crops and soil organisms at high concentrations. The study also found some of the soils contained sufficient plant nutrients (especially potassium, calcium and magnesium) to sustain high vegetable yields. This was attributed to continuous additions from irrigation water derived from sewage, which is high in nitrogen and P (Karanja *et al.*, 2010). However, the main challenge associated with the safe practice of urban agriculture is the possibility that continual use of polluted water sources, combined with poor agricultural practices, may lead to the accumulation of heavy metals in soils and uptake to crops.

Poor agriculture and land management practices have also impacted negatively on soils. The City of Nairobi (2009) noted that the biggest challenge has been deforestation in watershed areas, which is contributing to increased soil erosion. The loss of fertile soil, through erosion, is one of the factors that necessitates the use of fertilizers, which can negatively impact on soils. Sanchez and Swaminathan (2005) have observed that although synthetic fertilizers add necessary nutrients to cropland, they fail to restore organic matter to the soil as does manure, for example. Furthermore, regular use of synthetic fertilizers causes long-term depletion of organic matter and degradation of overall soil 'quality'. They further argue that such practices among farmers contribute to widespread nutrient mining. On the other hand, disposal of agro-chemicals, by industry and some farmers who grow flowers, has led to the contamination of soils; a practice that also leads to water pollution (City of Nairobi, 2009).

The overall impact of poor soil management is felt by the consumers of urban agricultural products. Traces of heavy metals, especially in leafy vegetables such as kale and spinach, have been detected, while infections with tapeworm, ascarids and protozoa are common among residents, especially those in informal settlements (Karanja *et al.*, 2010). In turn, this puts a substantial part of the urban population at risk, given that 50% of vegetables consumed in Nairobi are grown on such plots and are susceptible to these contaminants. However, regulating urban agriculture is not enough; there is a need for more research into the sites on which urban agriculture is practised to understand the impact on the area and the potential danger of eating produce grown within these spaces (cf. Cook *et al.*, 2005). It is hoped that since Kenya is a signatory to the Harare Declaration (of 2003), on Urban and Peri-Urban Agriculture in East and Southern Africa, policies will be put in place to create an environment enabling integration of urban agriculture into the urban economy.

7.3.3 The practice of urban agriculture in Johannesburg – South Africa

The city of Johannesburg is the largest urban complex in South Africa and one of the largest on the African continent. Though the city is the economic hub of South Africa, 33% of the city's population lives in poor housing characterized by backyard shacks and informal settlements (City of Johannesburg, 2003; Landau *et al.*, 2013). Generally, the practice of urban agriculture in Johannesburg is an *ad hoc* activity undertaken by households and communities on dumpsites, residential plots, riverbanks and other open spaces. Given the fluid nature of the activity, only 4.8% (164,500 ha) of the land area is under cultivation (City of Johannesburg, 2003).

There is no clear policy on land allocation for urban agriculture; the city does allow its own land to be used for food growing, but there is no provision relating to the allocation of land specifically for such activities, with gardeners 'invading' land on adjacent plots when they expand their gardens. Where some parcels of land are zoned for agriculture, they are far from residential areas. It has also been observed that land for urban agriculture is classified as part of the city's open space system, hence such land is temporary. A one-off grant of up to around US$5000 can be awarded to interested groups for equipment, and the amount can be increased to around US$15,000 if fencing and boreholes are required. Communities can also receive advice from Gauteng Department of Agriculture and Rural Development. The City of Johannesburg encourages food gardens institutionally by drafting formal constitutions for groups and sometimes land is provided, but preference is given to housing (Richards and Taylor, 2012).

Evidence from 'Orange Farm', a township located on the outskirts of the city, shows that gardeners are required to sign a lease agreement with the provincial government if they wish to use this land for gardening. Vacant land, including land designated for schools in Orange Farm, can be used until such a time that such land is designated for alternative development (Richards and Taylor, 2012). Yet despite these agreements, Orange Farm only has around 0.8% of households practising urban agriculture (City of Johannesburg, 2003; Richards and Taylor, 2012). On a wider level, and away from Orange Farm, there are a number of barriers preventing the practice across South Africa, ranging from financial to the general poor skills of urban farmers.

In most cases, urban agriculture subsists in an environment governed by a plethora of disjointed and uncoordinated legislation dealing with land use, health and the environment. Furthermore, Watson (1998) points out that prior to the 1994 democratic elections in South Africa, urban areas were home to elites; this, to a large extent, excluded poor people from urban areas.

Asomani-Boateng (2002) substantiates this argument by indicating that the majority of people who practise urban agriculture are the urban poor who cannot afford to pay for infrastructure and service provision. Above all, he emphasizes how the activity requires specialist skills and knowledge; urban farmers need to apply the appropriate production or farming techniques that are reconcilable with the urban situation. Usually urban farmers utilize traditional and inappropriate methods from rural areas. These individuals may lack the capacity to bring together their needs, know-how and activities in such a manner that it becomes a sustainable venture that contributes to their livelihoods (cf. Asomani-Boateng, 2002). Kroll and Rudolph (2010) argue that urban agriculture in Johannesburg remains a challenge, from the scarcity of land to erratic and torrential rains, degraded and polluted soils and other issues.

Soil management in Johannesburg

Much of South Africa has good agricultural soils characterized by loamy-clay soils (Bationo *et al.*, 2006). In addition, 35% of land in Johannesburg is not developed, partly due to degradation caused by large-scale mining activities that affected vast tracts of land. Mining activities have contributed to surface and groundwater pollution, soil pollution, high radiation levels, increased air-borne pollution and loss of land (City of Johannesburg, 2003). The situation is aggravated by small-scale mining and quarrying activities within the city that do not have adequate resources to invest in cost-effective and efficient technologies, and so generate large volumes of waste spoil. In addition, land degradation is also caused by industrial and commercial activities where inappropriate land and waste disposal practices are common – a factor that contributes to the removal of topsoil (City of Johannesburg, 2003).

Much of the land previously utilized for mining requires remediation and is consequently not fit for urban agricultural use. The City of Johannesburg (2003) observed that there was general evidence of escalating soil degradation, declining biological diversity and soil productivity mainly in open spaces caused by illegal clearing (for urban agriculture and informal settlements), dumping and uncontrolled burning. One could argue that the practice of urban agriculture in Johannesburg is a fragmented activity that is not well blended into the livelihoods of the people, despite support from government structures. Perhaps, most convincingly, it can also be argued that given the rapid rate of urbanization in Johannesburg (which stands at 2.6%) supported by the high concentration of economic activities (Turok, 2012; Landau *et al.*, 2013), indulgence in urban agriculture as a survival strategy is yet to gain importance as it has done in Harare and Nairobi.

7.4 Conclusions: The Future of Urban Agriculture and Soils in Africa

Soils support life, and commonality in the case study regions revolves around maximizing this valuable, non-renewable resource even where the inherent soil properties are far from optimal for supporting agriculture. We have seen that the practice of urban agriculture can be associated with both positive and negative impacts on soils. The use of organic fertilizers in Harare and Nairobi increased the fertility of the soils, but the contaminated composition of these organic inputs may eventually induce soil toxicity, decreasing yields and the passage of contaminants through crops to humans. Evidence from the City of Johannesburg shows that mining operations have polluted water sources and degraded land, thereby rendering it unsuitable for agriculture.

While this review provides evidence to suggest that urban agriculture is still a peripheral component of urban land use in many African cities, it does suggest that the continual search for food security among the urban poor is expanding the practice. However, unless major steps are taken, there will be continual contamination and degradation of soils that, in turn, will further constrain urban food security especially among poor households. Hence Zeeuw's (2003, cited in Haysom, 2007), observation that disjointed practices being experienced in African cities can only be resolved if the nexus between urban land-use planning, food security policies, urban environmental policies and urban health policies is achieved. Further research is required to explore the impact of this type of practice on the African continent: providing empirical material to suggest the potential for urban agriculture and whether populations are able to use this practice to become self-sufficient.

References

Angotti, T. (2013) *The New Century of the Metropolis: Urban Enclaves and Orientalism*. Routledge, Abingdon, UK.

Arku, G., Mkandawire, P., Aguda, N. and Kuuire, V. (2012) *Africa's Quest for Food Security: What is the Role of UA?* Occasional Paper No. 19.1. The African Capacity Building Foundation in Harare, Zimbabwe.

Asomani-Boateng, R. (2002) Urban cultivation in Accra: An examination of the nature, practices, problems, potentials and urban planning implications. *Habitat International* 26, 591–607.

Ayaga, G., Kibata, G., Lee-Smith, D., Njenga, M. and Rege, R. (2005) *Policy Prospects for Urban and Peri-UA in Kenya*. Urban Harvest – International Potato Centre, Lima, Peru.

Barber, B. (1991) Phosphate resources of Zimbabwe. *Fertilizer Research* 30, 247–278.

Bationo, A., Lungu, O., Naimi, M., Okoth, P., Smailing, E. and Thiombiano, L. (2006) African soils: Their productibility and profitablity of fertiliser use. Background paper prepared for the African Fertiliser Summit, 9–13 June 2006, Abuja, Nigeria.

Beesley, L. (2012) Carbon storage and fluxes in existing and newly created urban soils. *Journal of Environmental Management* 104, 158–165.

Bowyer, T.A.S and Tengbeh, G. (1995) The environmental implications of (illegal) UA in Harare, Zimbabwe. Working Paper 4. Paper presented at the ODA workshop on Environmental, Social and Economic (Illegal) UA in Harare, Zimbabwe 30th-31st August 1995.

Caputo, S. (2012) The purpose of urban food production in developed countries. In: Viljoen, A. and Wiskerke, J.S.C. (eds) *Sustainable Food Planning: Evolving Theory and Practice*. Wageningen Academic Publishers, Wageningen, the Netherlands.

Chipungu, L. (2011) Insights into urban development control challenges: A case study of Operation Murambatsvina/Restore Order in Zimbabwe. *Journal of the Built and Human Environment Review* 4(2), 17–32.

Choguill, C.L. (1995) Urban agriculture and cities in the developing world. *Habitat International* 19(2), 149–150.

City of Johannesburg (2003) *State of the Environment Report*. City of Johannesburg, Johannesburg, South Africa.

City of Nairobi (2009) *City of Nairobi Environmental Outlook*. City of Nairobi, Nairobi, Kenya.

Cook, H.F., Lee, H.C. and Perez-Vazquez, A. (2005) Allotments, plots and crops in Britain. In: Vijoen, A. (ed.) *CPULs: Continuous Productive Urban Landscapes*. Architectural Press, Oxford,UK.

Egziabher, A.G., Lee-Smith, D., Maxwell D.G., Memon P.A., Mougeot, L.J.A. and Sawio, C.J. (2006) Cities feeding people: An examination of UA in East Africa. International Development Research Centre, Ottawa, Canada.

Gumbo, B. (2000) UA in Harare: Integrating the past, present and the future perspectives – UZ/IAC Food and Nutrition Security in Urban Areas Regional Course: Strengthening Food and Nutrition Training in Southern Africa, Harare, 4–16 December 2000.

Hardman, M. and Larkham, P. (2014) *Informal Urban Agriculture: The Secret Lives of Guerrilla Gardeners*. Springer-Verlag, Dordrecht, the Netherlands.

Haysom, G. (2007) UA and food security. In: Braun, J.V. (ed.) *Towards a Healthy and Sustainable World Food Situation 2020–2050*. International Research Food policy Institute, Zurich, Switzerland.

Hursthouse, A., Tognarelli, D., Tucker, P., Marsan, F.A., Martini, C., Madrid, L. *et al.* (2004) Metal content of surface soils in parks and allotments from three European cities: Initial pilot study results. *Land Contamination and Reclamation* 12, 189–196.

IUSS Working Group WRB (2006) World reference base for soil resources 2006. World Soil Resources Reports No. 103. FAO, Rome.

Kamete, A.Y. (2007) Cold-hearted, negligent and spineless? Planning, planners and the (r)ejection of 'filth' in urban Zimbabwe. *International Planning Studies* 12(2), 153–171.

Kanyenze, G. (2004) Economic Structural Adjustment Programme (ESAP): Precursor to the Fast Track Resettlement. In: Friedrich, M. and Stifling, E. (eds) *Post-Independence Land Reform in Zimbabwe: Controversies and Impact on the Economy*. University of Zimbabwe, Harare, Zimbabwe.

Karanja, N., Njenga, M., Prain, G., Kang'ethe, E., Kironchi, G., Githuku, *et al.* (2010) Assessment of environmental and public health hazards in wastewater used for UA in Nairobi, Kenya. *Tropical and Subtropical Agro-Ecosystems* 12(1), 85–97.

Kisner, C. (2008) Green roofs for urban food security and environmental sustainability. Available at www.climate.org/topics/international-action/urban-agriculture.htm (accessed12 March 2015).

Kroll, F. and Rudolph M.T. (2010) *Urban Agriculture in Johannesburg – Risks, Constraints, Strategies based on Siyakhana Experience*. Paper presented at the Conference on 'Land Divided: Land and South African Society in 2013, in Comparative Perspective', University of Cape Town, 24th - 27th March 2013.

Lagerkvist, C.J. (2014) Economic drivers for urban and peri-urban agriculture. In: Magnusson, U. and Bergman, K.F. (eds) *Urban and Peri-urban Agriculture for Food Security in Low income Countries. Challenges and Knowledge Gaps*. SLU-Global Report, Uppsala, Sweden.

Landau, L.B., Segatti, A. and Freemantle, I. (2013) Urbanisation and migration: The governance of mobility and spatial inequalities. In: Pillay, U., Hagg, G., Nyanjoh, F. and Jansen, J. (ed) *State of the Nation, South Africa, Addressing poverty and Inequality*. Human Science Research Council Press, Cape Town, South Africa.

Lynch, K., Maconachie, R., Binns, T., Tengbe, P. and Bangura, K. (2013) Meeting the urban challenge? Urban agriculture and food security in post-conflict Freetown, Sierra Leone. *Applied Geography* 36, 31–39.

Madrid, F., Reinoso, R., Florido, M.C., Diaz-Barrientos, E., Ajmone-Marsan, F., Davidson, C.M. and Madrid, L. (2007) Estimating the extractability of potentially toxic metals in urban soils: A comparison of several extracting solutions. *Environmental Pollution* 147, 713–722.

Madrid, L. Diaz-Barrientos, E. and Madrid, F. (2002) Distribution of heavy metal contents of urban soils in parks in Seville. *Chemosphere* 49, 1301–1308.

Magidimisha, H.H., Chipungu, L. and Awuorh-Hayangah, R. (2012) Challenges and strategies among the poor: Focus on urban agriculture in KwaMashu, Durban, South Africa. *Journal of Agriculture, Food Systems and Community Development* 3(1), 109–126.

Mapanda, F., Mangwayana, E.M., Nyamangara, J. and Giller, K.E. (2005) The effect of long-term irrigation using wastewater on heavy metal contents of soils under vegetables in Harare, Zimbabwe. *Agriculture, Ecosystems and Environment* 107(2/3), 151–165.

Mbiba, B. (2000) Urban agriculture in Harare: Between suspicion and repression. In: Bakker, M., Dubbling, S., Guendel, U., Koschella, S. and de Zeeuw, H. (eds) (2000) *Growing Cities, Growing Food, Urban Agriculture on the Policy Agenda*. DSE, Feldafing, Germany, pp. 285–302.

MCN (1999) *Strategic Nakuru Structure Plan, Action Plan for Sustainable Urban Development of Nakuru Town and its Environs*. Nairobi. Kenya.

Mok, H.-F., Williamson, V.G., Grove, J.R., Burry, K., Barker, F. and Hamilton, A.J. (2014) Strawberry fields forever? Urban agriculture in developed countries: A review. *Agronomy for Sustainable Development* 34, 21–43.

Mougeot, L.A.J. (1999) For self-reliant cities: urban food production in a globalising south. In: Koc, M., MacRae, R., Mougeot, L.A.J. and Welsh, J. (eds) *For Hunger-Proof Cities: Sustainable Urban Food Systems*. IDRC, Ottawa.

Moustier, P. and Danso, G. (2006) Local economic development and marketing of urban produced food. In: van Veenhuisen, R. (ed.) *Cities Farming for the Future: Urban Agriculture for Green and Productive Cities*. RUAF Foundation. Ottawa.

Mutyavaviri, F. (2006) Impact of cultivation on soil and species composition of the Monavale Vlei, Harare. MSc thesis. University of Zimbabwe, Harare, Zimbabwe.

Nyamapfene, K. (1991) *Soils of Zimbabwe*. Nehanda Publishers, Harare, Zimbabwe.

Nyamasoka, B., Nyamugafata, P., Madyiwa, S. and Nyamangara, J. (2010) Effects of innovative farmer practices in soil fertility and water management on soil productivity and maize and leafy vegetable production in Harare, Zimbabwe. Second RUFORUM Biennial Meeting, 20–24 September 2010, Entebbe, Uganda.

Odendaal, N. (2012) Reality check: Planning education in the African urban century. *Cities* 29(3), 174–182.

Olima, W.H.A. (2006) UA constraints. University of Nairobi, Nairobi, Kenya.

Onyancha, C.K., Mathu, E.M., Mwea, S.K. and Ngecu, W.M. (2011) Dealing with sensitive and variable soils in Nairobi city. *International Journal of Research and Reviews in Applied Science* 9(2), 282–291.

Richards, R. and Taylor, S. (2012) Changing landuse on the periphery: A case study of UA and food gardening in Orange Farm. University of Witwatersrand. Johannesburg, South Africa.

Rossiter, D.G. (2007) Classification or urban and industrial soils in the World Reference Base for soil resources. *Journal of Soils and Sediments* 7, 96–100.

Sanchez, P.A. and Swaminathan M.S. (2005) Hunger in Africa: The link between unhealthy people and unhealthy soils. *The Lancet* 365, 442–444.

Saunyama, I.G.M. (2001) Food safety in crop post-harvest systems. Proceedings of an International Workshop sponsored by the Crop Post Harvests Programme of the United Kingdom Department for International Development, Harare, Zimbabwe, 20–21 September 2001.

Shumba, A., Nyamasoka, B., Nyamugafata, P., Madyiwa, S. and Nyamangara, J. (2010) Effects of optimised innovative soil fertility and water management technologies on maize production in two contrasting soils in Harare, Zimbabwe. Second RUFORUM Biennial Meeting, 20–24 September 2010, Entebbe, Uganda.

Tevera, D.S. (1996) Urban agriculture in Africa: A comparative analysis of findings from Zimbabwe, Kenya and Zambia. *African Urban Quarterly* 11(2/3), 181–187.

Thums, C.R., Farago, M.E. and Thornton, I. (2008) Bioavailability of trace metals in brownfield soils in an urban area in the UK. *Environmental Geochemistry and Health* 30, 549–563.

Tibaijuki, A. (2007) Cities can achieve more sustainable land use if municipalities combine urban planning and add development with environmental management. In: *Nairobi, Kenya: Atlas of Our Changing Environment*. UNEP.

Toriro, P. (2006) *Town Planning in Zimbabwe: History, Challenges and Urban Renewal Operation Murambatsvina*. RUAF foundation Harare, Zimbabwe.

Turok, I. (2012) *Urbanisation and Development in South Africa: Economic Imperatives, Spatial Distortions and Strategic Responses*. Human Settlements Group, International Institute for Environment and Development (IIED), London.

UN-FAO (2008) *The State of Food Insecurity in the World 2008: High Food Prices and Food Security – Threats and Opportunities*. FAO, Rome.

UN-Habitat (2003) *Challenges of Slums: Global Report on Human Settlements*. Earthscan, London.

UN-Habitat (2013) *Global Activities Report – Our Presence and Partnerships*. UN-Habitat, Nairobi, Kenya.

Viljoen, A. (2005) *CPULs: Continuous Productive Urban Landscapes*. Architectural Press, Oxford, UK.

Watson, V. (1998) Planning under political transition – Lessons from Cape Town's metropolitan planning Forum. *International Planning Studies* 3(3), 335–350.

Wiskerke, J.S.C. and Viljoen, A. (2012) Sustainable urban food provisioning: Challenges for scientists, policy-makers, planners and designers. In: Viljoen, A. and Wiskerke, J.S.C. (eds) *Sustainable Food Planning: Evolving Theory and Practice*. Wageningen Academic Publishers, Wageningen, the Netherlands.

Wopereis, M.C.S. and Maatman, A. (2002) Improving farming livelihoods in sub-Saharan Africa: The case for integrated soil fertility management. Strategy paper prepared for the International Training Program on Integrated Soil Fertility Management (ISFM) in the Tropics, IFDC-Africa Division, Lomé, Togo, 7–12 October 2002.

Wortman, S.E. and Lovell, S.T. (2013) environmental challenges threatening the growth of urban agriculture in the United States. *Journal of Environmental Quality* 42(5), 1283–1294.

Yoshikuni, T. (2007) *African Urban Experiences in Colonial Zimbabwe. A Social History of Harare Before 1925*. Weaver Press, Harare, Zimbabwe.

Zeeuw, H. (2004) The development of UA, some lessons learnt. Keynote paper for the International Conference 'UA Agro-tourism and City Regional Development', Beijing, 10th–14th October 2004.

Zeeuw, H. and Dubbeling, M. (2009) Cities, food and agriculture. Challenges and the way forward. RUAF Working Paper Series No. 3. RUAF Foundation, the Netherlands.

Zeeuw, H., Dubbeling, M., Veenhuizen, R. and Wilbers, J. (2009) Key issues and courses of action for municipal policy making on urban agriculture. Working Paper 2. RUAF Foundation, the Netherlands.

8 Urbanization and Soil Nutrient Challenges and Opportunities: Lessons from Malawian Cities

David D. Mkwambisi,[1]* Andrew J. Dougill,[2] Philip Antwi-Agyei[2,3] and Chikondi P. Chabvuta[4]

[1]Lilongwe University of Agriculture and Natural Resources, Bunda College Campus, Lilongwe, Malawi; [2]Sustainability Research Institute, School of Earth and Environment, University of Leeds, Leeds, UK; [3]Department of Environmental Science, College of Science, Kwame Nkrumah University of Science and Technology, Kumasi, Ghana; [4]Centre for Community Organization and Development, Lilongwe, Malawi

8.1 Introduction

It is estimated that the urban population in sub-Saharan Africa will double to about 600 million by 2030 (FAO, 2012). This urbanization has come with a recognition that strategies for meeting current and future urban food supply requirements are insufficient (Mougeot, 2005; Hovorka *et al.*, 2006; Zezza and Tasciotti, 2010) and that impacts on urban environmental conditions are poorly understood (Grimmond *et al.*, 2002; Lu *et al.*, 2010). In particular, the impact of urbanization on urban food production systems and soil nutrient cycles needs additional research. The need to consider urban land management practices that can reduce emissions of greenhouse gases and enhance carbon storage is increasingly being discussed at an international level (e.g. Rosenzweig *et al.*, 2011; Bulkeley and Castán Broto, 2013; Taylor and Peter, 2014). Indeed, this is now recognized by the Intergovernmental Panel on Climate Change, who included chapters on the urban dimensions of climate change in their Fifth Assessment Report (IPCC, 2013).

Many African cities are currently failing to realize the potential of urban agriculture in supporting households to meet their food requirements, as shown by previous studies in Dar es Salaam (Jacobi *et al.*, 2000), Kampala (Maxwell *et al.*, 1998) and Dakar (Mbaye and Moustier, 2000). Municipal solid waste and human waste from urban sanitation systems now offer the potential to address low soil nutrient concentrations and also to reduce environmental problems associated with poor waste management (Mkwambisi *et al.*, 2012). Better understanding of this relationship could help to address some of the challenges and provide opportunities for cities to meet both national and international development objectives.

Despite the important role of urban agriculture in reducing poverty and providing livelihoods for many households (Maxwell, 1999; Redwood, 2009), city authorities rarely provide formal space for it (Chipungu *et al.*, Chapter 7, this volume). There are typically no efforts put in place within city plans or policy support measures that can advance the role of urban agriculture.

* Corresponding author: david.mkwambisi@bunda.luanar.mw

The other sustainable food production challenge is that most urban African farmers cannot afford to purchase inorganic fertilizers, leading to year-on-year declines in levels of the essential plant macronutrients in soils used for agricultural production (Jones *et al.*, 2013). Failing to fully understand environmental problems and to link them to urban poverty and food systems is a pressing problem, as many urban poor people are resorting to unsustainable livelihoods such as charcoal production, moulding bricks and sand mining to raise income to enable the purchase of food to meet their household needs. Based on these challenges, several related research activities have been ongoing since 2005 in Malawi to improve understanding of the role of urban agriculture in contributing to urban food security, reducing municipal solid waste and their effects on urban soil properties. This chapter summarizes some of this research using findings from two Malawian cities (Lilongwe and Blantyre). The overall aim is to assess the potential for urban agriculture in Malawi to improve urban soil quality, crop yields and livelihood opportunities for urban residents. Specific research objectives are to:

1. Investigate the extent to which urban agriculture has improved the livelihoods and well-being of the urban poor, and other marginalized groups, in Lilongwe and Blantyre.
2. Assess farm-scale nutrient budgets for urban agricultural activities and how these are affected in Lilongwe by the availability of organic manure in the form of tobacco waste.
3. Evaluate the effectiveness of new ecosanitation organic manure products on soil quality and maize yields from recent pilot projects in Lilongwe.

8.2 Methods

8.2.1 Study sites

Research was undertaken with urban communities in Lilongwe and Blantyre in Malawi. These two cities are different in terms of economic, geographical and demographic structures, and capture many of the problems faced in African cities more widely.

Lilongwe lies in the fertile central region at an altitude of 1100 m above sea level and has ustalf (Alfisol) soils, which are the red-yellow soils of the Lilongwe plain (Lowole, 1965). These soils are typically highly fertile and the Lilongwe plain has extensive agricultural production for both commercial and subsistence needs of the nation. Lilongwe is an administrative and commercial centre with 669,100 residents, 25% of whom live below the nationally assessed poverty line (GOM, 2012). Residential plots in many medium- and high-income households were designed to incorporate crop production; however, government officials have regularly destroyed maize crops in poorer urban areas where no formal land ownership was assigned to local communities and livestock rearing has also been strictly prohibited in residential areas.

Blantyre, with a population of 711,200 in 2008 (GOM, 2008), is the largest commercial and industrial centre in Malawi. It covers a total area of 228 km^2 of hilly ground with soils typified by poorly developed Entisols, with their sandy texture making them less fertile and lower in soil organic matter (SOM) content. The majority (71%) of the city's residents live in unplanned settlements characterized by poor living conditions.

8.2.2 Urban agriculture and livelihoods

To assess food production and consumption and its determinants, a livelihood questionnaire was administered to a total of 330 households (stratified according to income, age, gender of household head and education level) engaged in urban agriculture in the two cities. In Lilongwe, 131 male-headed and 34 female-headed households were sampled, while in Blantyre 112 male-headed and 53 female-headed households were sampled. The sample was stratified into high-income, medium-income and low-income households based on a participatory wealth ranking exercise approach (Kebede, 2007) involving classification by traditional leaders of the relative wealth status of their community members (Mkwambisi, 2008). Focus groups with low-income residents in both cities were also held and provided an opportunity to elaborate on the results from the questionnaire. Finally, community workshops were held enabling feedback of key findings and the broader discussion of new opportunities for the use of domestic and municipal waste sources as a potential organic input to urban agriculture initiatives.

8.2.3 Soil nutrient budgets

We conducted a soil nutrient budget analysis of 27 farming households (21 male-headed and 6 female-headed households) in Lilongwe from different income groups, stratified by their proximity to an industrial site known as Kanengo, a major source of organic tobacco wastes. We hypothesized that households close to this site had improved access to a nutrient-rich organic waste compared to those located further away so the impact of distance from such an organic waste resource was studied across three regions with nine farm plots studied in each region. The first region covered farm plots located 0–5 km from Kanengo, the second region covered farm plots located 6–10 km from Kanengo and the final region was for farms over 10 km from the source. To determine nitrogen (N), phosphorus (P) and potassium (K) budgets, household grain yields were multiplied by nutrient concentrations of tropical crops as outlined by Kirchmann *et al.* (2005) to enable assessment of nutrient losses from harvested crops from both crop residues and the grain. We used indices that had been developed from studies in southern Ghana because of similarities in soil and agro-ecological characteristics in these areas (Stoorvogel and Smaling, 1990; FAO, 2004). Soil nutrient analyses used the following methods: modified macro-Kjeldahl digestion method for total N (Anderson and Ingram, 1993); modified Olsen's extraction procedures for extractable P (Olsen and Sommers, 1982); and exchangeable K following a sodium acetate extraction (Anderson and Ingram, 1993). These methods were also used to determine macro-nutrient concentrations in waste to calculate the amounts of nutrients available per year from this source.

Farmers were also asked (as part of semi-structured interviews) to explain the factors limiting their use of solid waste as an organic input. These factors were also discussed during community workshops and focus group discussions to gain a more detailed understanding of the decision-making followed by urban farmers in the two cities.

8.2.4 Assessing the effect of ecosanitation manure on maize yield and community perceptions

Ecological sanitation (also known as ecosanitation) is a holistic approach to sanitation and water management based on the systematic closure of local material flow cycles (Langergraber and Muellegger, 2005). This is a process whereby human faeces and urine are utilized as inputs for food production. In this case study, human faeces (also termed humanure) was applied as basal dressing fertilizer while urine was applied as top dressing liquid fertilizer for maize production.

The development aim of this new technological practice is principally to protect human health and reduce water pollution, but it also reduces the use of water in sanitation systems and recycles nutrients to help reduce the need for artificial fertilizers in agriculture. This project involved the use of demonstration plots (Fig. 8.1) to compare the performance of maize applied with various organic and inorganic fertilizers within study communities and at experimental plots at Bunda College of Agriculture, Lilongwe. In the experimental plots, there were four nutrient levels of humanure application (60, 75, 90 and 104 kg N ha^{-1}). Other plots were applied with cattle and goat manure at the same rates as that of humanure. Both humanure and livestock manures were applied one week before planting and once during planting. A questionnaire was also used to collect information from the households that participated in the utilization of the ecosanitation technology for providing humanure for soil treatments across Lilongwe.

8.3 Results

8.3.1 Urban agriculture and livelihoods

Overall, the households surveyed produced a mean maize yield of 802.5 ± 869.1 kg ha^{-1} (see Table 8.1 for breakdown by area and household type). At a household level, the mean maize production was 228 kg per capita of maize (or cereal equivalents), which is above the 181 kg per capita that the Government of Malawi recommends as an adequate food budget (GOM, 2012). Although considerable variation exists both within and between groups, more educated, wealthier

Fig. 8.1. Demonstration plot for ecosanitation products applied as fertilizers at Bunda College of Agriculture, Lilongwe, Malawi.

Table 8.1. Contribution of urban agriculture to household livelihoods in Lilongwe and Blantyre (Malawi) during the 2004/05 agricultural season with breakdown by household classifications.

Parameter	*n*	Maize yield (kg ha^{-1})	Total harvest consumed (kg person^{-1} year^{-1})	Amount of maize sold (kg)	Period taken to consume (months)
Lilongwe	142	855.4 ± 926.9	712.7 ± 820.4	122.8 ± 403.5	22.5 ± 39.1
Blantyre	152	758.5 ± 833.3	639.5 ± 765.5	109.7 ± 365.4	17.3 ± 34.6
Male-headed households	209	975.9 ± 943.3	825.7 ± 849.2	129.2 ± 435.1	21.1 ± 37.7
Female-headed households	85	385.1 ± 494.7	303.9 ± 454.0	80.2 ± 175.5	16.7± 34.7
Low-income households	120	297.5 ± 438.2	237.9 ± 424.6	59.6 ± 143.3	13.5 ± 31.2
High-income households	179	1155.5 ± 936.0	976.1 ± 845.2	148.6 ± 467.0	23.6 ± 39.5
Illiterate	17	183.3 ± 184.6	173.3 ± 189.8	10.0 ± 41.2	7.7 ± 23.6
Pre-school	4	312.5 ± 443.6	262.5 ± 343.8	50.0 ± 100.0	2.5 ± 1.7
Primary school	82	552.6 ± 697.8	475.7 ± 659.1	73.3 ± 205.2	15.9 ± 33.7
Secondary school	82	684.7 ± 830.6	559.6 ± 677.1	111.0 ± 351.5	23.5 ± 40.0
Post-secondary school	107	1203.4 ± 963.7	1002.7 ± 914.6	166.5 ± 507.9	22.5 ± 38.6

Note: Values are mean ± standard deviation.

and male-headed households typically obtained larger harvests than poorer, less educated and female-headed households (Table 8.1). For example, the 17 households where the household head was 'illiterate' only harvested an average of 183 ± 185 kg ha^{-1} from urban agriculture, while high-income households harvested 1156 ± 936 kg ha^{-1} from their plots.

Yields from female-headed households were lower (385 ± 495 kg ha^{-1}) when compared to that obtained by male-headed households (976 ± 943 kg ha^{-1}). Consultations with farmers and community focus groups revealed that these differences in harvest existed because male-headed households were typically wealthier and had access to large plots of land and better agricultural technology. Comments during stakeholder consultations noted that high-income households obtained higher maize yields than low-income households (1156 ± 936 kg ha^{-1} compared to 298 ± 438 kg ha^{-1}) due to level of education and their greater availability of skilled labour. This was also confirmed with community level workshops, and focus group discussions suggested that this was due to access to information, educational qualifications and resources such as good quality nutrient inputs. For example, a retired civil servant from Lilongwe city said:

> The availability of agricultural information in all the daily newspapers and information on packaging materials has allowed me to identify quality farm inputs and prevent some of the pest[s] attacking my crops.

This contrasted with the sentiments of farmers from low-income households. During the focus group discussions, the local chief complained:

> Most of the farmers here have no access to radios and are illiterate. This makes it difficult to understand instructions that are on the agricultural farm inputs. In addition, the instructions are mostly in English and are so brief to allow a farmer [to] practice the instructions.

The results show that households consumed around 80% of the maize they harvested from urban plots (Table 8.1). In terms of the period taken to consume the urban harvests, the results show that those farmers that have resources (e.g. high-income households) are able to depend on food from urban plots for a longer period of time. However, it is difficult to link this outcome to the size of the households as those from low-income and illiterate households generally have larger household sizes and hence their food will last for a shorter period of time.

8.3.2 Soil nutrient budgets

Farm scale results show that soils in Lilongwe require on average 77 kg N ha^{-1}, 19 kg P ha^{-1} and 28 kg K ha^{-1} per year to provide an annual nutrient balance given the nutrient outputs from the fields studied (Table 8.2). Farmers currently apply less inorganic fertilizers in farm plots located 0–5 km from the main source of tobacco wastes as compared to those located over 5 km from sources of tobacco waste.

Despite the high costs associated with managing municipal solid waste and increased prices of inorganic fertilizers, our survey data show a high demand for inorganic fertilizers with the majority of households (61%) reporting the usage of inorganic fertilizer. This compares with tobacco wastes (26%), livestock manure (9%) and recycled wastes (4%). For example, a female farmer aged

Table 8.2. Nutrient inputs and outputs from fields in Lilongwe in three regions with increasing distance from an industrial tobacco waste site.

Source and type of nutrient	Region 1 (0–5 km) kg ha^{-1} (*n* = 9)	Region 2 (6–10 km) kg ha^{-1} (*n* = 9)	Region 3 (over 10 km) kg ha^{-1} (*n* = 9)
Mineral fertilizer N	8.0 ± 11.4	64.5 ± 26.1	33.9 ± 10.3
Mineral fertilizer P	10.9 ± 6.4	19.3 ± 5.5	18.94 ± 7.1
Mineral fertilizer K	0 ± 0.0	0 ± 0.0	0 ± 0.0
Total crop output N	32.9 ± 19.4	58.1 ± 16.7	57.0 ± 21.3
Total crop output P	7.8 ± 4.6	13.8 ± 4.0	13.5 ± 5.0
Total crop output K	18.7 ± 11.0	33.1 ± 9.5	32.4 ± 12.1
N required	41.9 ± 26.0	98.5 ± 31.3	90.2 ± 29.5
P required	9.8 ± 6.5	24.3 ± 11.2	24.0 ± 12.1
K required	18.7 ± 11.0	33.1 ± 9.5	32.4 ± 12.1

Note: Values are mean ± standard deviation.
Mean nutrient deficits in Lilongwe (across all three regions) nitrogen = 77 kg ha^{-1}; phosphorus = 19 kg ha^{-1}; potassium = 28 kg ha^{-1}

48 years from Lilongwe city who has been practicing urban farming for over 10 years summarized the discussions with the following remark:

> We prefer tobacco wastes to other wastes as they are easy to apply, do not require recycling [saving time], have a high content of nutrients and they are available in good time [few months before the new agricultural season] for application. It is easy to transport the materials and they degrade very fast after application. On the other hand, solid organic wastes are difficult to collect and manage at household level; in addition, they are associated with low NPK content.

Soils have a mean N deficit of 77 kg N ha^{-1}, implying that farmers require 6.2 t of organic waste on average to meet the deficit for future sustainable maize production. Thus, it follows that 4000 ha of land could be supplied with sufficient nutrient inputs from the municipal solid waste that is currently delivered to dumping sites (Mkwambisi, 2008). The current tonnage of municipal solid waste taken to dumping sites (approximately 21,000 t) (Mkwambisi, 2008) is therefore not enough to fully meet requirements for potential land zoned for agriculture within and outside the city.

8.3.3 Ecological sanitation plots

Plots applied with ecosanitation manure produced higher maize yields (e.g. 2112 kg ha^{-1} for plots with 90 kg N ha^{-1} added) than those without fertilizers (1498 kg ha^{-1}). Interviews with urban farmers revealed that products from ecosanitation toilets have several uses. The results show that over 80% are using the manure for urban food production, while 43% are using the products to earn income. The main crops on which these products were applied are currently maize (*Zea mays* poaceae), vegetables such as spinach (*Spinacia oleracio* Chenopodiaceae) and kale (*Brassica oleracia* var. *acephala* Brassicaceae), beans (Leguminosae) and tomatoes (*Solanum lycopersicum*, Solanaceae).

8.4 Discussion

Urban agriculture activities in Malawi currently appear to favour educated, higher-income male-headed families who have retained access to the land and labour required for agricultural production within the confines of a city. The access to labour required for surface application of wastes is particularly important in areas such as Blantyre where soils are sandy, making water and nutrient retention reliant upon SOM levels. The observation of wealthier households being drivers for growth in urban food production corroborates findings from other studies (e.g. Maxwell, 1999), suggesting that more powerful urban interests have realized the value of under-utilized urban land and have converted it to agriculture. Land ownership for both rural and urban households can be used as a proxy for household wealth, and its relationship with favourable nutritional status shows that land enhances food security (Maxwell, 1999). However, the extent to which poor families used the economic opportunities presented by urban agriculture (either by selling food or through working for other urban farms) demonstrates that urban agriculture provides an important livelihood diversification strategy (Kwapata *et al.*, 2001; Mkwambisi *et al.*, 2011). It is pertinent to suggest that waste recycling can serve a key additional benefit to local communities and improved soil fertility as most poor communities are also faced with food security problems (Klemesu, 2000). Urban agricultural activities and their crop productivity could be enhanced if soil nutrient availability were to be increased through greater use of organic waste inputs. The findings corroborate the wider literature across sub-Saharan Africa suggesting that combined application of organic resources and inorganic mineral inputs (often termed as integrated soil fertility management approach) and maintaining SOM is a key to sustainable land-use management (e.g. Haiti *et al.*, 2007; Hilhorst and Muchena, 2000). Other important benefits resulting from the maintenance of SOM include improved retention of nutrients, increased buffering capacity in low acidity soils, and increased water-holding capacity. In many cases, this approach can also prevent the acidification problems that are commonly associated with the regular application of inorganic fertilizers to sandy soils of low inherent fertility (Dougill *et al.*, 2002).

Results on the utilization of organic waste as a source of plant nutrients confirm that cities in Malawi remain characterized by inadequate waste management policies and practices.

These findings corroborate other studies (e.g. Henry *et al.*, 2006; Oteng-Ababio *et al.*, 2013), suggesting that waste management constitutes one of the key challenges, but also an opportunity, for city authorities in sub-Saharan Africa. For instance, Henry *et al.* (2006) noted that municipal solid waste management posed enormous challenges to many cities in Kenya. The rates of waste generation in this study are broadly similar to those in other African cities (Beukering *et al.*, 1999). In addition, our household level consultations with urban food producers revealed that soil infertility was identified as a major cause of food insecurity. Opinions from key informants suggested that the nutritious food harvested from locations near rural areas is a result of agricultural produce from the organically enriched soils close to the location given the extra organic inputs that this waste source provides. High population densities can result in high waste generation and less waste removal allowing waste to decompose and thus provide an organic nutrient input where composting of municipal waste is taking place.

One potential problem for scaling up the use of new integrated waste management strategies is the perception that domestic waste management via urban agriculture creates health problems either because of the use of contaminated wastewater for irrigation that increases pathogen contamination leading to soil and water pollution (Mvena *et al.*, 1991; Afrane *et al.*, 2004; Redwood, 2007), or through pathogen concerns related to ecosanitation waste. For instance, studies by Khan *et al.* (2008) highlight the health risk of heavy metals in wastewater irrigated soils in Beijing. In Zimbabwe, Muchuweti *et al.* (2006) reported that soils irrigated with sewage sludge resulted in increased levels of heavy metals such as cadmium, copper, lead and zinc in maize. As yet, no studies have assessed these heavy metal risks in the Malawian context and further analysis to this end is required prior to more widespread uptake in the use of such composting technologies.

Despite large financial investments in handling waste, current waste management approaches lack integration of agriculture into city-level strategic plans. Even though concepts such as the 'waste minimization hierarchy' have been developed to achieve sustainable management of municipal waste (DEFRA, 2001), our work shows that such concepts are not being implemented in Malawian cities. Urban agriculture in sub-Saharan Africa can be valued as a strategy/opportunity to minimize waste delivered to dumping sites as it fits well into the waste hierarchy by moving waste from dumping sites to agricultural farms. Several studies in other parts of sub-Saharan Africa (Smit and Nasr, 1992; Jacobi *et al.*, 2000; Battersby and Marshak, 2013) have demonstrated the potential of urban agriculture to enhance food security while ensuring urban sustainability through the efficient use of resources. In particular, Smit and Nasr (1992) argued that urban wastes which would otherwise degrade the natural environments are converted into useful resources that conserve the environment through urban agriculture. Indeed, in recent years, various advanced technologies have been employed by African city authorities in an attempt to properly manage waste. However, wholesale importations of many of such advanced technologies have not yielded the desired results. This can be attributed to the incompatibility of advanced technology with the social, physical and economic realities prevailing in many sub-Saharan Africa cities, with our study noting the important controls of land and labour availability as being critical factors.

Ecosanitation technologies and their organic waste products are being accepted for producing maize with both communities and city health officials recognizing the role of the ecosanitation technology in reducing public health-related problems such as cholera and diarrhoea. There are now more permanent toilets in high-density areas, thereby saving land for other usages. Our preliminary plot-based field results showed that ecosanitation products can improve soil quality and maize yields. The efficacy of ecosanitation technology has been widely acknowledged (Simpson-Hebert, 2001; Sugden, 2006; Endale *et al.*, 2012). In using such approaches to achieve efficient waste management practices in cities across sub-Saharan Africa, ecosanitation technologies could be used to enhance urban food production and sustainability.

8.5 Conclusions

The Malawian urban agriculture and waste management case study research presented here provides preliminary evidence on how issues related to waste in urban areas can be used more effectively to address food insecurity, household poverty and soil nutrient benefits, especially in areas of sandy infertile soils as found around many cities in sub-Saharan Africa. Several constraints in waste management are identified and linked to the opportunity cost for land-use planning, soil fertility management and improved food security at household and city levels. City authorities need to more pro-actively utilize the opportunity to produce and use organic municipal waste for soil quality and agricultural production benefits within city boundaries. These initiatives offer 'win–win' benefits for soil fertility and water-holding capacity as well as reducing urban environmental problems associated with waste management and aquatic pollution.

Urban land management and planning debates at present typically ignore the role that urban farmers play, yet they are key actors in food production and affecting the cycling of nutrients within urban and peri-urban agricultural landscapes. Formalizing urban farming activities and supporting organic nutrient flows on to urban farmers' fields can drastically reduce solid waste management-related costs and improve food security through integrated soil nutrient management approaches. This study highlights project examples where it has been possible to implement strategies that not only improve the sanitation of the community but also improve household income and food security among low-income urban communities. Improved urban soil and land-use management has been achieved by empowering local communities with knowledge, skills and other resources that enable them to produce high quality compost manure for enhancing the nutrient and water-retention capacities of soils across urban landscapes.

Acknowledgements

The authors acknowledge the financial support provided by International Development Research Centre (IDRC) in Canada, the Centre for Community Organisation and Development (CCODE) in Malawi and Nippon International Cooperation for Community Development of Japan. We gratefully thank the editors for their input in improving this chapter.

References

Afrane, Y.A., Klinkenberg, E., Drechsel, P., Owusu-Daaku, K., Garms, R. and Kruppa, T. (2004) Does irrigated urban agriculture influence the transmission of malaria in the city of Kumasi, Ghana? *Acta Tropica* 89, 125–134.

Anderson, J.M. and Ingram, J.S.I. (1993) *Tropical Soil Biology and Fertility: A Handbook of Methods* (2nd edition). CAB International, Wallingford, UK.

Battersby, J. and Marshak, M. (2013) Growing communities: Integrating the social and economic benefits of urban agriculture in Cape Town. *Urban Forum* 24, 447–461.

Beukering, P., Sehker, M., Gerlagh, R. and Kumar, V. (1999) Analysing urban solid waste in developing countries: A perspective on Bangalore. India Collaborative Research in the Economics of Environment and Development Working Paper, 24.

Bulkeley, H. and Castán Broto, V. (2013) Government by experiment? Global cities and the governing of climate change. *Transactions of the Institute of British Geographers* 38, 361–375.

DEFRA (2001) *Guidance on Municipal Solid Waste Strategies*. Department of the Environment, Transport and the Regions, London.

Dougill, A.J., Twyman, C., Thomas, D.S.G. and Sporton, D. (2002) Soil degradation assessment in mixed farming systems of Southern Africa: Use of nutrient balance studies for participatory degradation monitoring. *The Geographical Journal* 168, 195–210.

Endale, Y.T., Yirsaw, B.D. and Asfaw, S.L. (2012) Pathogen reduction efficiency of on-site treatment processes in eco-sanitation system. *Waste Management and Research* 30, 750–754.

FAO (2004) The state of food insecurity in the world. Monitoring progress towards the World Food Summit and Millennium Development Goals. Food and Agriculture Organization, Rome.

FAO (2012) Growing greener cities in Africa. First status report of urban and peri-urban horticulture in Africa. Available at www.fao.org/docrep/016/i3002e/i3002e.pdf (accessed 1 April 2015).

GOM (2012) Integrated household survey. National Statistics Office, Zomba, Malawi.

GOM (2008) Malawi census report 2008. National Statistics Office, Zomba, Malawi.

Grimmond, C., King, T., Cropley, F., Nowak, D. and Souch, C. (2002) Local-scale fluxes of carbon dioxide in urban environments: Methodological challenges and results from Chicago. *Environmental Pollution* 116, S243–S254.

Haiti, K.M., Swarup, A., Dwivedi, A.K., Misra, A.K. and Bandyopadhyay, K.K. (2007) Changes in soil physical properties and organic carbon status at the topsoil horizon of a vertisol of central India after 28 years of continuous cropping, fertilization and manuring. *Agriculture, Ecosystems and Environment* 119, 127–134.

Henry, R.K., Yongsheng, Z. and Jun, D. (2006) Municipal solid waste management challenges in developing countries – Kenyan case study. *Waste Management* 26, 92–100.

Hilhorst, T. and Muchena, F.M. (eds) (2000) *Nutrients on the Move – Soil Fertility Dynamics in African Farming Systems*. International Institute for Environment and Development, London.

Hovorka, A.J. (2006) Urban agriculture: Addressing practical and strategic gender needs. *Development Practice* 16, 51–61.

IPCC (2013) Climate change 2013: The physical science basis. Contribution of working group I to the IPCC fifth assessment report. Cambridge University Press, Cambridge, UK.

Jacobi, P., Amend, J. and Kiango, S. (2000). Urban agriculture in Dar es Salaam: Providing for an indispensable part of the diet. In: Bakker, N. (ed.) *Growing Cities, Growing Food: Urban Agriculture on the Policy Agenda*. German Foundation for International Development, Feldafing, Germany.

Jones, A., Breuning-Madsen, H., Brossard, M., Dampha, A., Deckers, J., Dewitte, O. *et al.* (eds) (2013) *Soil Atlas of Africa*. European Commission, Publications Office of the European Union, Luxembourg..

Kebede, B. (2007) Community wealth-ranking and household surveys: An integrative approach. Q-Squared Working Paper No. 38. Available at www.sarpn.org/documents/d0002826/Household_surveys_Kebede_Jun2007.pdf (accessed 1 April 2015).

Khan, S., Cao, Q., Zheng, Y.M., Huang, Y.Z. and Zhu, Y.G. (2008) Health risks of heavy metals in contaminated soils and food crops irrigated with wastewater in Beijing, China. *Environmental Pollution* 152, 686–692.

Kirchmann, H., Nyamangara, J. and Cohen, Y. (2005) Recycling municipal wastes in the future: From organic to inorganic forms? *Soil Use and Management* 21, 152–159.

Klemesu, M.A. (2000) Urban agriculture and food security, nutrition and health. In: Barker, N., Dubbeling, M., Gindel, S., Sabel-Koschella, U. and de Zeeuw, H. (eds) *Growing Cities, Growing Food – Urban Agriculture on the Policy Agenda*. DSE, Feldafing, Germany, pp. 99–118.

Kwapata, M.B., Nkanaunena, G.A., Maliro, M.F.A., Kamwendo, P.M., Dawa, G. and Bokosi, J. (2001) *The Current Status and Contribution of Urban Agriculture and Identification of Information, Communication and Training Needs in Malawi*. University of Malawi, Lilongwe, Malawi.

Langergraber, G. and Muellegger, E. (2005) Ecological sanitation – A way to global sanitation problems? *Habitat International* 31, 433–444.

Lowole, M.W. (1965) *Soil Map of Malawi*. Department of Agricultural Research, Lilongwe, Malawi.

Lu, D., Xu, X., Tian, H., Moran, E., Zhao, M. and Running, S. (2010) The effects of urbanisation on net primary productivity in south-eastern China. *Environmental Management* 46, 404–410.

Maxwell, D. (1999) The political economy of urban food security in Sub-Saharan Africa. *World Development* 27, 1939–1953.

Maxwell, D., Levin, C. and Csete, J. (1998) Does urban agriculture help prevent malnutrition? Evidence from Kampala. *Food Policy* 23, 411–424.

Mbaye, A. and Moustier, P. (2000). Market-oriented urban agricultural production in Dakar. *Growing Cities, Growing Food: Urban Agriculture on the Policy Agenda*, pp. 235–256.

Mkwambisi, D.D. (2008) Urban agriculture in Malawi: Poverty, reduction, waste management and institutional barriers. Unpublished PhD Thesis, University of Leeds, UK.

Mkwambisi, D.D., Fraser, E.D.G. and Dougill, A.J. (2011) Urban agriculture and poverty reduction: Evaluating how food production in cities contributes to food security, employment and income in Malawi. *Journal of International Development* 23, 181–203.

Mkwambisi, D.D., Fraser, E.D.G. and Dougill, A.J. (2012) Urban agriculture, institutions and development. In: Jacquet, P., Pachauri, R. and Tubiana, L. (eds) *A Planet of Life: Towards Agricultural Change?* TERI Press, New Delhi, India, pp. 73–76.

Mougeot, L.J.A. (2005) Introduction. In: Mougeot, L.J.A. (ed.) *AGROPOLIS: The Social, Political and Environmental Dimensions of Urban Agriculture*. Earthscan, London, pp. 1–30.

Muchuweti, M., Birkett J.W., Chinyanga E., Zvauya, R., Scrimshaw, M.D. and Lester, J.N. (2006) Heavy metal content of vegetables irrigated with mixtures of wastewater and sewage sludge in Zimbabwe: Implications for human health. *Agriculture, Ecosystems and Environment* 112, 41–48.

Mvena, Z., Lupanga, I. and Mlozi, M. (1991) Urban agriculture in Tanzania: A study of six towns. A Draft Report submitted to the IDRC. Sokoine University of Agriculture, Morogoro, Tanzania.

Olsen, S.R. and Sommers, L.E. (1982) Phosphorus. In: Page, A.L. (ed.) *Methods of Soil Analysis: Chemical and Microbiological Properties*. American Society of Agronomy, Madison, Wisconsin, pp. 403–430.

Oteng-Ababio, M., Melara Arguello, J.E. and Gabbay, O. (2013) Solid waste management in African cities: Sorting the facts from the fads in Accra, Ghana. *Habitat International* 39, 96–104.

Redwood, M. (2007) Urban agriculture and changing food markets. *Exporting Good Governance: Temptations and Challenges in Canada's Aid Program* 3, 205.

Redwood, M. (ed.) (2009) *Agriculture in Urban Planning. Generating Livelihoods and Food Security*. Earthscan, London.

Rosenzweig, C., Solecki, B., Hammer, S. and Mehrota, S. (2011) *Climate Change and Cities: First Assessment Report of the Urban Climate Change Research Network*. Cambridge University Press, Cambridge, UK.

Simpson-Hebert, M. (2001) Ecological sanitation and urban sustainability. *First International Conference on Ecological Sanitation* 5–8.

Smit, J. and Nasr, J. (1992) Urban agriculture for sustainable cities: Using wastes and idle land and water bodies as resources. *Environment and Urbanisation* 4, 141–152.

Sugden, S. (2006) Malawi – One step closer to sustainable sanitation: Experiences of an eco-sanitation project. *The Value of Environmental Sanitation – Case Studies* 13.

Stoorvogel, J.J. and Smaling, E.M.A. (1990) Assessment of soil nutrient depletion in sub-Saharan Africa: 1983–2000. Report 28. Winand Staring Centre, Wageningen, the Netherlands.

Taylor, A. and Peter, C. (2014) Strengthening climate resilience in African cities: A framework for working with informality. CDKN Working Paper, May 2014.

Zezza, A. and Tasciotti, L. (2010) Urban agriculture, poverty and food security: Empirical evidence from a sample of developing countries. *Food Policy* 35, 265–273.

9 Impact of Gold Mining on Mercury Contamination and Soil Degradation in Amazonian Ecosystems of French Guiana

Michel Grimaldi,[1]* Stéphane Guédron[2] and Catherine Grimaldi[3]

[1]*Institut de Recherche pour le Développement, Institute of Ecology and Environmental Sciences of Paris, Bondy, France;* [2]*ISTerre, Université Grenoble, Grenoble, France;* [3]*INRA, Rennes, France*

9.1 Introduction

As early as the 16th century, major expeditions were searching for gold and other precious metals and ores in the 'New World' (Nriagu, 1994; Tandeter, 2006). Later, in the middle of the 19th century, gold rushes that started in California spread to South America, as well as Australia and South Africa (Nriagu, 1994; Ali, 2006). Their main goal was to extract alluvial gold, occurring as fine particles in sediments derived from soils and weathered rocks. Another more current strategy of artisanal mining groups is to extract eluvial gold concentrated in the soil as fine particles and nuggets, originating from the *in situ* weathering of rocks. The extent of such artisanal small-scale gold mining (ASGM) activities varies with the price of gold (Malm, 1998). According to the United Nations Industrial Development Organization, 10–15 million people worked on ASGM sites throughout the world in 2004 (Veiga and Baker, 2004) in more than 160 mainly developing countries (Telmer and Veiga, 2009). Unfortunately, there is little information about the size and type of ASGM operations (Telmer and Veiga, 2009) or the surface area of soil and sediment worked; thus, the exact spatial extent and magnitude of impacts are difficult to determine.

Gold mining changes land use drastically, with a considerable impact on the entire ecosystem, including soil. First, extracting eluvial deposits requires removing vegetation and then using powerful water jets to strip off surface soil horizons and reach the gold-bearing layer (Fig. 9.1). Besides the large loss of topsoil, these practices affect the turbidity of watercourses (Fig. 9.1) and sometimes their drainage patterns (Telmer *et al.*, 2006; Hammond *et al.*, 2007) and thus have impacts far beyond the local site of operation. Second, fine gold particles in sediments and soils are recovered by amalgamating them with mercury (Hg) (Nriagu, 1994; Malm, 1998), which is subsequently burnt to evaporate the Hg. The amalgamation process releases 650–1350 t Hg into the global environment annually (mean annual release of 1000 t, of which 350 t are emitted to the atmosphere through burning of amalgam and 650 t are discharged into the hydrosphere as elemental and particulate Hg (Telmer and Veiga, 2009). Mercury contamination of aquatic trophic chains is the most alarming health risk for local populations, whose protein diet is largely provided by carnivorous fish and

* Corresponding author: michel.grimaldi@ird.fr

 Land-use Change Impacts on Soil Processes: Tropical and Savannah Ecosystems (eds F.Q. Brearley and A.D. Thomas)

Fig. 9.1. Illustration of artisanal or semi-industrial gold mining activities in French Guiana. From top left to bottom right: (a) mechanical removal of surface soil horizons, (b) stripping off of gold-bearing material with water jets to recover gold particles and nuggets, (c) digging of a derivation canal, (d) settling basin, and water turbidity in Combat Creek (e) before and (f) during artisanal small-scale gold mining activities.

their predators (Lebel *et al.*, 1996, 1997; Fréry *et al.*, 2001; Laffont *et al.*, 2011). These top-chain predators concentrate Hg in its most toxic and bio-accumulative form, monomethyl-Hg (Agaki *et al.*, 1995; Mason *et al.*, 1995; Barbosa *et al.*, 2003; Coquery *et al.*, 2003).

In this chapter, we focus on the impact of ASGM on the Hg contamination of tropical soils, with special attention on how the change in land use (i.e. gold extraction practices) alters the soil processes that affect Hg fate and behaviour. We explain the soil formation and transformation processes that determine the accumulation or mobilization of Hg, reflecting the dual role of soils as both 'sink' and 'source' of Hg. First, we compare the fate and behaviour of Hg from ASGM with that from the natural geochemical background, considering the nature and abundance of carrier phases and soil functioning. Second, we describe how former and current ASGM activities intensify the soil processes that favour Hg mobilization by surface and near-surface runoff to watercourses, Hg emission to the atmosphere and production of monomethyl-Hg.

This chapter combines research carried out in French Guiana, a territory located in northern South America (Tessier *et al.*, 2003; Grimaldi *et al.*, 2004, 2008; Guédron *et al.*, 2006, 2009, 2011a,b), with studies from Brazilian Amazonia. French Guiana, like much of Amazonia, is greatly impacted by ancient and current, legal and illegal ASGM activities. The impact on tropical soils, as well as on ecosystems as a whole, has been well studied in these areas. Furthermore, as the environmental legislation in force since 2006 in French Guiana prohibits the use of Hg at gold mining sites, it is relevant to compare ASGM with or without the amalgamation practice. For a better understanding of the soil processes involved, the main research issue is to promote mining practices that can reduce health and environmental risks.

9.2 Origins and Carrier Phases of Mercury Accumulated in Soils

Mercury inputs from ASGM activities add to Hg already accumulated in the soil from two distinct origins: (i) weathering of parent rocks (lithogenic Hg); and (ii) atmospheric deposition derived from degassing of the Earth's crust and oceans, along with long-term human activities since the industrial era.

The magmatic and metamorphic rocks making up the Guiana and Brazilian shields, which border the Amazon basin to the north-east and south, have low concentrations of Hg, ranging from 4 to 7 ng g^{-1} for granites, basalts and andesites in French Guiana (Guédron *et al.*, 2006). These concentrations are lower than, or equal to, those obtained for rocks of a comparable nature found in Brazil (Fiorentino *et al.*, 2011) and on other continents (Gao *et al.*, 1998). Except for certain minerals, which are generally sulfur (S)-bearing (cinnabar: HgS), Mercury is present as impurities in the crystalline lattices of minerals (Andersson, 1979). Weathered materials (alterites) have higher Hg concentrations than parent rocks. The weathering zone is often 20–40 m thick in humid tropical regions where rock exposures are limited to inselbergs. Analysis of several alterites derived from the same parent rock shows that the Hg concentration increases progressively with the degree of weathering, up to approximately 50 ng g^{-1} (Oliveira *et al.*, 2001; Guédron *et al.*, 2006). Mercury concentrations in alterites are themselves lower than those from the soil horizons above them, in particular the clayey and micro-aggregated Oxisol horizons that are a few metres thick. The Hg concentration in Oxisols can reach 800 ng g^{-1} in the fine fraction (<2 mm) and sometimes more than 1 μg g^{-1} in the ferruginous coarse fraction (>2 mm), which often represents more than half of the bulk soil fraction (Oliveira *et al.*, 2001; Grimaldi *et al.*, 2008).

The strong contrast between the parent material and the Oxisols can be explained by the long period of intense weathering under a hot and humid climate. During weathering, Hg can concentrate in Oxisols, along with elements such as iron (Fe), aluminium (Al), manganese and titanium, while other elements, including calcium, magnesium and silica, are more mobile (Nahon, 1991). The proportion of lithogenic Hg has been estimated using refractory elements (Fe, niobium, uranium and zinc), which are known to be relatively immobile, not derived from an atmospheric source, and are indicators of the degree of weathering (Oliveira *et al.*, 2001). It appears that rocks are of minor importance as a source of Hg (in French Guiana, 10–25% of total Hg depending on soil type) compared to long-term natural atmospheric deposition and anthropogenic inputs (Guédron *et al.*, 2006; Grimaldi *et al.*, 2008) (Fig. 9.2). The contribution of anthropogenic Hg increases even more near gold mining sites using amalgamation, where the total Hg concentrations in the soil can reach several micrograms per gram locally (Guédron *et al.*, 2006, 2009).

While most of the Hg accumulated in soils is not derived from their parent rocks, these materials have an indirect influence on the Hg enrichment, from natural or anthropogenic sources, due to the nature and abundance of secondary minerals produced by weathering, which are likely to fix Hg. The Hg carrier phases identified in Amazonian soils by selective extraction methods (Kostka and Luther, 1994; Bloom *et al.*, 2003) are: (i) organic matter; and (ii) amorphous and crystallized Fe-oxides (Al-substituted goethite and haematite), both in the fine fraction (<2 mm) and in ferruginous nodules (Roulet and Lucotte, 1995; Roulet *et al.*, 1998; Guédron *et al.*, 2009). Hg has a strong affinity for the S groups of organic molecules (Schuster, 1991; Skyllberg *et al.*, 2000; Manceau and Nagy, 2008), which explains its commonly observed correlation with S and organic carbon (C) (Guédron *et al.*, 2009). Owing to the rapid mineralization of organic matter in Oxisols, oxy(hydr)oxides (whether associated with residual organic matter or not) take over in importance for Hg retention. The ability of Fe-oxides to retain Hg in its divalent form (Hg^{2+}) is even higher when these oxides are amorphous or poorly crystallized, because of their large surface area and the atomic ordering defects induced by a high rate of substitution of

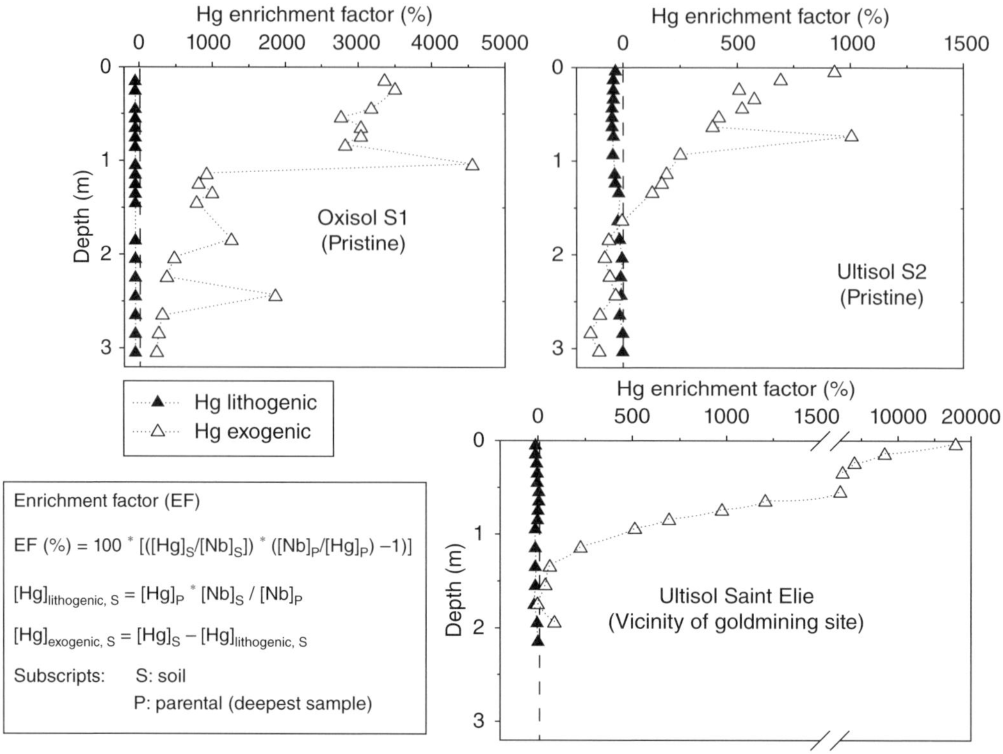

Fig. 9.2. Lithogenic and exogenic mercury (Hg) enrichment factors (%) in pristine Oxisol and Ultisol from the Leblond toposequence (upper panel) and in an Ultisol sampled near the Saint Elie goldmine camp (French Guiana). Niobium (Nb) was selected as an immobile element, uranium (U) as proxy for Hg behaviour during weathering and pedogenesis, and the mean composition of the three lowermost samples of both Ultisols' alterite as reference material. (Modified from Guédron *et al.*, 2006.)

Fe by Al (Schwertmann and Latham, 1986; Schwertmann and Cornell, 2000). These various organic and mineral phases are associated with kaolinite, forming clayey micro-aggregates in the upper soil horizons (Tandy *et al.*, 1990), so that the abundance of clay particles (<2 μm fraction) accounts for the Hg content of the soil, in the same geological setting. Highly clayey Oxisols developed on sedimentary formations of the central Amazonian basin (Roulet *et al.*, 1998) have lower Hg concentrations (<200 ng g^{-1}) than Oxisols on the Guiana Shield (350–810 ng g^{-1} – Richard *et al.*, 2000; Guédron *et al.*, 2006, 2009; Grimaldi *et al.*, 2008) and Brazilian Shield (80–640 ng g^{-1} – Oliveira *et al.*, 2001). Indeed, the Oxisols of the central Amazonian basin are richer in clay but poorer in oxides, which is an indirect result of the nature of their parent rocks.

9.3 Soil Processes Controlling Mercury Concentrations in Soils

Soil processes influence the fate of Hg not only by determining the nature and abundance of carrier phases, but also more directly by controlling hydric and biological soil functioning, which affect Hg speciation and migration. There is a high degree of pedological diversity in Amazonia (composition and functioning) across a wide range of geological formations and climatic conditions at the regional scale (Richter and Babbar, 1991; Quesada *et al.*, 2011). Even at the local scale, soils formed from the same bedrock can display distinctively different properties (Chauvel *et al.*, 1987; Boulet *et al.*, 1993). As a consequence, away from gold mining sites the pedological diversity associated with past and present soil formation processes explains the spatial variability of Hg content observed for the same parent rock and under the same conditions of atmospheric input (Roulet *et al.*, 1998; Oliveira *et al.*, 2001; do Valle *et al.*, 2005; Guédron *et al.*, 2006; Grimaldi *et al.*, 2008).

This pedological diversity results from the transformation of Oxisols by two active natural processes (i.e. erosion and hydromorphy) under present-day tectonic and climatic conditions. First, erosion, diffuse under forest cover, produces a gradual thinning of the soil, bringing poorly structured and more impermeable alterites closer to the surface, thus leading to the development of Ultisols that are superficially depleted in clay by dissolution and eluviation (Grimaldi *et al.*, 2004). Second, hydromorphy is favoured by the presence of these alterites near the surface on slight slopes, where vertical or lateral water drainage is limited. Therefore, under long-lasting conditions of excess water, this process results in the chemical and physical degradation of the soil (depletion of Fe and clay; disappearance of micro-aggregated structure).

The micro-aggregated structure of Oxisols, well developed in thicknesses varying from one to a few metres, leads to an effective porosity that facilitates water and gas transfers; thus, water drainage is primarily vertical and occurs under oxidizing conditions. The high Hg content of Oxisols, sometimes down to a depth of 3 m, can be explained by the abundance of carrier phases, as well as by water transfer and biological activity, which redistribute Hg of atmospheric origin in the soil profile (Guédron *et al.*, 2006; Grimaldi *et al.*, 2008) (Fig. 9.3). The Hg can migrate with other substances transported by water, in particular organo-mineral complexes such as fulvic acids associated with Fe and Al (Ravichandran, 2004). Bioturbation of the soil, i.e. structural rearrangement of soil particles by invertebrates such as termites, worms and ants, is active in Oxisols, also contributing to the penetration of particulate-bound Hg deep into the soil, while ensuring aggregation of the soil and thus maintaining its permeability (Lavelle *et al.*, 2006).

In contrast, in the Ultisols, drainage during heavy rains is essentially lateral and superficial, due to the presence of alterites at depths of less than 1 m (Molicova *et al.*, 1997; Grimaldi *et al.*, 2004). The Hg carrier phases become less abundant in the surface layers of the Ultisols due to frequent water excess and reducing conditions (Guédron *et al.*, 2006; Grimaldi *et al.*, 2008) (Fig. 9.3). Moreover, the poorly structured alterites are depleted in Hg, which is primarily lithogenic and preferentially integrated into the lattice of crystallized Al- and Fe-oxides (Guédron *et al.*, 2009).

Hydromorphic soils are characterized by poor water drainage and reducing conditions (do Nascimento *et al.*, 2004; Fritsch *et al.*, 2006). These soils are located in areas of slight or flat slopes of the watershed and in depressions close to the riverbed, and are sometimes developed on summit shoulders or mid-hillslope areas that are hydrologically more isolated (Fig. 9.3). In the hydromorphic soils, reducing conditions favour dissolution of amorphous and crystallized Fe-oxides (Patel-Sorrentino *et al.*, 2007), as well as emission of elemental gaseous Hg into the atmosphere by reduction of its ionic divalent species (Tessier *et al.*, 2003). Consequently, the two natural processes of transformation of Oxisols into Ultisols and hydromorphic soils lead to a redistribution of natural or anthropogenic Hg in the soil or, more often, to its export, as discussed later.

9.4 Relative Importance of Mercury Anthropogenic Inputs versus Geochemical Background

The last great gold rush in Amazonia released approximately 2000 t Hg into the environment (Malm, 1998), reaching 200 t year^{-1} in Brazil during the 1980s, of which 60% was released into the atmosphere due to burning of gold-Hg amalgams (Lacerda and Salomons, 1998). Among other human activities, burning associated with Amazonian deforestation also releases considerable quantities of Hg: 6–9 t year^{-1} (Roulet *et al.*, 1998).

Consequently, Hg fluxes reaching the soil via direct wet and dry atmospheric deposition (above the forest canopy) were estimated at 151 μg m^{-2} year^{-1} at less than 10 km from a mining site in Brazil (Tumpling *et al.*, 1996), decreasing rapidly with distance to 18–25 μg m^{-2} year^{-1} at 100 km (Fostier *et al.*, 2000; Fadini and Jardim, 2001). At this distance, the contribution of gold mining to wet atmospheric fall out was estimated at 7–17 μg m^{-2} year^{-1}. Hg stocks were evaluated at 0.65 and 2 g m^{-2} in the uppermost

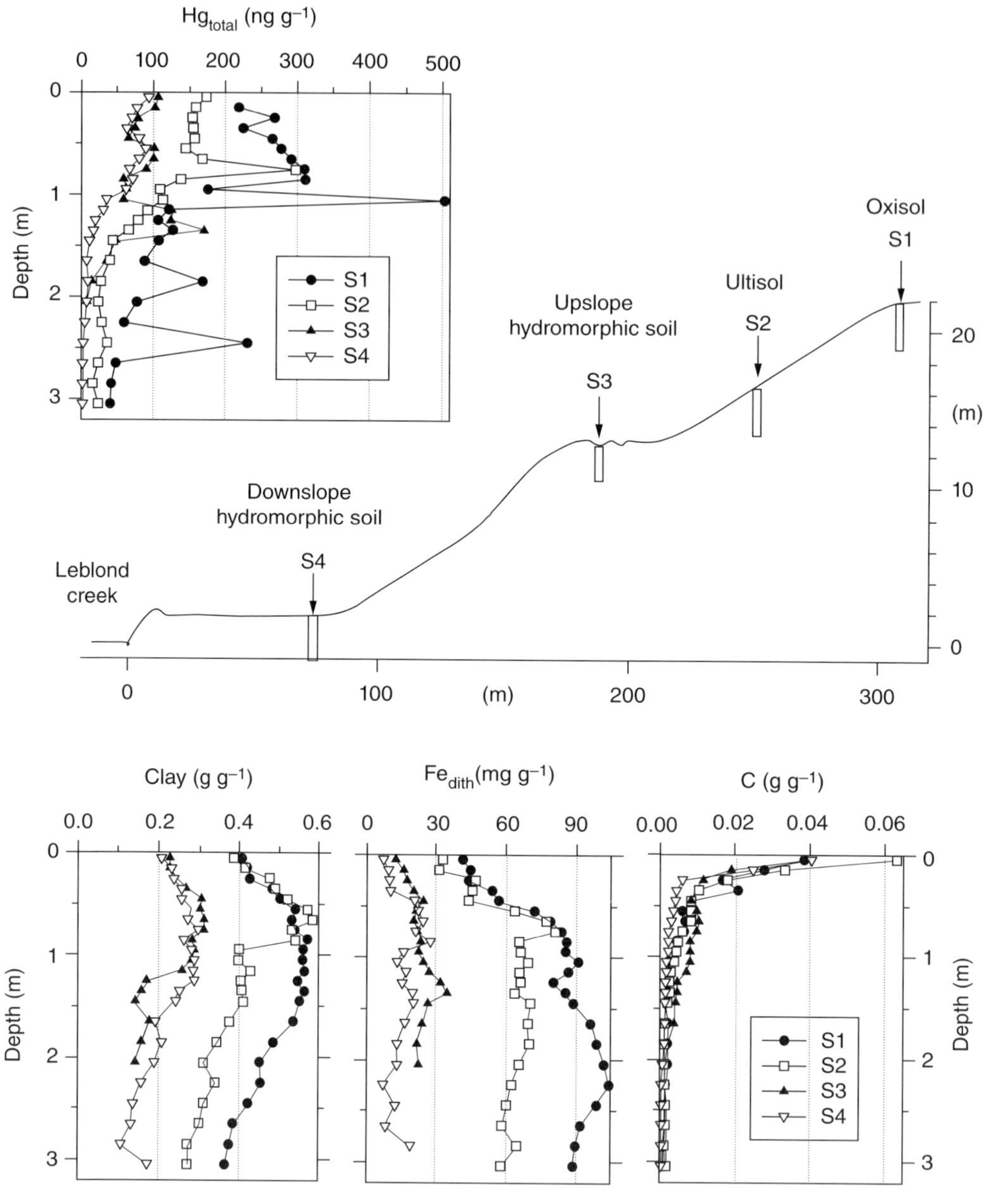

Fig. 9.3. Total mercury (Hg$_{total}$), clay, crystalline iron extracted with dithionite (Fe$_{dith}$) and total carbon (C) contents in four soil profiles (S1–S4) of the pristine Leblond toposequence (French Guiana). (Modified from Guédron *et al.*, 2006 and unpublished data.)

2 m of two Oxisols of French Guiana, respectively, without including the Hg accumulated in ferruginous nodules (Guédron *et al.*, 2009). A Hg flux of 17 μg m^{-2} year^{-1} over the last 150 years would have supplied 2.6 mg m^{-2}, that is to say, only 0.39 and 0.13% of the Hg stock, respectively, in both Oxisols. In the hydromorphic soils, which are markedly less rich in Hg, the same Hg flux would account for 1.7% of a stock of 0.15 g m^{-2}. These calculations do not take into account dry atmospheric deposition (Hg^{2+} bound to aerosols as well as elemental Hg^0),

which, in forests, is adsorbed on to the leaves of the canopy before being washed off by rains or reaching the ground with the fall of leaf litter (Fostier *et al.*, 2000; Mélières *et al.*, 2003; Guédron *et al.*, 2013). Mining activities at the beginning of the European colonization of South America, before the 150 years assumed here for calculating fluxes, released large quantities of Hg (Nriagu, 1993, 1994), which could also contribute to the Hg stock in the soils. Despite these approximations, it appears that anthropogenic activities, in particular gold mining, only make a minor contribution to the Hg stock of soils compared to other natural atmospheric sources over long periods. Away from gold mining sites, the geochemical background of Hg is thus predominant in controlling its content in Amazonian soils.

However, Hg pollution is clearly apparent in soils intensely reworked by gold prospecting activities. The Hg concentrations at such sites are an order of magnitude higher, on average, than those from pristine soils, irrespective of their texture (Guédron *et al.*, 2009) (Fig. 9.2). Their spatial variability over short distances is high (Guédron *et al.*, 2009). Exceptionally high concentrations were observed in an undisturbed soil of a gold mining camp, probably caused by the input of Hg^0 in liquid (waste or accidental pollution) and gaseous phases (amalgam burning) (Guédron *et al.*, 2006). It is thus noteworthy that a major proportion of Hg is present in the form of isolated droplets of elemental Hg (Hg^0), quantifiable by pyrolysis (Bollen *et al.*, 2008) that are sometimes amalgamated with particles of gold (Guédron *et al.*, 2009).

9.5 Mobilization of Mercury from Soils

Natural or anthropogenic Hg stored in soil can be mobilized by water fluxes on the surface and in the soil, or emitted to the atmosphere. In Amazonia, intense rain events play a major role in leaching Hg from the atmosphere and the canopy to the soil and watercourses. Mercury is mobilized in its various chemical forms, either dissolved or associated with solid suspended particles. In streams draining Oxisols, concentrations of dissolved total Hg are lower (around 1 ng l^{-1}) than in both rainwater and throughfall (4 ng l^{-1}), confirming the role played by Oxisols as a sink for Hg (Guédron *et al.*, 2011a) (Fig. 9.4a). In contrast, surface runoff and the temporary near-surface groundwater formed during strong rainfalls in the Ultisols mobilize Hg, both associated with particles and in soluble form. A large part of the mobilized Hg is in particulate form during the peaks of discharge, while the proportion of dissolved Hg tends to increase during the discharge recession (Tessier *et al.*, 2003; Guédron *et al.*, 2011a). Due to the considerable soil erosion produced by ASGM practices, particulate Hg fluxes into streams include not only anthropogenic Hg but also long-term accumulated Hg from the natural background.

Long-lasting water saturation in hydromorphic soils after intense rainfall favours the reduction of Hg^{2+} and its emission into the atmosphere as a gaseous phase. This other mechanism for the mobilization of Hg in the soil was demonstrated by experiments in the laboratory and in the field (Tessier *et al.*, 2003; Peretyazhko *et al.*, 2006). The concentrations of gaseous elemental Hg in the atmosphere range from 2–9 ng m^{-3} above well-drained soils and from 14–95 ng m^{-3} in pristine areas where the drainage is less efficient (Tessier *et al.*, 2003). In French Guiana, the gold rushes occurring since the middle of the 19th century affected mostly flat areas along the river network. Reducing conditions often occur in this environment, accentuated by disturbance of the prospected area's geomorphology.

9.6 Methylation of Mercury in Former Gold-mined Sites

In former gold mine flats, soil disturbance from ASGM activities has impeded soil drainage, which favours methylation of Hg. Guédron *et al.* (2011a) observed an increase in the concentration of various Hg species in a stream, between the entry and outlet of a former mined site, especially for both dissolved and particulate monomethyl-Hg (Fig. 9.4a). Total Hg concentration of suspended sediments increased along with increasing Hg concentrations of ASGM contaminated riparian soils. Although monomethyl-Hg represented less than 1% of the total Hg dissolved

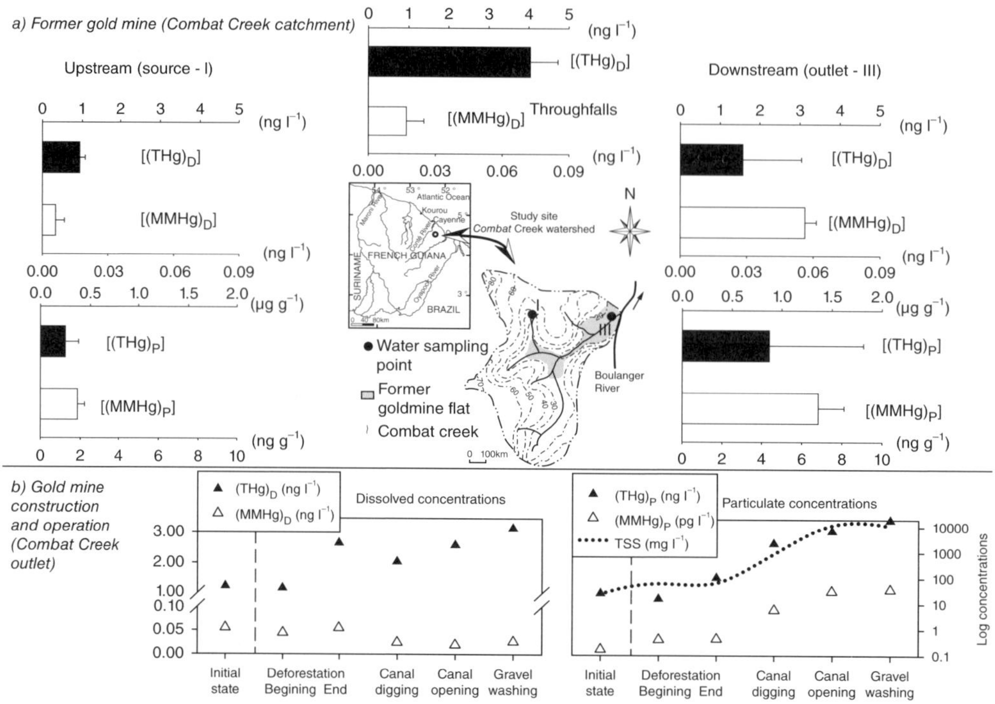

Fig. 9.4. (a) Mean and standard deviation of dissolved (D) and particulate (P) total mercury (THg) and monomethylmercury (MMHg) concentrations in throughfalls, the stream source and the stream outlet of the former goldmine flat of the Combat Creek catchment (French Guiana) sampled between and during 14 storm events at the beginning and in the middle of the 2005 and 2006 rainy seasons. (Modified from Guédron *et al.*, 2011a.) (b) Dissolved (D) and particulate (P) total mercury (THg), monomethylmercury (MMHg) and total suspended solid (TSS) concentrations monitored at the outlet of Combat Creek (French Guiana) during construction and operation of an experimental gold mine in the former goldmine flat of the Combat Creek catchment in December 2006. (Unpublished data.)

in rainwater, throughfall and water draining the pristine Oxisols, it reached 2–3% in water draining the exploited soils. A comparable trend was observed for particulate monomethyl-Hg. Unlike Oxisols, which tend to retain or transform the monomethyl-Hg originating from the canopy and litter, soils contaminated by former eluvial gold mining are sources of both dissolved and particulate monomethyl-Hg (Fig. 9.4a). Between the entry and outlet of the former mined site, the specific flux in the stream (flux per unit area of the catchment) of dissolved and particulate monomethyl-Hg increased six-fold and four-fold, respectively, while total Hg specific flux remained stable for the dissolved fraction and doubled for the particulate fraction (Guédron *et al.*, 2011a).

Former gold mining activities dramatically disturb the geomorphology of prospected areas. Many zones of stagnant water remain even several decades after the mining activity, leading to the accumulation of organic matter under anoxic conditions. These stagnant waters and, to a lesser extent, the pore waters of water-logged soils, are highly enriched in monomethyl-Hg (up to 0.8 ng l^{-1}), containing 10–100 times the concentrations in waters draining the Oxisols (Guédron *et al.*, 2011b). Because of the abundance of sulfide and reduced Fe in these waters, methylation, favoured by the high Hg and organic C, can be attributed to sulfate and Fe-reducing bacteria (Ullrich *et al.*, 2001; Benoit *et al.*, 2003). Dissolved and particulate monomethyl-Hg is mobilized when production zones are connected to the drainage network during intense rainfall events. This mobilization is expressed by sudden peaks of monomethyl-Hg concentration in the stream, the strongest ones being observed at the beginning of the rainy season, before the areas of stagnant water are replenished or diluted (Guédron *et al.*, 2011a).

In many Amazonian regions, eluvial gold mining is currently practised on formerly mined sites. Far more than deforestation and the removal of superficial soil layers, the highest particulate emissions are caused by excavation of diversion canals, which are subsequently used to supply water to a site's installations and to wash auriferous gravels, before digging a settling pond. A full-scale simulation of a gold mining operation at a site at Combat Creek (French Guiana, unpublished data), without use of Hg, pointed to large increases in Hg fluxes at the start of the operation: approximately 10 times more total Hg and 50 times more particulate monomethyl-Hg were released downstream of the mining site (Fig. 9.4b). Dissolved Hg fluxes increased less markedly, and dissolved monomethyl-Hg fluxes even decreased temporarily (Fig. 9.4b). Particulate Hg emissions remained high during rain events, because of erosion of deforested areas on the mining site. In the long term, these cumulative Hg emissions to the drainage network can be higher than inputs due to the opening of the sites. In particular, it is advisable to avoid overflowing of settling basins into rivers, because these basins act as zones of water stagnation. The highest concentrations of dissolved monomethyl-Hg observed in two basins studied (1.4 and 2.5 ng l^{-1}) (Guédron *et al.*, 2011b) are similar to those recorded in storage dams (Boudou *et al.*, 2005; Muresan *et al.*, 2008), at the sediment–water interface. In the settling basins, high concentrations of dissolved organic matter, sulfates and Fe-oxides indeed favour the activity of Fe- and sulfate-reducing bacteria.

9.7 Recent Legislative Attempts to Reduce Mercury Emissions and Methylation

Since 2006, legislation in French Guiana has improved semi-industrial gold mining practices by prohibiting the use of Hg and replacing it with techniques that combine concentration by gravity and centrifugation. This legislation also requires construction of closed water circuits to strip off the gold-bearing layers of soil or sediment, thus decreasing emissions of fine Hg-rich particles to rivers during mining operations. Besides improving mining practices, the most critical points for miners are to reforest sites to avoid Hg methylation after mining activities. Rapid drainage of the settling basins, followed by rapid revegetation, appears to be the best method to decrease post-mining Hg and monomethyl-Hg emissions (Guédron *et al.*, 2011b). The success of rehabilitation depends on human intervention to restore soil fertility, which includes reconstituting physico-chemical characteristics and recovering biological functions (Loubry, 2002; Schimann *et al.*, 2007). These interventions have to be prepared before mining operations by sampling local representative plant species and building a nursery. Next, revegetation requires covering the worked materials with an organic matter layer (stored during excavation of the organic horizon, the first step after deforestation), in which pioneer species and then the local species grown in the nursery are planted.

Since such promising practices are recent, not enough time has elapsed to evaluate their short- and medium-term impacts on Hg emissions and methylation. In particular, the fate of Hg methylation in deep and sub-surface groundwater of rehabilitated soils must be studied further. Although little to no information about the drainage capacity of these soils is available, it is probable that before they recover an aggregated structure that allows drainage, they will become saturated during rainfall events. Therefore, this disturbed material, consisting of an Fe-rich soil horizon and fine Hg-rich particles from settling ponds, might become a sub-surface source of monomethyl-Hg in groundwater under anoxic conditions.

9.8 Conclusions

Artisanal small-scale gold mining, both at present and in the past, has had a major impact on soils and ecosystems. The most crucial risk identified is the use of Hg to amalgamate gold. The impacts are not only local, as in areas directly affected by the working of auriferous deposits, or downstream from mining sites due to increased release of fine Hg-rich particles into the drainage network and food-chain contamination, but also global, via atmospheric emissions.

The Hg used in eluvial gold mining is added to the already important geochemical background of tropical soils. Indeed, soils of humid tropical regions are naturally rich in Hg, because of their high contents of carrier phases, in particular Al- and Fe-oxyhydroxides. Over long periods of time, these soils have accumulated Hg resulting from (i) the intense weathering of rocks, which nevertheless have low concentrations of Hg, and, more particularly, (ii) the input of atmospheric Hg derived from degassing of the Earth's crust and oceans, or from long-term anthropogenic emissions. This accumulation took place simultaneously with pedogenic processes that have shaped soil development, thus leading to great variability in Hg content linked to the diversity of tropical soils. The Hg used in ASGM activities can be stored in soils under its divalent form, as natural Hg, because tropical soils have high Hg-storage capacities. On gold mining sites, large quantities of anthropogenic Hg can be found in the form of droplets of elemental Hg or gold–Hg amalgams.

Although soils generally behave as a sink for Hg, from natural or anthropogenic origins, they can also become a source. By destroying the vegetation and soil cover, gold mining increases the erosion and transfer of soil particles enriched in Hg towards rivers. Contamination of the aquatic food chain by Hg results, above all, from erosive practices that export both natural and anthropogenic Hg.

Artisanal small-scale gold mining affects most flat areas along river networks, where reducing conditions in hydromorphic soils favour atmospheric Hg emission. Above all, by reworking the soil, gold mining disturbs local hydrological and geochemical conditions, which then become favourable for Hg methylation. In this way, settling basins of present-day sites, but also flat riparian areas affected by former gold mining, act as biochemical reactors producing monomethyl-Hg, the most toxic bio-accumulative form of Hg.

Although the use of Hg has been officially prohibited in French Guiana since 2006, being replaced by techniques combining concentration by gravity and centrifugation, its use is common in other parts of the developing world. However, even when Hg amalgamation is not used, gold mining without any precautions creates conditions that favour the transport of Hg already present in the soils to river systems and its transformation into monomethyl-Hg. The risk of Hg contamination for humans thus tends to eclipse the other problems arising from ASGM operations. In fact, ASGM also causes complete deforestation of the areas exploited and consequently a loss of biodiversity and ecosystem functions, landscape modifications and soil degradation.

References

Agaki, H., Malm, O., Kinjo, Y., Harada, M., Branches, F.J.P., Pfeiffer, W.C. and Kato, H. (1995) Methylmercury pollution in the Amazon, Brazil. *The Science of the Total Environment* 175(2), 85–95.

Ali, S.H. (2006) Gold mining and the golden rule: A challenge for producers and consumers in developing countries. *Journal of Cleaner Production* 14, 455–462.

Andersson, A. (1979) Mercury in soils. In: Nriagu, J.O. (ed.) *The Biogeochemistry of Mercury in the Environment.* Elsevier/North-Holland Biomedical Press, Amsterdam, pp. 79–112.

Barbosa, A.C., Souza, J.D., Dorea, J.G., Jardim, W.F. and Fadini, P.S. (2003) Mercury biomagnification in a tropical black water, Rio Negro, Brazil. *Archives of Environmental Contamination and Toxicology* 45, 235–246.

Benoit, J.M., Gilmour, C.C., Heyes, A., Mason, R.P. and Miller, C.L. (2003) Geochemical and biological controls over methylmercury production and degradation in aquatic ecosystems. In: Cai, Y. and Braids, O.C. (eds) *Biogeochemistry of Environmentally Important Trace Elements.* American Chemical Society, ACS Symposium Series, Washington, DC, pp. 262–297.

Bloom, N.S., Preus, E., Katon, J. and Hiltner, M. (2003) Selective extractions to assess the biogeochemically relevant fractionation of inorganic mercury in sediments and soils. *Analytica Chimica Acta* 479, 233–248.

Bollen, A., Wenke, A. and Biester, H. (2008) Mercury speciation analyses in $HgCl_2$-contaminated soils and groundwater – Implications for risk assessment and remediation strategies. *Water Research* 42, 91–100.

Boudou, A., Maury-Brachet, R., Coquery, M., Durrieu, G. and Cossa, D. (2005) Synergic effect of goldmining and damming on mercury contamination in fish. *Environmental Science and Technology* 39, 2448–2454.

Boulet, R., Lucas, Y., Fritsch, E. and Paquet, H. (1997) Geochemical processes in tropical landscapes: role of soil covers. In: Paquet, H. and Clauer, N. (eds) *Soils and Sediments: Mineralogy and Geochemistry*. Springer-Verlag, Berlin, pp. 67–96.

Chauvel, A., Lucas, Y. and Boulet, R. (1987) On the genesis of soil mantle of the region of Manaus, Central Amazonia, Brazil. *Experientia* 43, 234–241.

Coquery, M., Cossa, D., Azemard, S., Peretyazhko, T. and Charlet, L. (2003) Methylmercury formation in the anoxic waters of the Petit-Saut reservoir (French Guiana) and its spreading in the adjacent Sinnamary river. *Journal de Physique IV* 107, 327–331.

do Nascimento, N.R., Bueno, G.T., Fritsch, E., Herbillon, A.J., Allard, T., Melfi, A.J. *et al.* (2004) Podzolization as a deferralitization process: A study of an Acrisol-Podzol sequence derived from Palaeozoic sandstones in the northern upper Amazon Basin. *European Journal of Soil Science* 55, 523–538.

do Valle, C.M., Santana, G.P., Augusti, R., Filho, F.B.E. and Windmöller, C.C. (2005) Speciation and quantification of mercury in Oxisol, Ultisol, and Spodosol from Amazon (Manaus, Brazil). *Chemosphere* 58, 779–792.

Fadini, P.S. and Jardim, W.F. (2001) Is the Negro River basin (Amazon) impacted by naturally occuring mercury? *The Science of the Total Environment* 275, 71–82.

Fiorentino, J.C., Enzweiler, J. and Angélica, R.S., 2011. Geochemistry of mercury along a soil profile compared to other elements and to the parental rock: Evidence of external input. *Water, Air and Soil Pollution* 221, 63–75.

Fostier, A.-H., Forti, M.C., Guimaraes, J.R.D., Melfi, A.J., Boulet, R., Espirito Santo, C.M. and Krug, F.J. (2000) Mercury fluxes in a natural forested Amazonian catchment (Serra do Navio, Amapa State, Brazil). *The Science of the Total Environment* 260, 201–211.

Fréry, N., Maury-Brachet, R., Maillot, E., Deheeger, M., Mérona de, B. and Boudou, A. (2001) Gold-mining activities and mercury contamination of native Amerindian communities in French Guiana: Key role of fish in dietary uptake. *Environmental Health Perspectives* 109, 449–456.

Fritsch, E., Herbillon, A.J., do Nascimento, N.R., Grimaldi, M. and Melfi, A.J. (2006) From Plinthic Acrisols to Plinthosols and Gleysols: Iron and groundwater dynamics in the tertiary sediments of the upper Amazon basin. *European Journal of Soil Science* 58, 989–1006.

Gao, S., Luo, T.C., Zhang, B.R., Zhang, H.F., Han, Y.W., Zhao, Z.D. and Hu, Y.K. (1998) Chemical composition of the continental crust as revealed by studies in East China. *Geochimica Cosmochimica Acta* 62, 1959–1975.

Grimaldi, C., Grimaldi, M. and Guédron, S. (2008) Mercury distribution in tropical soil profiles related to origin of mercury and soil processes. *The Science of the Total Environment* 401, 121–129.

Grimaldi, C., Grimaldi, M., Millet, A., Bariac, T. and Boulègue, J. (2004) Behavior of chemical solutes during a storm in a rainforested headwater catchment. *Hydrological Processes* 18, 93–106.

Guédron, S., Grimaldi, C., Chauvel, C., Spadini, L. and Grimaldi, M. (2006) Weathering versus atmospheric contributions to mercury concentrations in French Guiana soils. *Applied Geochemistry* 21, 2010–2022.

Guédron, S., Grangeon, S., Lanson, B. and Grimaldi, M. (2009) Mercury speciation in a tropical soil association: Consequence of gold mining on Hg distribution in French Guiana. *Geoderma* 153, 331–346.

Guédron, S., Grimaldi, M., Grimaldi, C., Cossa, D., Tisserand D. and Charlet L. (2011a) Amazonian former gold mined soils as a source of methylmercury: Evidence from a small scale watershed in French Guiana. *Water Research* 45, 2659–2669.

Guédron, S., Cossa, D., Grimaldi, M. and Charlet, L. (2011b) Methylmercury in tailings ponds of Amazonian gold mines (French Guiana): Field observations and an experimental flocculation method for in situ remediation. *Applied Geochemistry* 26(2), 222–229.

Guédron S., Grangeon S., Jouravel G., Charlet L. and Sarret G. (2013) Atmospheric mercury incorporation in soils of an area impacted by a chlor-alkali plant (Grenoble, France): Contribution of canopy uptake. *The Science of the Total Environment* 445–446, 356–364.

Hammond, D.S., Gond, V., de Thoisy, B., Forget, P.-M. and DeDijn, B.P.E. (2007) Causes and consequences of a tropical forest gold rush in the Guiana Shield, South America. *Ambio: A Journal of the Human Environment* 36, 661–670.

Kostka, J.E. and Luther III, G.W. (1994) Partitioning and speciation of solid phase iron in saltmarsh sediments. *Geochimica Cosmochimica Acta* 58, 1701–1710.

Lacerda, L.D. and Salomons, W. (1998) *Mercury in the Amazon. A Chemical Time Bomb?* Springer-Verlag, Heidelberg, Germany.

Laffont, L., Sonke, J.E., Maurice, L., Luna-Monroy, S., Chincherros, J., Amouroux, D. and Behra, P. (2011) Hg isotope signatures of human hair: A comparative study between fish-eating people and gold-miners in Bolivia. *Environmental Science and Technology* 45, 9910–9916.

Lavelle, P., Decaens, T., Aubert, M., Barot, S., Blouin, M., Bureau, F. *et al.* (2006) Soil invertebrates and ecosystem services. *European Journal of Soil Biology* 42, S3–S15.

Lebel, J., Mergler, D., Lucotte, M., Amorin, M., Dolbec, J.,Miranda, D. *et al.* (1996) Evidence of early nervous system dysfunctions in Amazonian populations exposed to low levels of methylmercury. *Neurotoxicology* 17, 157–168.

Lebel, J., Roulet, M., Mergler, D., Lucotte, M. and Larribe, F. (1997) Fish diet and mercury exposure in a riparian Amazonian population. *Water Air and Soil Pollution* 97, 31–44.

Loubry, D. (2002) *Livret technique pour la conduite de la revégétalisation sur les surfaces minières alluvionnaires de Guyane*. IRD, Cayenne, France.

Malm, O. (1998) Gold mining as a source of mercury exposure in the Brazilian Amazon. *Environmental Research*, Section A, 77, 73–78.

Manceau, A. and Nagy, K.L. (2008) Relationships between Hg(II)-S bond distance and Hg(II) coordination in thiolates. *Dalton Transactions* 11, 1385–1508.

Mason, R.P., Reinfelder, J.R. and Morel, F.M.M. (1995) Bioaccumulation of mercury and methylmercury. *Water, Air and Soil Pollution* 80, 915–921.

Mélières, M.-A., Pourchet, M., Charles-Dominique, P. and Gaucher, P. (2003) Mercury in canopy leaves of French Guiana in remote areas. *The Science of the Total Environment* 311, 261–267.

Molicova, H., Grimaldi, M., Bonell, M. and Hubert, P. (1997) Using TOPMODEL towards identifying and modelling the hydrological patterns within a headwater humid, tropical catchment. *Hydrological Processes* 11, 1169–1196.

Muresan, B., Cossa, D., Richard, S. and Dominique, Y. (2008) Monomethylmercury sources in a tropical artificial reservoir. *Applied Geochemistry* 23, 1101–1126.

Nahon, D.B. (1991) *Introduction to the Petrology of Soils and Chemical Weathering*. John Wiley and Sons, Inc., New York.

Nriagu, J.O. (1993) Legacy of mercury pollution. *Nature* 363, 589.

Nriagu, J.O. (1994) Mercury pollution from the past mining of gold and silver in the Americas. *The Science of the Total Environment* 149, 167–181.

Oliveira de, S.M.B., Melfi, A.J., Fostier, A.H., Forti, M.C., Favaro, D.I.T. and Boulet, R. (2001) Soils as an important sink for mercury in the Amazon. *Water, Air and Soil Pollution* 26, 321–337.

Patel-Sorrentino, N., Lucas, Y., Eyrolle, F. and Melfi, A.J. (2007) Fe, Al and Si species and organic matter leached off a ferrallitic and podzolic soil system from Central Amazonia. *Geoderma* 137, 444–454.

Peretyazhko, T., Charlet, L. and Grimaldi, M. (2006) Production of gaseous mercury in hydromorphic soils in the presence of ferrous iron: A laboratory study. *European Journal of Soil Science* 57, 190–199.

Quesada, C.A., Lloyd, J., Anderson, L.O., Fyllas, N.M., Schwarz, M. and Czimczik, C.I. (2011) Soils of Amazonia with particular reference to the RAINFOR sites. *Biogeosciences* 8, 1415–1440.

Ravichandran, M. (2004) Interactions between mercury and dissolved organic matter – A review. *Chemosphere* 55(3), 319–331.

Richard, S., Arnoux, A., Cerdan, P., Reynouard, C. and Horeau, V. (2000) Mercury levels of soils, sediments and fish in French Guiana, South America. *Water, Air and Soil Pollution* 124, 221–244.

Richter, D.D. and Babbar, L.I. (1991) Soil diversity in the tropics. *Advances in Ecological Research* 21, 315–389.

Roulet, M. and Lucotte, M. (1995) Geochemistry of mercury in pristine and flooded ferralitic soils of a dense tropical rain forest in French Guiana, South America. *Water, Air and Soil Pollution* 80, 1079–1088.

Roulet, M., Lucotte, M., Saint-Aubin, A., Tran, S., Rheault, I., Farella, N. *et al.* (1998) The geochemistry of mercury in central Amazonian soils developed on the Alter-do-Chão formation of the lower Tapajos River Valley, Pará state, Brazil. *The Science of the Total Environment* 223(1), 1–24.

Schimann, H., Joffre, R., Roggy, J.-C., Lensi, R. and Domenach, A.-M. (2007) Evaluation of the recovery of microbial functions during soil restoration using near-infrared spectroscopy. *Applied Soil Ecology* 37, 223–232.

Schuster, E. (1991) The behavior of mercury in the soil with special emphasis on complexation and adsorption processes – A review of the literature. *Water, Air and Soil Pollution* 56, 667–680.

Schwertmann, U. and Cornell, R.M. (2000) *Iron Oxides in the Laboratory: Preparation and Characterization*. Wiley-VCH, Weinheim, Germany.

Schwertmann, U. and Latham, M. (1986) Properties of iron oxides in some new Caledonian oxisols. *Geoderma* 39, 105–123.

Skyllberg, U., Xia, K., Bloom, P.R., Nater, E.A. and Bleam, W.F. (2000) Binding of mercury (II) to reduced sulfur in soil organic matter along upland-peat soil transects. *Journal of Environmental Quality* 29, 855–865.

Tandeter, E. (2006) The mining industry. In: Bulmer-Thomas, V., Coatsworth, J.H. and Cortés Conde, R. (eds) *The Cambridge Economic History of Latin America. Vol. 1, The Colonial Era and the Short Nineteenth Century*. Cambridge University Press, Cambridge, UK, pp. 315–356.

Tandy, J.C., Grimaldi, M., Grimaldi, C. and Tessier, D. (1990) Mineralogical and textural changes in French Guiana oxisols and their relation with microaggregation. In: Douglas, L.A. (ed.) *Soil Micromorphology*. Elsevier, Amsterdam, pp. 191–198.

Telmer, K. and Veiga, M.M. (2009) World emissions of mercury from artisanal and small scale gold mining. In: Pirrone, N. and Mason, R. (eds) *Mercury Fate and Transport in the Global Atmosphere Emissions, Measurements and Models*. Springer-Verlag, Heidelberg, Germany, pp. 131–172.

Telmer, K., Costa, M., Angélica, R.S., Araujo, E.S. and Maurice, Y. (2006) The source and fate of sediment and mercury in the Tapajos River, Pará, Brazilian Amazon: Ground-and space-based evidence. *Journal of Environmental Management* 81, 101–113.

Tessier, E., Amouroux, D., Grimaldi, M., Stoichev, T., Grimaldi, C., Dutin, G. and Donard, O.F.X. (2003) Mercury mobilization in soil from a rainfall event in a tropical forest (French Guyana). *Journal de Physique IV* 107, 1301–1304.

Tumpling, W.V., Wilken, R.R. and Einax, J. (1995) Mercury contamination in the northern Pantanal region, Mato Grosso, Brazil. *Journal of Geochemical Exploration* 52, 127–134.

Ullrich, S.M., Tanton, T.W. and Abdrashitova, S.A. (2001) Mercury in the aquatic environment: A review of factors affecting methylation. *Critical Reviews in Environmental Science and Technology* 31, 241–293.

Veiga, M. and Baker, R. (2004) Protocols for environmental and health assessment of mercury released by artisanal and small scale miners. Report to the Global mercury project: Removal of barriers to introduction of cleaner artisanal gold mining and extraction technologies, GEF/UNDP/UNIDO, Vienna, Austria.

10 Erosion and Sedimentation Effects on Soil Organic Carbon Redistribution in a Complex Landscape in Western Ecuador

Marife D. Corre,[1]* Jeroen M. Schoorl,[2] Free de Koning,[3] Magdalena López-Ulloa[4] and Edzo Veldkamp[1]

[1]Büsgen Institute – Soil Science of Tropical and Subtropical Ecosystems, Georg-August University Göttingen, Göttingen, Germany; [2]Soil Geography and Landscape, Wageningen University, Wageningen, the Netherlands; [3]Conservation International Ecuador, Quito, Ecuador; [4]Environmental Engineering, Universidad de las Americas, Quito, Ecuador

10.1 Introduction

Soil organic carbon (SOC) contains a large proportion of the nutrient-holding capacity of most soils and contributes to important structural properties such as aggregate stability, fertility, erodibility and water-holding capacity. In recent years, losses of SOC due to land-cover change and agricultural practices have contributed about 12 to 15% of the anthropogenic CO_2 emissions to the atmosphere (~1.2 Pg C year^{-1}), the bulk of which is released from tropical regions (Le Quéré *et al.*, 2009, Van der Werf *et al.*, 2009). Agricultural practices lead to a reduction in ecosystem carbon (C) stocks due to the removal of aboveground biomass with subsequent burning (Mutuo *et al.*, 2005) and accelerated loss of SOC as a result of decomposition (Powers *et al.*, 2011) and soil erosion (Yoo, 2005). A recent framework for improving prediction of SOC response to global warming calls for research approaches that represent spatial gradients in landscapes in order to exemplify the spatial patterns of factors (e.g. soil type and topographic position) that regulate SOC stabilization and loss (Schmidt *et al.*, 2011). These authors also pointed out that one of the major weaknesses of current models is the lack of representation of the edaphic factors that regulate SOC at the landscape scale, which could better reflect observations and inform policies. While the impact of land-use change on SOC stocks has been investigated at a landscape scale (e.g. de Koning *et al.*, 2003; Powers, 2004), research efforts on the effects of soil erosion on changes of SOC are still limited (e.g. Chaplot and Poesen, 2012).

Soil erosion is a complex process, involving four phases: detachment, breakdown of aggregates, transport/redistribution and sedimentation. These four phases of erosion strongly influence the distribution and quantity of SOC in a landscape (Fang *et al.*, 2006). Typically, erosion reduces the SOC storage at upper slope positions and sedimentation increases the SOC storage at footslopes and depressions compared to areas without erosion processes (Liu *et al.*, 2003). The four phases of soil erosion depend strongly on land cover and use, parent material,

* Corresponding author: mcorre@gwdg.de

soil texture, landscape position and climate. Vegetation cover enhances infiltration and reduces surface runoff and consequently erosion (Francis and Thornes, 1990; Kosmas *et al.*, 2000). For example, a study on soil erosion in eastern Spain demonstrated that the interaction between vegetation and parent material prevented high runoff and erosion rates on marl soils, which developed on Cretaceous lime-rich mudstone, when plant cover increased (Cerda, 1999). Parent material and its position in the landscape influence transport-limited and detachment-limited erosion, and hence the spatial patterns of soil redistribution (Schoorl *et al.*, 2002, 2006). Soil texture affects transport and redistribution of soil as clay minerals and the light fractions of organic matter are more easily removed and redistributed over the landscape than the heavier silt and sand fractions (Lal, 2003). Landscape position and landform shape influence transfer of water within and between landscapes which, in turn, controls soil redistribution and sediment deposition (Swanson *et al.*, 1988). Finally, precipitation intensity strongly influences soil erosion and sedimentation processes.

Erosion and sedimentation affect SOC distribution in two ways. First, erosion and sedimentation redistribute considerable amounts of SOC at a variety of scales within landscapes. Therefore, accurate estimates of soil redistribution in the landscape are needed to quantify the relative contribution of erosion and sedimentation to landscape-scale changes in SOC storage. Second, erosion and sedimentation may alter the biological process of SOC mineralization by burying surface soil layers and by indirectly affecting vegetation growth, soil depth and water availability. In the present study, we evaluated how land-use changes affect the distribution of SOC within a complex tropical landscape through the processes of erosion and sedimentation. Our objectives were: (i) to estimate the present SOC storage at a landscape scale using predictors such as slope, elevation, texture, land-use type and landscape position; (ii) to estimate soil redistribution under the present land-use conditions and under different land-use change scenarios using an erosion–sedimentation model; and (iii) to estimate the redistribution of SOC caused by erosion–sedimentation processes and its effect on landscape-scale SOC stocks. Implications for land-use policy options for the study area are also discussed.

10.2 Materials and Methods

10.2.1 Study site

The study area (Fig. 10.1) is located in the southern part of Manabí province in western Ecuador (1° 16′–01° 37′ S, 80° 22′–80° 28′ E). This area is within the Chocó biogeographical region, which is one of the world's hotspots of biodiversity (Myers *et al.*, 2000) and a priority area for implementation of conservation measures. Mean annual precipitation varies from 800–1300 mm (INAMHI, 2002) and mean annual temperature is 25°C. The main agricultural land uses are coffee–agroforestry systems (Fig. 10.2), pastures and upland rice fields. Coffee–agroforestry systems, on which more than 60% of the farmers in the area depend for their income (INEC, 2001), were established 30–40 years ago by the replacement of the forest understorey vegetation with coffee plants (mainly *Coffea arabica*; Rubiaceae) and planting of fruit trees and timber species with a multi-layered canopy structure. Pastures (consisting mainly of *Panicum maximum*; Poaceae) were established approximately 30 years ago after slashing and burning of the original vegetation. Upland rice (*Oriza sativa*; Poaceae) fields were also established after slashing and burning of original vegetation and/or agroforestry systems and are rotations of 4–5 years rice and 2–3 years fallow.

The geology and landscape genesis of the Noboa-Pajan area dates back to the Oligocene and two physiographic units characterize the study area (Fig. 10.1): sloping areas situated north-west, and river valleys with alluvial and colluvial deposits situated south-south-east of Noboa and Pajan villages (Schoorl *et al.*, 2006). Parent material of the upland areas mainly consist of sediments deposited by deep-sea currents (turbidites), resulting in soil layers from fine clays to silts and fine sands. In general, Lithic Udorthents (Entisols) are formed on the higher landscape positions and sloping areas (hereafter called the upland soils). Mollic Udifluvents (Entisols) are developed on depositional lower landscape positions and river valleys (hereafter called the lowland soils).

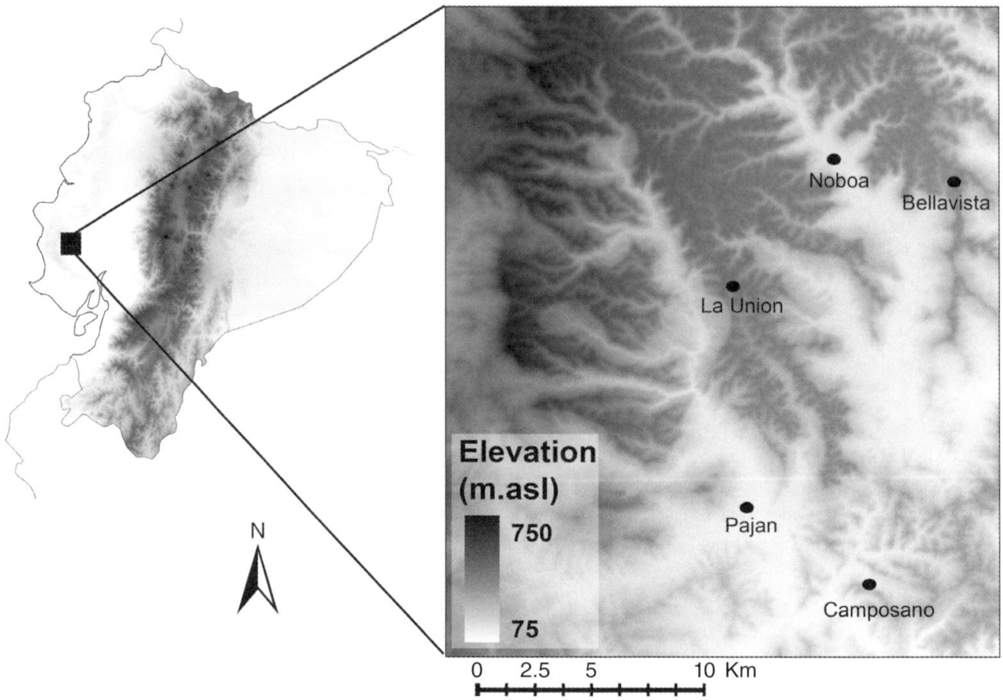

Fig. 10.1. Digital Elevation Model of the Noboa-Pajan study area, located in the southern part of Manabí province, western Ecuador (UTM coordinates: 545.5 and 580.5 km (E, W) and 9819.5 and 9850.5 km (S, N)). The total landscape area is 812.55 km^2.

Fig. 10.2. Coffee–agroforestry system in the Noboa-Pajan study area, western Ecuador.

10.2.2 Site selection, soil sampling and measurement of soil organic carbon

We selected 12 sites in each of the three land-use systems (36 sites in total) to represent the two major physiographic soil units. Six sites of each land use were located on alluvial fans and valley bottoms with lowland soils. The other six sites of each land use were located on sloping areas with upland soils. At each geo-referenced site (within an area of 50 m × 50 m), a 3 m × 3 m regular grid was laid out and at each grid point (total of nine sampling points) a soil sample was collected to 25 cm depth. Soil samples from the nine sampling points of each plot were air-dried, sieved through a 2-mm sieve, ground and analysed for SOC using a CNS Elemental Analyser (Elementar Vario EL, Hanau, Germany). Soil bulk density was determined for each sampling point using a core of 250 cm^3. From each soil core, all roots were removed by hand, stones (if present) were weighed to correct for the volume, and the soil was oven-dried at 105°C for 24 h. The SOC stock was calculated for each sampling point from the bulk density, SOC concentration and soil depth, and the nine sampling points in a plot were averaged to represent a replicate plot (Table 10.1).

10.2.3 Modelling and statistical analyses

Determination of actual soil organic carbon stocks using terrain-based regulatory factors

Using a general linear model with backward stepwise elimination, we developed a model for predicting SOC stocks (as the dependent variable) using the following regulatory factors (independent variables): elevation, slope, texture (as continuous variables), land-use type and soil-landform class (as categorical variables). Furthermore, we completed a Geographic Information System (GIS) database of these variables for our study area, allowing extrapolation of the model developed from the plot scale (Table 10.1) to the landscape scale. We examined the explanatory power of significant effects using a Type II Likelihood Ratio test. Assumptions of normality and homogeneity of variance were tested on all variables prior to developing the statistical model. Using the model, we calculated the SOC stocks at the landscape scale and generated the SOC map for baseline condition (Fig. 10.3). For all statistical analyses, the general regression model of Statistica 6.0 was used (StatSoft Inc., 2003).

Soil redistribution under four scenarios of land-use changes

To calculate the amount of soil redistribution at the landscape scale, we used the LAPSUS (LandscApe ProcesS modelling at mUlti dimensions and Scales) modelling framework (Schoorl and Veldkamp, 2001; Schoorl *et al.*, 2000, 2002; Claessens *et al.*, 2005). This model requires data on topography (generated for the study area using a Digital Elevation Model (DEM) with 29 m resolution; Fig. 10.1), land-use type, geology, soil type, soil depth, rainfall,

Table 10.1. Concentrations and stocks of soil organic carbon (SOC) in the top 25 cm depth in three land-use types in the Manabí province in western Ecuador.

Land use	Soil – physiographic units	SOC (mg C g^{-1})	Bulk density (g cm^{-3})	SOC (mg C ha^{-1})	Slope (°)	Elevation (m)
		Mean ± SE, n = 6			Range	
Coffee–agroforestry	Udifluvents on river valleys	28 ± 2	1.07 ± 0.02	74 ± 5	0–0	125–148
	Udorthents on sloping areas	26 ± 2	1.12 ± 0.02	72 ± 7	2–30	190–471
Pasture	Udifluvents on river valleys	23 ± 3	1.18 ± 0.04	67 ± 6	0–16	121–160
	Udorthents on sloping areas	26 ± 3	1.17 ± 0.05	73 ± 8	5–34	180–477
Upland rice	Udifluvents on river valleys	19 ± 1	1.14 ± 0.05	54 ± 4	0–25	105–156
	Udorthents on sloping areas	23 ± 1	1.16 ± 0.02	67 ± 4	2–20	175–471

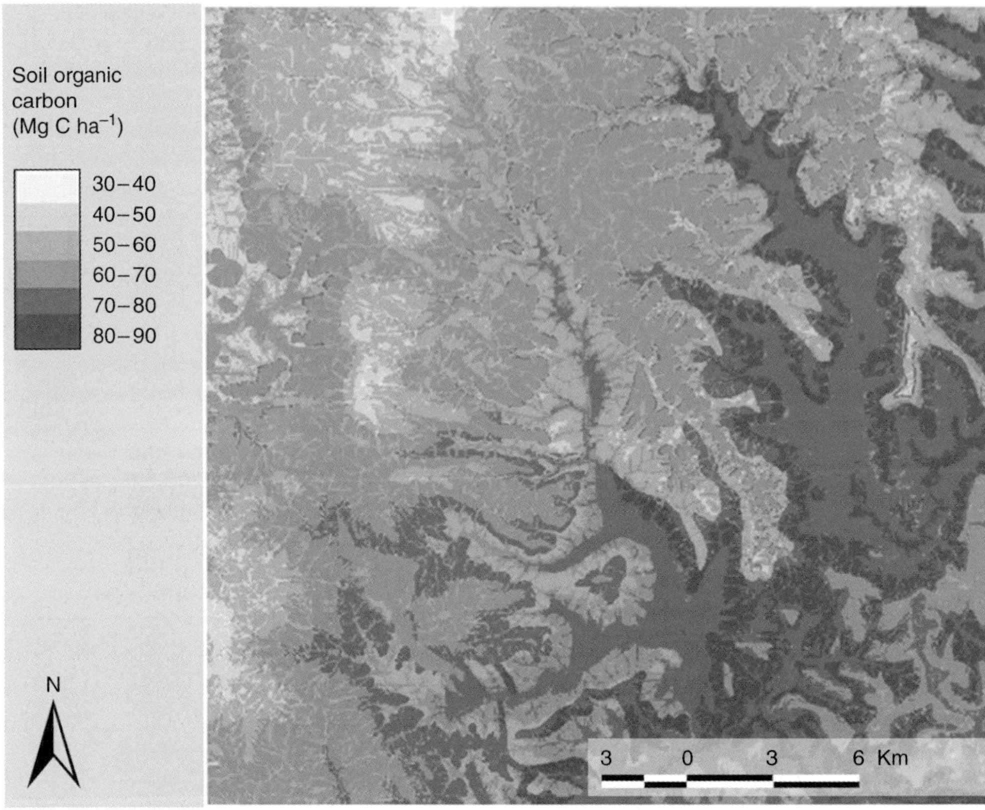

Fig. 10.3. Landscape-scale soil organic carbon stocks under the present land-use types in the Manabí province in western Ecuador. Values were calculated using a statistical model between SOC stocks and regulatory factors (see Section 10.2.3) combined with a GIS database of regulatory factors.

erodibility and soil permeability. Attributes derived from the DEM are the local slope and the area of the upslope contributing to the drainage area, which are calculated using the algorithm of multiple downslope flow (Quinn *et al.*, 1991). Some input variables such as erodibility, soil depth and soil permeability were not measured but estimated in the field based on soil profile surveys. For the model-calculated soil redistribution, we stratified the landscape into four soil textural classes: sand, clay, loam soils and soils located in river terraces and river/stream valley areas (Fig. 10.4). The rate of soil redistribution was then calculated by soil textural classes.

First, we ran LAPSUS for the present land-use types, topography and soil conditions in the area. This model-run corresponds to scenario '0' or the baseline conditions with which all subsequent scenarios were compared. To explore the possible changes in SOC storage in the area as a result of erosion and sedimentation, we defined four 'extreme' scenarios of land-use changes: scenario 1 is conversion of all coffee–agroforestry systems to pastures; scenario 2 is conversion of all coffee–agroforestry systems to rice fields; scenario 3 is conversion of all rice fields to coffee–agroforestry systems; and scenario 4 is conversion of all pastures to coffee–agroforestry systems. For each scenario, the model was run to calculate soil erosion and sedimentation rates over a period of 1 year. Each scenario produced new spatial patterns of total amounts of erosion (Fig. 10.5a) and sedimentation (Fig. 10.5b) throughout the landscape. To determine the fraction of eroded soil lost from the study area, we calculated the Sediment Delivery Ratio (*SDR*), which is the sediment fraction exported out of the landscape.

$$SDR = 1 - (Tot\_sed/(Tot\_ero \times (-1))) \quad (10.1)$$

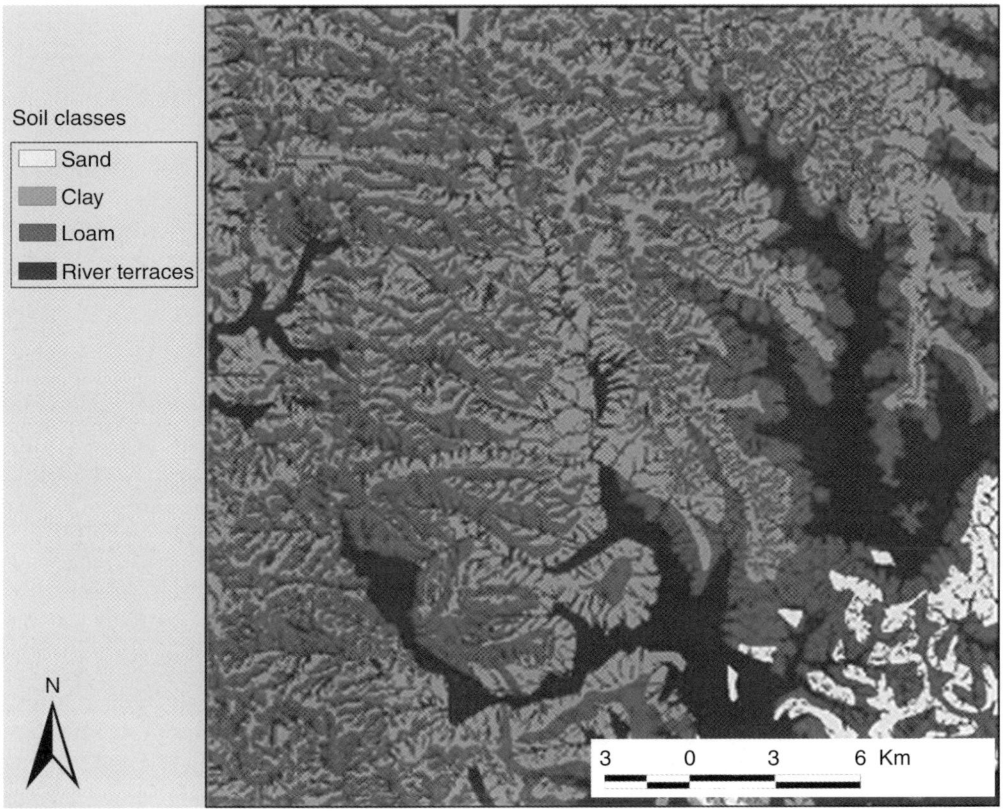

Fig. 10.4. Soil stratification by textural classes based on combined information of field soil surveys, geology and Digital Elevation Model (see Section 10.2.3) in the Manabí province in western Ecuador.

Where *Tot_sed* is the total amount of re-sedimentation (Mg ha^{-1} $year^{-1}$) and *Tot_ero* is the total amount of erosion (Mg ha^{-1} $year^{-1}$). A *SDR* value of 1 means all eroded soil is exported out of the whole landscape whereas a *SDR* value of 0 indicates all eroded soil is deposited within the landscape. We used the same climatic conditions for all land-use change scenarios, which were the average monthly values (e.g. rainfall, temperature) during 1990–2002 taken from the weather stations (i.e. Colimes de Pajan and Camposano 2) near our study area.

Soil organic carbon balance in the landscape

The LAPSUS model-calculated erosion (Fig. 10.5a) and sedimentation rates (Fig. 10.5b) for each land-use change scenario and soil textural class were combined with the SOC map (Fig. 10.3) to calculate changes in SOC storage at the landscape level. We used the following equations to calculate SOC losses and gains for each land-use change scenario and soil textural class:

$$SOC_{losses} = (C_{soil} \times Tot\_ero)/\text{depth} \quad (10.2)$$

$$SOC_{gains} = (C_{soil} \times Tot\_sed)/\text{depth} \quad (10.3)$$

Where SOC_{losses} is area-weighted average of SOC losses by erosion processes (corresponding to a negative value in Mg C ha^{-1} $year^{-1}$); SOC_{gains} is area-weighted average of SOC gains by sedimentation processes (corresponding to a positive value in Mg C ha^{-1} $year^{-1}$); C_{soil} is the area-weighted average of SOC stock (Mg C ha^{-1}) (Fig. 10.3); *Tot_ero* is area-weighted average of soil erosion losses (m $year^{-1}$); *Tot_sed* is area-weighted average of soil sedimentation gains (m $year^{-1}$) (Fig. 10.5; data expressed as m $year^{-1}$ = Mg ha^{-1} $year^{-1}$ ÷

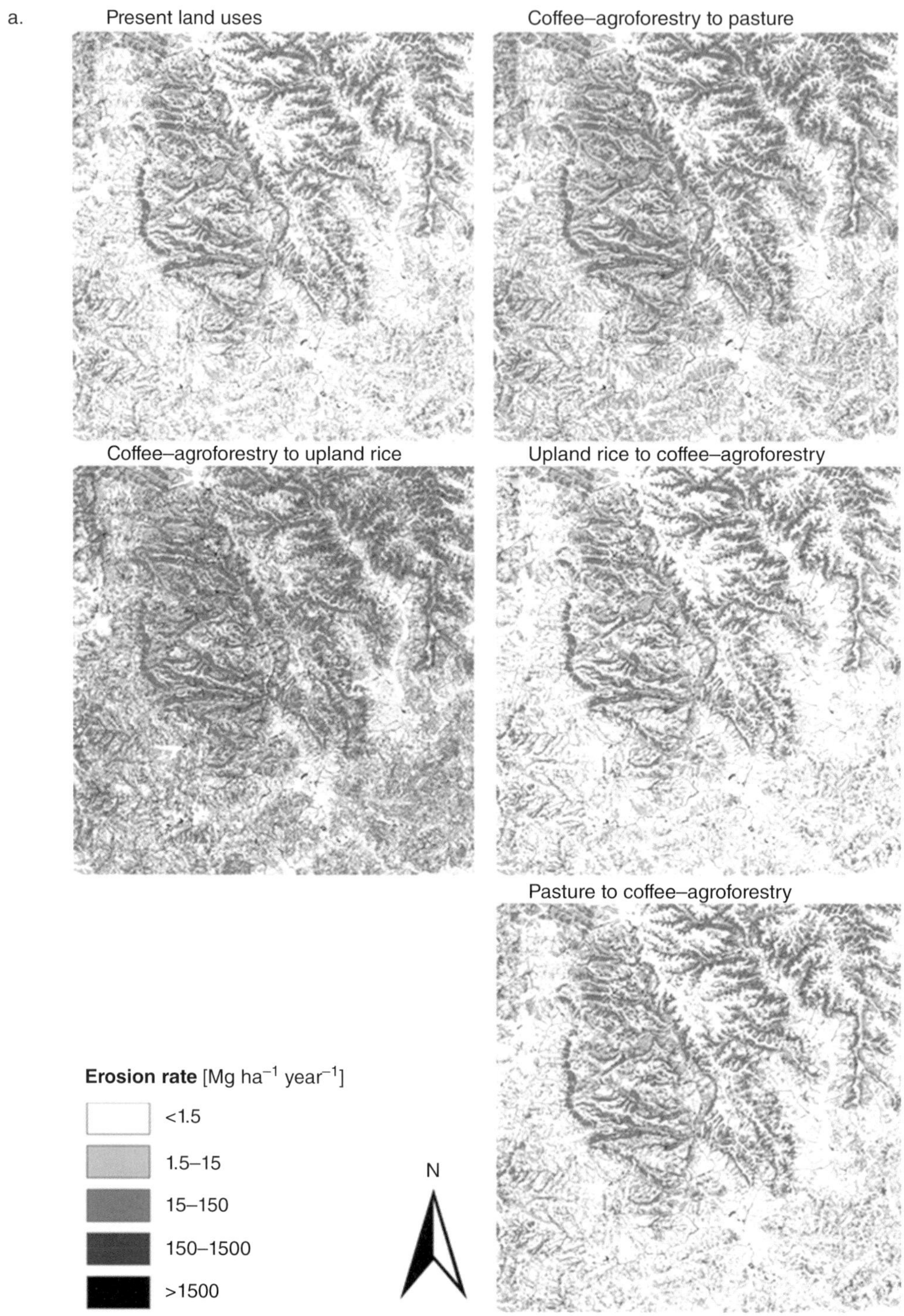

Fig. 10.5. (a) LAPSUS model-estimated erosion rates in the landscape area of 812.55 km^2 under different land-use change scenarios in the Manabí province in western Ecuador.

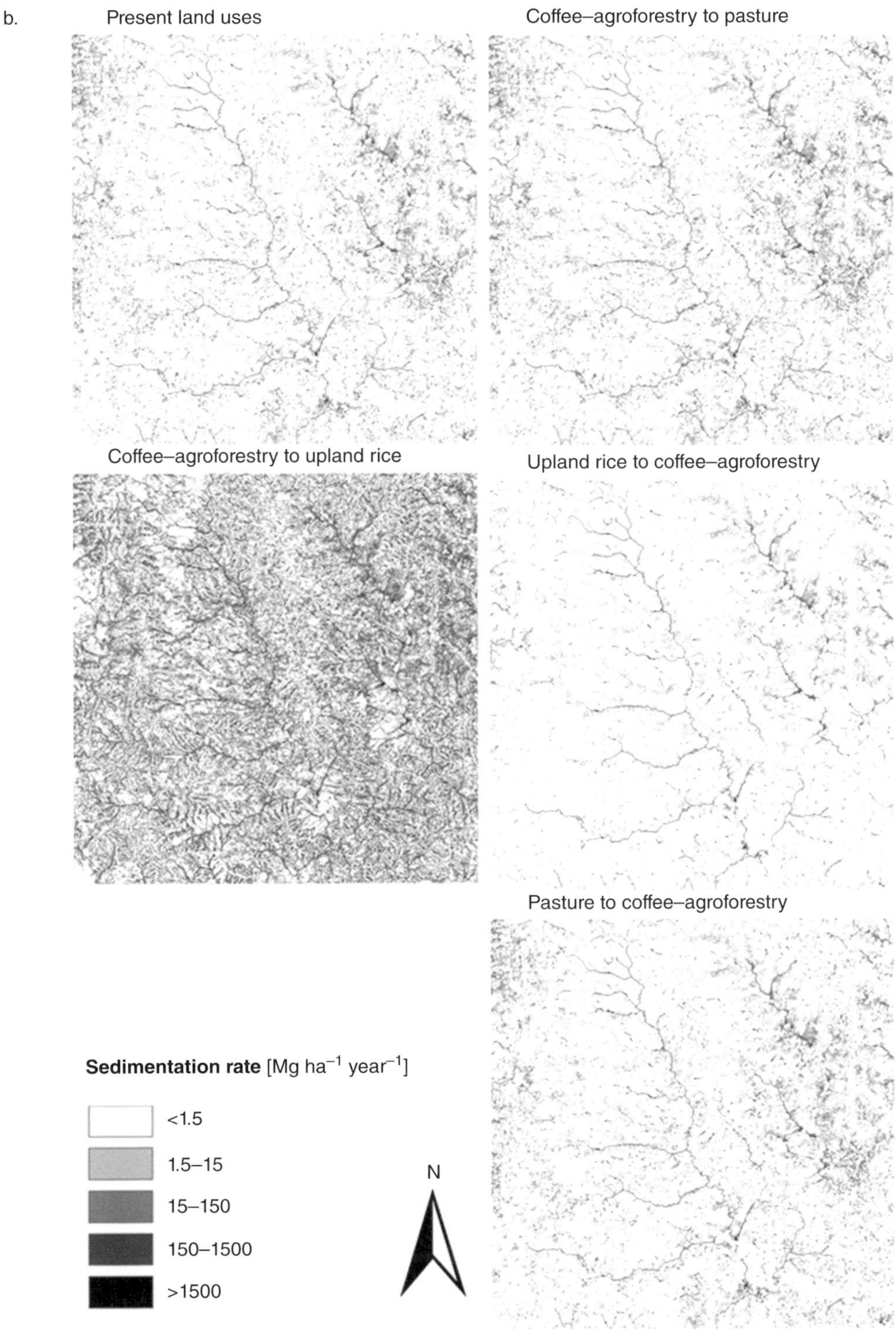

Fig. 10.5. (b) LAPSUS model-estimated sedimentation rates in the landscape area of 812.55 km^2 under different land-use change scenarios in the Manabí province in western Ecuador.

$10{,}000\ m^2\ ha^{-1} \div$ bulk density Mg m^{-3}); depth is the soil sampling layer (m).

To determine the net SOC redistribution at the landscape scale, we calculated the SOC balance between SOC losses and SOC gains:

$$\text{SOC balance} = SOC_{gains} - SOC_{losses} \qquad (10.4)$$

A negative SOC balance corresponds to SOC exported via erosion out of the landscape (loss), whereas a positive SOC balance corresponds to SOC deposited through sedimentation into the landscape (gain). SOC balance was calculated for all four land-use change scenarios.

10.3 Results

10.3.1 Soil organic carbon baseline

The significant variables that explained SOC stocks at the landscape scale were: elevation ($P < 0.01$), texture (sand) ($P < 0.05$), land-use type (LU1 = coffee–agroforestry; LU2 = pasture) ($P < 0.05$), and soil-landform class (SL1 = lowland soils) ($P < 0.01$), as reflected in the regression model:

$$\begin{aligned}\text{SOC}\ (\text{Mg C ha}^{-1}) &= 87.4 - 0.06 \times \text{elevation} \\ &- 10.9 \times \text{sand} + 4.8 \times \text{LU1} + 3.7 \times \text{LU2} \\ &- 8.3 \times \text{SL1} \qquad (10.5)\end{aligned}$$

This model explained 46% of the variability in SOC stocks of the 36 measured sites ($P < 0.01$). Application of this model to the landscape scale using the GIS database generated the map of SOC stocks of the study area (Fig. 10.3). The highest SOC stocks (in the south-east corner of the area) were found in lowland soils on river valleys, river terraces and lower hills, whereas lower values were found in upland soils on higher landscape positions (north-west corner of the area). SOC stocks in the top 25 cm depth ranged from 30–87 Mg C ha^{-1} and the area-weighted mean was 63.6 Mg C ha^{-1}. The SOC map illustrates that the actual SOC stocks were strongly related to topography and topography-related soil textural classes (Figs. 10.1, 10.3 and 10.4), suggesting that topography-driven water erosion and sedimentation processes play an important role in this landscape.

10.3.2 Impacts of erosion and sedimentation by land-use change on SOC redistribution

Soil erosion losses (Fig. 10.5a) and sedimentation gains (Fig. 10.5b) showed stark contrasts among the four land-use change scenarios (Table 10.2). Scenario 2 (all coffee–agroforestry converted to rice) resulted in the highest erosion losses and highest sedimentation gains within the landscape, whereas scenario 3 (all rice converted to coffee–agroforestry) resulted in the lowest erosion and sedimentation rates within the landscape. For all scenarios, the area affected by sedimentation (positive values) was much smaller (Fig. 10.5b) compared to the area affected by erosion (negative values) (Fig. 10.5a). Compared to the baseline condition, scenarios 3 and 4 (all pasture converted to coffee–agroforestry) showed comparable amount of soil loss from the landscape (*Tot_out*) whereas scenarios 1

Table 10.2. Model-estimated erosion and sedimentation rates for the entire landscape area of 81,255 ha under different land-use change scenarios in the Manabí province in western Ecuador.

Scenario	*Tot_ero* (Mg ha^{-1} $year^{-1}$)	*Tot_sed* (Mg ha^{-1} $year^{-1}$)	*Tot_out* (Mg ha^{-1} $year^{-1}$)	SDR
0	–124.2	34.8	–89.4	0.72
1	–161.0	39.2	–121.8	0.76
2	–311.4	88.5	–223.0	0.71
3	–113.4	29.8	–83.6	0.74
4	–120.0	34.1	–86.0	0.72

Values are area-weighted means of the four soil textural stratifications in Figure 10.4.
Scenario: 0 = present land uses (baseline conditions); 1 = conversion from coffee–agroforestry to pasture; 2 = conversion from coffee–agroforestry to upland rice; 3 = conversion from upland rice to coffee–agroforestry; 4 = conversion from pasture to coffee-agroforestry; *Tot_ero* = area-weighted average of soil erosion loss from the landscape; *Tot_sed* = area-weighted average of soil sedimentation gain in the landscape; *Tot_out* = area-weighted average of net soil erosion loss from the landscape (*Tot_ero* – *Tot_sed*).
SDR = sediment delivery ratio = 1 – (*Tot_sed*/(*Tot_ero* × (–1))).

(all coffee–agroforestry converted to pasture) and 2 showed considerably more soil exported out of the entire landscape (Table 10.2).

SOC redistribution in the landscape, caused by land-use change effects on erosion and sedimentation, showed the highest impact in clay soil zones on depositional lower landscape positions and in lowland soils on river terraces, whereas the lowest impact was found in sand and loam soils on upper landscape positions (Fig. 10.6). Among land-use change scenarios, the strongest impacts of erosion and sedimentation on SOC redistribution occur in scenario 2. For the landscape as a whole, SOC losses (negative values) ranged from 2.5 Mg C ha^{-1} in scenario 3 to 7.0 Mg C ha^{-1} in scenario 2 (Fig. 10.6a), whereas SOC gains only ranged from 1.0 to 2.0 Mg C ha^{-1}, respectively (Fig. 10.6b). The SOC balance was negative at the landscape level for all scenarios (Fig. 10.6c), indicating that the net effect of erosion and sedimentation processes was the export of SOC from the entire landscape. However, the

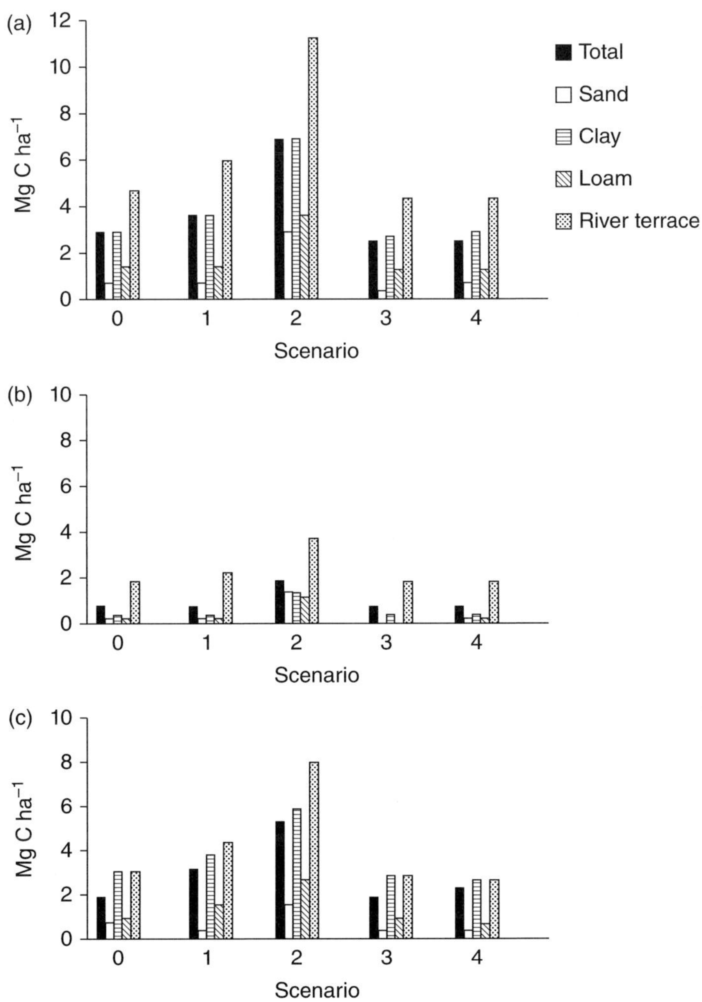

Fig. 10.6. Area-weighted averages of (a) soil organic carbon (SOC) losses by erosion; (b) SOC gains by sedimentation; and (c) SOC balance ($SOC_{gains} - SOC_{losses}$) in the landscape under different land-use change scenarios in the Manabí province in western Ecuador: 0 = present land uses, conversions from 1 = coffee–agroforestry to pasture; 2 = coffee–agroforestry to upland rice; 3 = upland rice to coffee–agroforestry; and 4 = pasture to coffee–agroforestry.

conversion from coffee–agroforestry systems to upland rice fields resulted in twice as great a loss of SOC relative to the present land-use configuration (scenario 0 or baseline conditions).

10.4 Discussion

10.4.1 Soil organic carbon baseline

Spatial variation of SOC in the landscape normally hinders generating landscape-scale estimates of SOC stocks. Our study illustrates that a considerable part (46%) of the spatial variation in SOC stocks can be predicted using soil-topographic land-use attributes. Such attributes have been shown to regulate spatial patterns of SOC at a large scale (de Koning *et al.*, 2003; Powers and Veldkamp, 2005), but application of predictive relationships between SOC and spatially based regulatory factors to generate landscape-scale estimates of SOC stock is seldom done (Pennock *et al.*, 1994). The upper landscape positions with sandy to loamy soils showed lower SOC stocks than the depositional lower landscape positions with clay soils. Our results agree well with earlier studies that also found decreasing SOC stocks with increasing elevation (Bergstrom *et al.*, 2001; Chaplot *et al.*, 2001). The effects of topography and topography-induced water erosion on landscape-scale SOC storage have also been demonstrated earlier (Pennock *et al.*, 1994; Pennock and van Kessel, 1997).

10.4.2 Soil redistribution and SOC balance under different land-use change scenarios

Our results showed that the overall erosion susceptibility of the study area was highly dependent on land-use type. Additionally, erosion and runoff rates from upstream located land uses may also have an impact on erosion and sedimentation rates downslope (Schoorl and Veldkamp, 2001). Land-use types that do not have permanent vegetation cover (e.g. upland rice) result in larger soil redistribution rates than land-use types with permanent cover (e.g. pastures or coffee–agroforestry systems). It should be kept in mind that our calculated soil losses were not calibrated against an independent dataset. However, the range of soil losses (83–223 Mg ha^{-1} year^{-1}) are comparable with other studies that found maximum erosion losses of 109 and 250 Mg ha^{-1} year^{-1} in Cameroon and Spain, respectively (Ambassa-Kiki and Nill, 1999; Schoorl *et al.*, 2002). Yet these studies were done in drier areas than our study area, where the annual precipitation is greater and with more intense events.

Our approach only concentrated on the redistribution of SOC on the basis of currently estimated SOC stocks. We did not consider the complete SOC balance, particularly SOC inputs and mineralization. All land-use systems have C input through root and leaf litter and C losses through mineralization, which will strongly influence the SOC budget. Although all the land-use change scenarios resulted in negative landscape-scale SOC balances, this does not necessarily mean that the land-use systems are not inputting and stabilizing SOC locally. Instead, this only means that the net effect of soil erosion and sedimentation is dominant and resulted in a removal of SOC from the studied landscape. This modelling exercise also provides an insight that if quantification of SOC is only based on a local scale and not on the entire landscape, changes in SOC stocks with land-use change could be negative or positive, depending on whether erosion or sedimentation is dominant.

Soil organic carbon balance was strongly influenced by changes in land use and soil types. Conversions of pasture and upland rice to coffee–agroforestry systems, particularly in sandy to loamy soils, have the lowest losses of SOC. This land use/soil type association may therefore be considered for soil conservation and mitigation of CO_2 emissions. In contrast, conversion from coffee–agroforestry systems to upland rice, as is presently occurring in parts of the study area due to low coffee prices (de Koning *et al.*, 2007), is the worst scenario in terms of SOC conservation. Soil organic carbon losses from sand and loamy soils were approximately eight times lower than clay soils and lowland soils on river terraces. A similar trend of soil erosion differences due to soil texture, which is related to parent material, was reported from a study in eastern Spain with lower erosion rates on sandy soils than on clay soils under different vegetation covers (Cerda, 1999). High infiltration rates in sandy soils probably lead to much lower soil erosion rates than in clay soils. Loamy soils, which are mainly located in mid-slope and footslope areas, suffered less from erosion processes. Conversely, lowland soils on river terraces and river/stream valley areas were most susceptible to erosion processes. This demonstrates that the offsite effects on lowland soils were more important than the onsite consequences of land-use change.

Although our simple mass balance approach probably results in a reasonable estimate of short-term effects of soil redistribution on the SOC balance in the landscape, the results should be interpreted with care if long-term effects are considered. For example, we did not consider the effects of erosion and sedimentation on SOC mineralization, nor do we know what happened to the SOC exported from the study area. Part of the exported SOC may be deposited in terrestrial deposits and another part may be exported to aquatic ecosystems (Hope *et al.*, 1994; Stallard, 1998). In addition, as much as 70% of the exported SOC may be decomposed during transportation and sedimentation (Beyer *et al.*, 1993), although a lower estimate of about 20% was reported by others (Jacinthe *et al.*, 2001). Also, we used the average monthly climatic values in the LAPSUS model, which may not reflect erosion and sedimentation events occurring during pulses of heavy rainfall. While these varied estimates show that presently not enough is known about the fate of eroded SOC, generating a SOC budget at the landscape level as we did in the present study will be a first step towards model development in which processes affecting SOC redistribution, e.g. erosion, transport, sedimentation and mineralization, should be integrated.

10.4.3 Land-use policy options

Our approach can directly support land-use policies at the landscape level. As the conversion of coffee–agroforestry systems to pastures or rice leads to the largest soil redistribution and SOC loss from the landscape, policies that aim at reducing SOC loss should implement incentives that make agroforestry a more competitive land-use option as compared to other alternatives. In the same landscape studied, de Koning *et al.* (2007) applied a land-use allocation model to analyse different policy options to support coffee–agroforestry. They conclude that direct payment per hectare is more efficient to maintain or increase coffee–agroforestry areas than payment per kilogram of coffee. Price buffering is possibly the most cost-effective way to support coffee–agroforestry production, especially when farmers are highly averse. Also, relatively low payments per hectare of forest can increase forested land in the studied landscape.

The present methodology can also be used to inform policies and methodologies for REDD+ (Reducing Emissions from Deforestation and forest Degradation + conservation, sustainable management of forests, and enhancement of forest carbon stocks). Set-up, implementation and monitoring costs of REDD+ projects are highly variable (Rendón Thompson *et al.*, 2013) and improved methodologies for the estimation of SOC redistribution at the landscape scale can contribute to reducing costs and increasing benefits. The Government of Ecuador is implementing its national REDD+ programme and is part of the UN-REDD initiative (MAE, 2011). Since 2008, the government has run the Socio Program, which provides direct economic incentives to families and indigenous and local communities to conserve native forests (de Koning *et al.*, 2011), and has currently over 1.1 million hectares of native ecosystems enrolled in the programme. For such initiatives, it is necessary to understand the importance of SOC stocks in the C balance and the effect of land-use change on these stocks. This will help inform their design and the quantification of the benefits of reduced deforestation and forest restoration in terms of CO_2 emissions.

Acknowledgements

This study was funded by the Tropical Ecology Support Program of the German International Cooperation and the German Federal Ministry of Education and Research. We thank the many landowners who allowed us to collect field data. M.D.C. also acknowledges funding from the Robert Bosch Foundation (Germany) and the German Research Foundation (Co 749/1-1).

References

Ambassa-Kiki, R. and Nill, D. (1999) Effects of different land management techniques on selected topsoil properties of a forest Ferralsol. *Soil and Tillage Research* 52, 259–264.

Bergstrom, D.W., Monreal, C.M. and St Jacques, E. (2001) Spatial dependence of soil organic carbon mass and its relationship to soil series and topography. *Canadian Journal of Soil Science* 81, 53–62.

Beyer, L., Frund, R., Schleuss, U. and Wachendorf, C. (1993) Colluvisols under cultivation in Schleswig-Holstein. 2. Carbon distribution and soil organic-matter composition. *Zeitschrift für Pflanzenernährung und Bodenkunde* 156, 213–217.

Cerda, A. (1999) Parent material and vegetation affect soil erosion in eastern Spain. *Soil Science Society of America Journal* 63, 362–368.

Chaplot, V. and Poesen, J. (2012) Sediment, soil organic carbon and runoff delivery at various spatial scales. *Catena* 88, 46–56.

Chaplot, V., Bernoux, M., Walter, C., Curmi, P. and Herpin, U. (2001) Soil carbon storage prediction in temperate hydromorphic soils using a morphologic index and digital elevation model. *Soil Science* 166, 48–60.

Claessens, L., Heuvelink, G.B.M., Schoorl, J.M. and Veldkamp, A. (2005) DEM resolution effects on shallow landslide hazard and soil redistribution modelling. *Earth Surface Processes and Landforms* 30, 461–477.

de Koning, F., Aguiñaga, M., Bravo, M., Chiu, M., Lascano, M., Lozada, T. and Suarez, L. (2011) Bridging the gap between forest conservation and poverty alleviation: The Ecuadorian Socio Bosque program. *Environmental Science and Policy* 14, 531–542.

de Koning, G.H.J., Veldkamp, E. and López-Ulloa, M. (2003) Quantification of carbon sequestration in soils following pasture to forest conversion in northwestern Ecuador. *Global Biogeochemical Cycles* 17.4.

de Koning, G.H.J., Benítez, P., Muñoz, F. and Olschewski, R. (2007) Modelling the impacts of payments for biodiversity conservation on regional land-use patterns. *Landscape and Urban Planning* 83, 255–267.

Fang, H.J., Cheng, S.L., Zhang, X.P., Liang, A.Z., Yang, X. M. and Drury, C. F. (2006) Impact of soil redistribution in a sloping landscape on carbon sequestration in Northeast China. *Land Degradation and Development* 17, 89–96.

Francis, C.F. and Thornes, J.B. (1990) Runoff hydrographs from three Mediterranean vegetation cover types. In: Thornes, J.B. (ed.) *Vegetation and Erosion, Processes and Environments*. Wiley, Chichester, UK, pp. 363–384.

Hope, D., Billett, M.F. and Cresser, M.S. (1994) A review of the export of carbon in river water – Fluxes and processes. *Environmental Pollution* 84, 301–324.

INAMHI (2002) *Instituto Nacional de Meteorología e Hidrología. Anuarios Meteorológicos (1995–2002)*. INAMHI, Quito, Ecuador.

INEC, M. a. S (2001) II Censo nacional agropecuario: resultados provinciales y cantonales – Manabí. Instituto Ecuatoriano de Estadísticas y Censos, Quito, Ecuador.

Jacinthe, P.A., Lal, R. and Kimble, J. (2001) Assessing water erosion impacts on soil carbon pools and fluxes. In: Lal, R., Kimble, J.M., Follett, R.F. and Stewart, B.A. (eds) *Assessment Methods for Soil Carbon Pools*. CRC/Lewis Publishers, Boca Raton, Florida, pp. 427–450.

Kosmas, C., Gerontidis, S. and Marathianou, M. (2000) The effect of land use change on soils and vegetation over various lithological formations on Lesvos (Greece). *Catena* 40, 51–68.

Lal, R. (2003) Soil erosion and the global carbon budget. *Environment International* 29, 437–450.

Le Quéré, C., Raupach, M.R., Canadell, J.G., Marland, G. *et al.* (2009) Trends in the sources and sinks of carbon dioxide. *Nature Geoscience* 2, 831–836.

Liu, S.G., Bliss, N., Sundquist, E. and Huntington, T.G. (2003) Modeling carbon dynamics in vegetation and soil under the impact of soil erosion and deposition. *Global Biogeochemical Cycles* 17.

MAE (Ministry of Environment of Ecuador) (2011) National programme document – Ecuador UN-REDD Programme Sixth Policy Board Meeting, 21–22 March 2011 Da Lat, Vietnam. Available at: www.un-redd.org (accessed 1 April 2015).

Mutuo, P.K., Cadisch, G., Albrecht, A., Palm, C.A. and Verchot, L. (2005) Potential of agroforestry for carbon sequestration and mitigation of greenhouse gas emissions from soils in the tropics. *Nutrient Cycling in Agroecosystems* 71, 43–54.

Myers, N., Mittermeier, R.A., Mittermeier, C.G., da Fonseca, G.A.B. and Kent, J. (2000) Biodiversity hotspots for conservation priorities. *Nature* 403, 853–858.

Pennock, D.J., Anderson, D.W. and Dejong, E. (1994) Landscape-scale changes in indicators of soil quality due to cultivation in Saskatchewan, Canada. *Geoderma* 64, 1–19.

Pennock, D.J. and van Kessel, C. (1997) Effect of agriculture and of clear-cut forest harvest on landscape-scale soil organic carbon storage in Saskatchewan. *Canadian Journal of Soil Science* 77, 211–218.

Powers, J.S (2004) Changes in soil carbon and nitrogen after contrasting land-use transitions in Northeastern Costa Rica. *Ecosystems* 7, 134–146.

Powers, J.S. and Veldkamp, E. (2005) Regional variation in soil carbon and delta C-13 in forests and pastures of northeastern Costa Rica. *Biogeochemistry* 72, 315–336.

Powers, J.S., Corre, M.D., Twine, T.E. and Veldkamp, E. (2011) Geographic bias of field observations of soil carbon stocks with tropical land-use changes precludes spatial extrapolation. *Proceedings of the National Academy of Sciences, USA* 108, 6318–6322.

Quinn, P., Beven, K., Chevallier, P. and Planchon, O. (1991) The prediction of hillslope flow paths for distributed hydrological modeling using digital terrain models. *Hydrological Processes* 5, 59–79.

Rendón Thompson, O.R., Paavola, J., Healey, J.R., Jones, J.P.G., Baker, T.R. and Torres, J. (2013) Reducing emissions from deforestation and forest degradation (REDD+): Transaction costs of six Peruvian projects. *Ecology and Society* 18, http://dx.doi.org/10.5751/ES-05239-180117.

Schmidt, M.W.I., Torn, M.S., Abiven, S. , Dittmar, T, Guggenberger, G., Janssens, I.A., Kleber, M., Kogel-Knabner, I., Lehmann, J., Manning, D.A.C., Nannipieri, P., Rasse, D.P., Weiner, S. and Trumbore, S.E. (2011) Persistence of soil organic matter as an ecosystem property. *Nature* 478, 49–56.

Schoorl, J.M. and Veldkamp, A. (2001) Linking land use and landscape process modelling: A case study for the Alora region (south Spain). *Agriculture Ecosystems and Environment* 85, 281–292.

Schoorl, J.M., Sonneveld, M.P.W. and Veldkamp, A. (2000) Three-dimensional landscape process modelling: The effect of DEM resolution. *Earth Surface Processes and Landforms* 25, 1025–1034.

Schoorl, J.M., Veldkamp, A. and Bouma, J. (2002) Modeling water and soil redistribution in a dynamic landscape context. *Soil Science Society of America Journal* 66, 1610–1619.

Schoorl, J.M., Claessens, L., Lopez-Ulloa, M., de Koning, G.H.J. and Veldkamp, A. (2006) Geomorphological analysis and scenario modelling in the Noboa-Pajan area, Manabi Province, Ecuador. *Zeitschrift für Geomorphologie* 145, 105–118.

Stallard, R.F. (1998) Terrestrial sedimentation and the carbon cycle: Coupling weathering and erosion to carbon burial. *Global Biogeochemical Cycles* 12, 231–257.

StatSoft_Inc (2003) STATISTICA (data analysis software system), version 6.

Swanson, F.J., Kratz, T.K., Caine, N. and Woodmansee, R.G. (1988) Landform effects on ecosystem patterns and processes. *BioScience* 38, 92–98.

Van der Werf, G.R., Morton, D.C., DeFries, R.S., Olivier, J.G.J., Kasibhatla, P.S., Jackson, R.B. *et al.* (2009) CO_2 emissions from forest loss. *Nature Geoscience* 2, 737–738.

Yoo, K., Amundson, R., Heimsath, A.M. and Dietrich, W.E. (2005) Erosion of upland hillslope soil organic carbon: Coupling field measurements with a sediment transport model. *Global Biogeochemical Cycles* 19, GB3003.

11 Pastoralism and Kalahari Rangeland Soils

Andrew D. Thomas,[1]* David R. Elliott,[2] Tasmin N.L. Griffith[2] and Helen Mairs[2]
[1]Department of Geography and Earth Sciences, Aberystwyth University, Aberystwyth, UK; [2]School of Science and the Environment, Manchester Metropolitan University, UK

11.1 Introduction

Grazing lands cover almost half the global land area and an estimated 70% of the world's poorest billion people rely on income generated from pastoralism (FAO/IIASA/ISRIC/ISS-CAS/JRC, 2009). In rural dry sub-humid environments such as the Kalahari, pastoral farming is the only viable livelihood for most people, and cattle are central to the Tswana way of life. Cattle not only provide a major source of household income, but confer prestige to families within their communities (Campbell, 1990). The vast majority of livestock are reared on communal land where fences are absent and grazing resources are shared. The absence of surface water in the Kalahari means that animals are dependent on groundwater from boreholes (Perkins and Thomas, 1993). As more boreholes have been established, livestock grazing has expanded across the Kalahari (Rohde *et al.*, 2006), leading to spatially defined areas of impact around the watering points (Perkins and Thomas, 1993). The location of boreholes therefore controls the distribution and intensity of grazing impacts in the landscape. The most frequently disturbed areas, or sacrifice zones, are characterized by unconsolidated sand and an absence of vascular plant cover. With increasing distance from the borehole, there is typically a zone of less palatable annual grasses where the more nutritious perennial species have been preferentially grazed (Dougill *et al.*, 1999). These zones have little biological soil crust (BSC) cover as the fragile crusts are highly susceptible to breakage by trampling (Dougill and Thomas, 2004; Berkeley *et al.*, 2005; Thomas and Dougill, 2007). BSCs form from the association of soil particles with cyanobacteria, algae, lichens and fungi (Belnap and Lange, 2003). They bind sand grains into aggregates, reducing erodibility (McKenna-Neuman and Maxwell, 2002; Thomas and Dougill, 2007) and add carbon (C) and nitrogen (N) to soils (Belnap and Lange, 2003; Elbert *et al.*, 2009). They are widespread in drylands regions and the sub-tropics, and are such an important part of the landscape they have been called ecosystem engineers (Viles, 2008). A reduction in BSC cover is synonymous with increased erodibility (McKenna-Neuman and Maxwell, 2002; Langston and McKenna-Neuman, 2005) and a reduction in soil nutrient content (Belnap and Lange, 2003).

Areas that have been grazed intensively for long periods commonly have a higher density of woody shrubs than areas that have not been grazed (Moleele and Mainah, 2003; Gil-Romera, 2010; Eldridge *et al.*, 2011). This is because the sustained selective removal of grasses gives a

* Corresponding author: ant23@aber.ac.uk

competitive advantage to shrubs that gradually increase in density and area. Many studies consider this change in vegetation community structure to be the single most important process of degradation affecting the rangelands of southern Africa because of the associated reduction in grazing potential (Moleele and Mainah, 2003; MEA, 2005). However, shrub encroachment has been shown to induce a series of alterations to the soil microclimate, microbial populations and nutrient cycles that make the relationship with degradation much less certain (Eldridge *et al.*, 2011).

Grazing will inevitably lead to changes in soils and vegetation and this has led to the view that, in Botswana, communal pastoralism is a degradation-causing, low-output system. This, in turn, has encouraged the privatization of land through the Tribal Grazing Lands Policy (TGLP) that was seen as a way of increasing productivity and reducing degradation (Dougill *et al.*, 1999). However, there is little evidence that the TGLP has met its environmental, pastoral production or societal objectives (Thomas *et al.*, 2000). Rohde *et al.* (2006) have also argued, very powerfully, that rangeland management policy across the whole of southern Africa is the result of a flawed model of development. A clearer understanding of the impact of grazing on soil properties and the likely consequences of policies affecting grazing lands is therefore of utmost importance. This requires an approach that quantifies the impact of different grazing regimes on soil properties as well as changes associated with shrub encroachment. Importantly, the findings must be evaluated in terms of realistic livelihood options for pastoral farmers.

This chapter presents field-based evidence of the changes in soil properties resulting from grazing-related disturbance and shrub encroachment in the south-west Kalahari. The objectives were: (i) to determine the impact of different grazing intensities on soil CO_2 efflux, soil organic carbon (SOC) and chlorophyll *a* (a proxy for photosynthetic biomass); and (ii) to quantify differences in microclimate, SOC and soil CO_2 efflux between open-grass sites, and the shrub/tree canopy. The results from the field experiments are discussed and interpreted within the broader context of grazing land management and policy in the Kalahari.

11.2 Methods

11.2.1 Study site

Experiments were conducted in an open fine-leaf savannah on Kalahari Sand soils in south-west Botswana (25° 56'51" S, 22° 25'40" E). Vegetation cover is a mix of perennial (*Eragrostis* sp.; Poaceae) and annual (*Schmidtia* spp.; Poaceae) grasses, woody shrubs (*Grewia flava*; Tiliaceae and *Acacia mellifera*; Leguminosae) and trees (predominantly *Acacia erioloba*; Leguminosae). *Grewia flava* is a common encroaching species in the Kalahari as well as East Africa (Gil-Romera, 2010). Mean annual precipitation between 1996 and 2012 was 314 mm, with a low of 130 mm in 2002–2003 and a high of 532 mm in 2001–2002. Seasonal variations in air temperature are extreme, with summer maxima frequently in excess of 40°C and winter below 0°C.

Soils are weakly acidic psamments (Entisols), with little or no horizon development. In lightly grazed areas, around 80% of the surface is covered in a 3–4-mm deep BSC, composed of cyanobacteria (dominated by *Phormidium*), bacteria (mostly Proteobacteria) and fungi (mostly Ascomycota) (Thomas, 2012). Crusts are enriched in ammonium, total N and organic C compared to the mineral soil (Thomas and Dougill, 2007; Thomas *et al.*, 2011; Thomas, 2012). The stock of SOC to 1 m was 39.4 ± 4.1 t C ha^{-1}, and >10% of this was contained within the upper 2 cm (5.4 t ha^{-1}) (Thomas *et al.*, 2012).

11.2.2 Field experiment 1: Grazing disturbance simulation

Four 1 m × 1 m plots, approximately 0.5 m apart, were laid out and enclosed within a fence in July 2009. In February 2010, treatments were applied to the soils within each plot to simulate different levels of livestock disturbance. To simulate the effects of intense grazing, the BSC was removed to a depth of 1 cm in one plot and buried in a 1-cm layer of unconsolidated sand in a second plot (together, both plots are classified as heavily disturbed). Soil burial by wind-blown material is a common, albeit localized occurrence in heavily grazed areas where the breakup of the BSC promotes sand transport. To simulate light grazing impacts, the BSC was broken into small fragments, approximately 1–2 cm in length. The fourth plot was left untreated as a control. After treatment, measurements were made on four occasions in contrasting seasons over the following 15 months. Monthly rainfall and mean air temperatures during the sampling months were: February (48.7 mm; 27.4°C); July (0 mm; 12.4°C); November (103.8 mm; 28.9°C) and April (55.8 mm; 20.5°C). For the purposes of analysis, data were grouped into three sets: pre-treatment (February 2010), first year after disturbance (February 2010 in the week following treatment, July and November 2010) and second year after disturbance (April 2011).

The assumption that differences between plots are due to the treatments is only valid if the natural variability in soil properties does not increase between sites 1 m and 5 m apart (the distance over which the plots were located). This assumption was tested by quantifying the variability in soil CO_2 efflux between respiration chambers as a function of distance between them (Thomas, 2012). There are two main sources of variability in Kalahari Sand properties: (i) islands of fertility around shrubs and trees (Wang *et al.*, 2009); and (ii) the extent and degree of BSC development (Dougill and Thomas, 2004; Thomas and Dougill, 2007). The plots in this study were located in plant interspaces and the construction of the fences around the plots in June 2009 meant that grazing disturbance was minimized for 7 months prior to the start of the experiment. Thus, when treatments were applied to the plots there was a relatively uniform cover of BSCs at each site. Two factors, (i) selecting sites in plant interspaces, and (ii) allowing BSCs to develop evenly across the sites prior to the start of the experiments, maximized homogeneity in BSC and soil properties before treatments were applied. Experimental data confirmed that differences in CO_2 efflux between plots did not increase with the distance between them (Thomas, 2012). Differences in soil CO_2 efflux were just as likely to be the same between chambers 1 m apart as those that were 5 m apart. This result validates the sampling and replication strategy adopted in this grazing experiment.

Soil CO_2 efflux was measured using static closed respiration chambers, the design of which is outlined in Thomas *et al.* (2011) and Thomas (2012). Three replicate chambers were used within each plot and soil CO_2 efflux determined seven times per 24 h. Within each measurement cycle, a syringe was used to extract two 10 ml air samples from the chambers at 15 min intervals. The gas was injected into 6 ml pre-evacuated glass vials and CO_2 concentrations determined using an Agilent GC 3000 gas chromatograph. Soil CO_2 efflux data from each chamber were integrated over the measurement period and the resulting cumulative change used to determine mean CO_2 efflux rates.

All BSC samples were collected in triplicate after insertion of 50 mm diameter sterile Petri dishes into the surface to remove the crust. Subsurface soils were also collected in triplicate from a soil pit. The chlorophyll *a* content of the surface soil/BSC samples was determined within 12 h of collection by heating samples to 60°C in the dark in high performance liquid chromatography (HPLC)-grade 100% methanol. Concentrations in filtered extracts were determined from absorbance values at 652 nm, 665.2 nm and 750 nm (Porra, 2006) using an Ocean Optics HR4000 field spectrometer, responsive at 200 nm to 1120 nm with a mean sampling interval of 0.27 nm. The remaining samples were air dried, bagged and returned to the UK for determination of total C and N using a Leco TruSpec CN elemental analyser and organic C, following the method of Walkley and Black (1934).

11.2.3 Field experiment 2: Soil properties and microclimate under vegetation and in open-grass sites

Experiments were undertaken at the same study site to determine whether a suite of soil and micrometeorological properties were significantly different under shrub/tree canopies compared to open-grass sites. Data were collected in November 2011, February 2012 and July 2012 from three *rewia flava* and *cacia erioloba* sub-canopy sites, three *Eragrostis* and three *Schmidtia* grass sites.

Soil CO_2 efflux was measured using the same static closed respiration chambers as in the grazing experiment. Three replicate shrub and three tree sites were sampled, each with two chambers under the east and west side of the canopy. Six different open-grass sites were sampled, each with a single chamber. Soil CO_2 efflux was determined at each site three to four times per 24 h over each measurement period. CO_2 concentrations were determined onsite using an infra-red gas analyser. Simultaneous soil moisture (integrated over the top 5 cm soil) measurements were made with a Delta T ML2x theta probe and soil surface temperature readings using a TN1 infra-red sensor. Incoming solar radiation was measured continuously 2 m above the ground surface in an open site as well as underneath a *rewia flava* canopy using a Skye sensor and logger. Soil and BSC samples were collected from soil pits at each site at 0–1 cm; 1–2 cm; 2–5 cm; 5–10 cm and 10–20 cm depths for subsequent determination of total C and N.

11.2.4 Statistical analyses

After checking for normality, a two-way analysis of variance (ANOVA) was undertaken using SPSS 19 to test the significance of differences in soil CO_2 efflux and soil properties due to the effects of treatment and season. To test for the effects of treatment, mean soil CO_2 efflux rates of each chamber in each season, were compared. When the analysis was significant ($P < 0.05$), a post-hoc Tukey's test was undertaken to evaluate differences.

11.3 Results

11.3.1 The impact of intense and light grazing on SOC, chlorophyll *a* and CO_2 efflux

Soil organic carbon in the control plot did not change significantly during the experiment, ranging from 0.35% ± 0.08 over the first year to 0.41% ± 0.02 in the second year of measurement (Table 11.1). Light grazing had no significant impact on the SOC content of the soil surface. Heavy grazing disturbances, however, led to a significant decline in SOC that persisted for the entire 2 years of the experiment (Table 11.1).

Chlorophyll *a* concentrations in the surface soil of the control plot were naturally variable between seasons and increased significantly in years 1 and 2 compared to the pre-treatment conditions (Table 11.1). Light grazing did not significantly alter this pattern, despite the decline in chlorophyll *a* in year 1. Heavy grazing, however, led to a significant decline in chlorophyll *a* during the first year after treatment.

Soil CO_2 efflux was significantly affected by seasonal changes in moisture and temperature (see Thomas, 2012 for more details). Simulated grazing intensity also had a significant effect on soil CO_2 efflux ($P < 0.01$) but *post hoc* tests reveal these effects were only apparent in February and April, and not when soils were dry in November and July (Fig. 11.1). In February, soil CO_2 efflux was significantly greater from soil where the BSC was heavily disturbed compared to the control ($P < 0.05$) and from where the BSC was lightly disturbed ($P < 0.05$). In April, soil CO_2 efflux from the heavily disturbed soil was also significantly higher than from the control soil ($P < 0.05$). CO_2 efflux was significantly higher from soil where the BSC was lightly and heavily disturbed than the control ($P < 0.01$). There were no significant differences between any of the other treatments.

Table 11.1. Soil organic carbon (SOC) and chlorophyll *a* (Chl *a*) concentrations in the control, lightly and heavily disturbed grazing plots in the south-west Kalahari.

	Control			Light disturbance			Heavy disturbance		
	Pre-treatment	First year	Second year	Pre-treatment	First year	Second year	Pre-treatment	First year	Second year
SOC (%)	0.39 ± 0.02	0.35 ± 0.08	0.41 ± 0.02	0.27 ± 0.01	0.23 ± 0.01	0.24 ± 0.01	0.39 ± 0.06	0.16 ± 0.05[a]	0.14 ± 0.09[b]
Chl a (μg g^{-1})	0.94 ± 0.04	1.55 ± 1.15[a]	4.43 ± 0.28[b]	1.72 ± 0.47	1.36 ± 0.43	3.52 ± 0.13[b]	1.78 ± 0.50	0.60 ± 0.46[a]	1.55 ± 1.57

Note: Means with standard deviations are given for the pre-treatment period and the first and second years after treatment. Significant differences ($P < 0.05$) are indicated by letters.

[a]Significantly different between pre-treatment and year 1 ($P < 0.05$).

[b]Significantly different between pre-treatment and year 2 ($P < 0.05$).

11.3.2 Differences in microclimate, SOC and soil CO_2 efflux in open-grass and vegetation canopy sites

Soil CO_2 efflux was highest in March and lowest in November when soil moisture was very low (below detection limits) (Fig. 11.2). When dry, there were no significant differences between sites, but when soil moisture increased, differences between the sites became apparent. Efflux was significantly greater at the *cacia erioloba* and *rewia flava* sites compared to the open-grass sites in March and June ($P < 0.05$) but there were no significant differences in November.

Total C and N concentrations in the soils underneath the *cacia erioloba* trees were significantly greater than those under *rewia flava* shrubs and in open-grass sites ($P < 0.05$) (Table 11.2). Total C and N concentrations were also significantly greater under *rewia flava* than in the grass sites ($P < 0.05$) (Table 11.2). Soil C:N ratios were consistent between sites, ranging from 9:1 to 10.5:1.

In the hot summer and autumn months of November and March, air temperatures at the grass site were higher than those under *rewia flava* (Table 11.3). This reflects the shading provided to the soil surface by the shrub canopy. Conversely, in the relatively cooler winter month of June, air temperatures at the grass sites were less than locations under shrubs. This is reflected in the incoming solar radiation at each site (Fig. 11.3). November 2011 and March 2012 were relatively cloud free and the diurnal changes in open site solar radiation are described by sine curves. Peak incoming solar radiation was similar in November and March. June 2012 was cloudy,

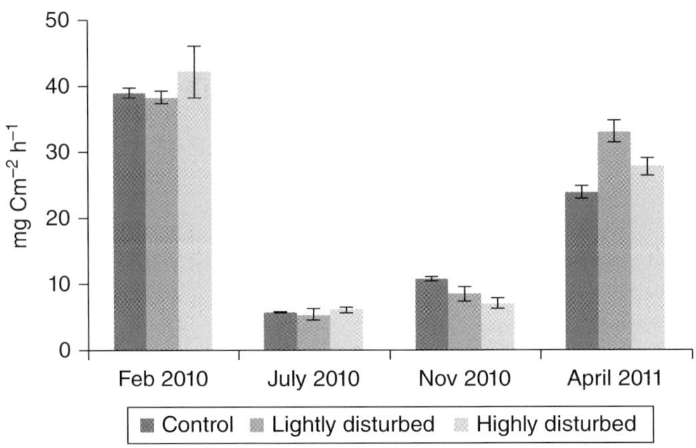

Fig. 11.1. Mean soil CO_2 efflux from control, lightly and heavily disturbed grazing plots in February, July, November 2010 and April 2011 in the south-west Kalahari. Error bars represent the standard error.

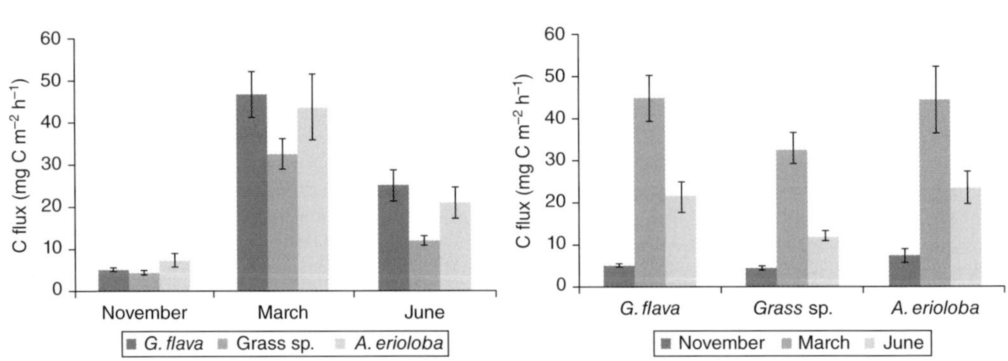

Fig. 11.2. Seasonal variations in mean soil respiration in soils under *Grewia flava*, grass and *Acacia erioloba* in the south-west Kalahari. Error bars represent the standard error.

Table 11.2. Total carbon (C) and nitrogen (N) concentrations under *Grewia flava, Acacia erioloba* and in open-grass sites in the south-west Kalahari.

	C (%)	N (%)	C:N ratio
Open-grass sites	0.18 ± 0.01[a]	0.02 ± 0.01[a]	9:1[a]
Grewia flava	0.39 ± 0.07[b]	0.04 ± 0.01[b]	9.8:1[a]
Acacia erioloba	0.84 ± 0.10[c]	0.08 ± 0.01[c]	10.5:1[a]

Note: Data are means of values from November, March and June with standard errors. Letters indicate significant differences ($P < 0.05$) between sites.

Table 11.3. Mean, maximum and minimum air temperatures (°C) in the open-grass, *Grewia flava* and *Acacia erioloba* sites in the south-west Kalahari during three monitoring periods.

Temp. (°C)	November 2011			March 2012			June 2012		
	Grass	*G. flava*	*A. erioloba*	Grass	*G. flava*	*A. erioloba*	Grass	*G. flava*	*A. erioloba*
Mean	33.0	32.2	29.3	30.8	29.8	28.5	12.9	13.4	13.6
Max.	59.5	59.5	60.5	55.0	54.5	48.5	33.0	34.5	39.0
Min.	4.5	5.5	5.5	9.5	11.5	9.5	–3.0	–1.0	–2.5

particularly over the first 4 days of the study and peak radiation was much lower than in the previous study periods. The radiation reaching the soil surface underneath the *rewia flava* canopy was substantially reduced in all measurement periods (Fig. 11.3). This was most apparent in June when, despite the lack of leaf cover, the woody branches and slightly lower sun angle meant that much less light penetrated the canopy.

11.4 Discussion

11.4.1 Grazing and soils in rangelands

Understanding the impact of grazing on soils and SOC is particularly important in sub-humid savannah rangelands because SOC stocks are low and even small changes can have a large impact on ecosystem function. Soil organic matter and SOC are essential to biological productivity and underpin numerous ecosystem functions in terrestrial environments (Thomas, 2012). Land-use change, largely deforestation and conversion to agricultural land, was responsible for 20% of the global anthropogenic emissions of CO_2 in the 1990s (IPCC, 2007) and is the main cause of net C release in Africa (Henry *et al.*, 2009). Our data suggest that intense grazing, which breaks up existing BSCs, needs to be considered as an additional factor in accelerating land-atmosphere CO_2 emissions.

Belnap and Lange (2003, and references therein) and Elbert *et al.* (2009) demonstrate that in some environments, primary production by BSCs can be the major source of SOC. The field-based experimental data presented here clearly demonstrate that intense grazing has a detrimental impact on SOC, at least over a period of 2 years. Compared to lightly grazed and ungrazed soils, C losses as CO_2 efflux are significantly greater in intensively disturbed soils, and C and N concentrations significantly lower. There are likely to be several reasons why destruction of BSCs due to grazing leads to a reduction in SOC. First, BSCs affect the boundary layer between the atmosphere and soil, influencing gas diffusivity, water infiltration and retention. A reduction in the ability of soils to retain moisture and therefore sustain primary productivity from shallow rooting grasses and BSCs may be particularly important in reducing SOC. Second, autotrophic organisms in the BSC will utilize CO_2 generated from the mineral soil below during photosynthesis, simultaneously reducing CO_2 losses and adding to the SOC pool.

Light grazing, where disturbance to the soil surface is relatively minor and discrete bits of BSC remain, has no discernible negative effect on soil properties. There is some evidence that

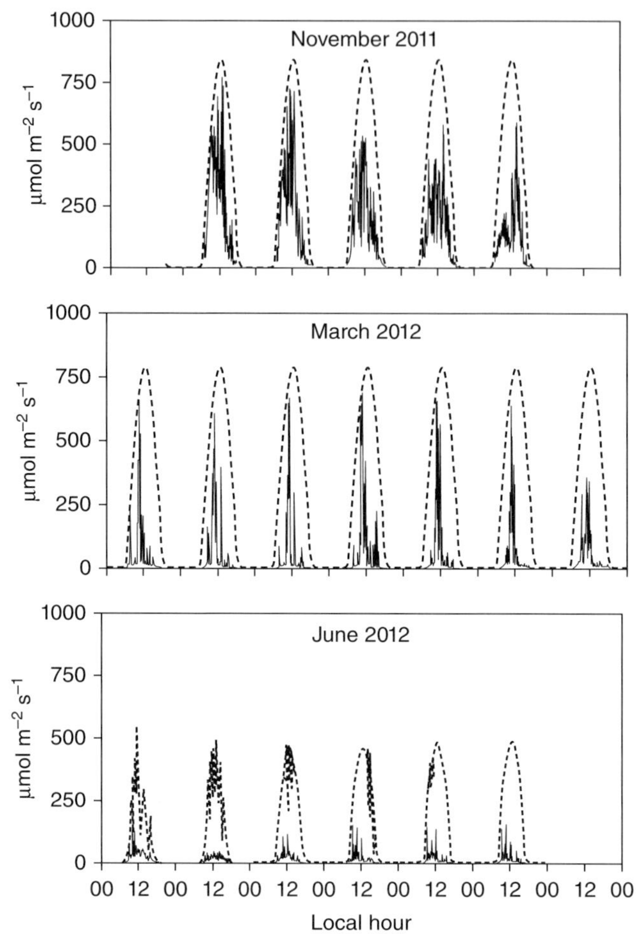

Fig. 11.3. Diurnal variation in open-grass site (thick dashed line) and below *G. flava* canopy (thin solid line) incoming solar radiation (630 nm) in November 2011 and March and June 2012.

light grazing actually increases soil concentrations of C and chlorophyll *a*. This is likely to be due to the increased surface roughness created by the disturbed blocks of BSC, which create areas of shade, reducing mean soil surface temperatures and prolonging autotrophic microbial activity of BSC organisms. The time to recover BSC photosynthetic function after disturbance will be important for management decisions, and our data suggest this may be at least 2 years for heavily grazed sites. The good news is that, despite severe disturbance whether from complete removal or burial, BSCs do reform and start to generate chlorophyll and organic C. What is much less certain, and subject to intense debate, is whether changes will result in a long-term reduction in productivity and what, if anything, should be done about it.

The seemingly logical conclusion is to recommend that stocking densities are determined by carrying capacity assessments, based on grass production during average rainfall years. Carrying capacity is defined as the maximum livestock density that can be supported on a sustainable basis (FAO, 1988), and in Botswana this typically equates to one cow per 10–20 ha (Stocking and Murnaghan, 2001). The rationale is that if livestock density is kept below this number then long-term degradation of ecosystem resources will be avoided, even in dry years when grass production is low. Rotational grazing systems can be used to ensure no areas of land and soil are disturbed for too long and grazed land has time to recover. Furthermore, because this will ensure CO_2 emissions are reduced, it opens up the possibility of rewarding pastoral farmers through carbon credit payment schemes such as REDD+ (United Nations, 2011; Stringer *et al.*, 2012). However, the situation may not be this straightforward, and our data hint at some of the complicating factors that call into question these assumptions and conclusions. At the very least, rangeland health assessments should include information on the extent and type of BSCs in the area and take into account recovery times after disturbance. Rohde *et al.* (2006) have also questioned the assumptions upon which many grazing management decisions are based in southern Africa, namely that: (i) they are stable systems in equilibrium; (ii) improper land-use practices will destabilize them; and (iii) a reduction in livestock numbers, or a change in the tenure system, is needed to return to a more productive and less degraded state.

The inter-annual variability of precipitation in the Kalahari (and all drylands) means that primary production, including that from BSCs, is inherently temporally and spatially variable (NoyMeir, 1973; De Leeuw and Tothill, 1990; Behnke *et al.*, 1993; Rohde *et al.*, 2006). A carrying capacity set in the drought years of 2002–2005 in the southern Kalahari will be considerably lower that one set in any year since 2007 when rainfall has been average or above average for 5 consecutive years. Importantly, grazing strategies that do not take advantage of the additional resources provided in the productive years (as a carrying capacity strategy would dictate), are not appropriate, practical or even sustainable. Under these conditions, opportunistic pastoral strategies are the rational approach, rather than one that adheres to fixed-stocking rates (White, 1993; Stocking and Murnaghan, 2001; Sullivan and Rohde, 2002). Increasing stocking densities to take advantage of rains provides immediate gain and increases the chance of at least some of the herd surviving when drought returns and rangeland productivity drops (Stocking and Murnaghan, 2001). The following excerpt from an interview with a communal farmer from the Kalahari illustrates the point (from Campbell, 1990, p. 52):

> Of course, I knew the drought must come. The more cattle I have when it comes, the more chance I have that some of them will survive.... Mostly, we try to keep our cattle during a drought knowing that if we sell them when thin, we get little for them and sometimes they are condemned.

Interestingly, Campbell (1990) demonstrates these issues are not confined to communal areas, albeit for subtly different reasons, by quoting another farmer who purchased and fenced land:

> If I am honest, I must say I have too many cattle on the ranch, especially during the drought when grass is poor. But I have problems. I had to drill three boreholes before striking water. I got a rebate, but it still costs much more than I expected. If I kept less cattle, I would be unable to cover repayments to the bank. Then, my relatives have cattle. As the eldest, I have to help them with grazing and water. Even the people I employ have a few cattle and I find it impossible to refuse to keep these here.
>
> Campbell (1990, p. 52)

The apparent disconnect between recommendations derived from scientific studies of soil and ecological processes and what the people need and, indeed, should, do in order to farm in this environment, may be illusory. BSCs dominated by cyanobacteria, typical of much of the Kalahari and elsewhere, recover relatively quickly from even intense disturbance. Similarly, annual grasses followed by perennials will return after intense grazing within 2–3 years in the Kalahari. Maintaining land in pristine condition is not a priority for semi-subsistence livestock farmers in an environment where rainfall is highly variable. This does not mean to say that an ever-increasing number of livestock on a decreasing amount of communal grazing land is sustainable or advisable. It will eventually lead to a coalescing of highly disturbed areas and an elimination of any opportunity to let the land lie fallow and grass and BSC cover to re-establish. Decisions on grazing land management should take into account natural variations in productivity and be based on a holistic assessment of the grass and BSC resource.

11.4.2 The spatial distribution of resources and links to degradation

Additional complicating factors relate to the spatial distribution of soil resources in the landscape.

It is well established that the spatial distribution of vascular plants in drylands is heterogeneous, in response to low and variable precipitation (Noy-Meir, 1973, 1985). Our data clearly show that C, N and soil microbial activity (evidenced by CO_2 efflux) are greater under shrubs and trees compared to open areas. The processes leading to these 'islands of fertility' can relate to organic matter additions and faunal activity (e.g. Dean *et al.*, 1999) and/or deposition of enriched eroded material (e.g. Tongway and Ludwig, 1994). Given the lack of major wind erosion in the Kalahari (Thomas and Dougill, 2007), it is likely that any sub-plant canopy enrichment will occur through the former mechanism (Dougill and Thomas, 2004; Berkeley *et al.*, 2005). The implications of this spatial heterogeneity are contested. Schlesinger *et al.* (1990) associated the development of spatial heterogeneity in soil and water resources in the south-west USA with land degradation. They argued that intensive grazing reduced grass cover and resulted in an invasion of woody shrub species that, once established, was difficult to reverse. However, Ludwig and Tongway (1996) concluded that patchiness increased ecosystem productivity and that it was vital to ecosystem functioning and biodiversity. There is a further apparent contradiction, in that intense grazing will lead to reductions in SOC in the short term, but shrub encroachment, the result of long-term grazing pressures, will increase SOC (Eldridge *et al.*, 2011). This highlights the need to be very careful in the use of terms such as degradation and desertification. Feedback mechanisms between climate, land use, soils and vegetation are complex, and do not lend themselves easily to terms that oversimplify multiple ecosystem processes.

What is clear, however, is that shrub encroachment will lead to an increase in the amount of C (and N) contained in soils and vegetation compared to grassland dominated areas (Eldridge *et al.*, 2011). Long-term, intensively grazed areas in the Kalahari are likely to have a higher shrub density than lightly grazed areas (Moleele and Mainah, 2003). Changes in land use and management practices to store and sequester C are becoming integral to global efforts to address climate change (Dougill *et al.*, 2012; Stringer *et al.*, 2012). However, this may not be simple in the rangelands of the Kalahari where there is a complex relationship between land tenure, grazing intensity, climate and soil impacts. Increases in the C stocks due to shrub encroachment are likely to be higher than those from carefully managed grazing lands and so there is a possibility that this system could reward practices that reduce the grazing potential of the rangelands. Stringer *et al.* (2012) stress the importance of a collaborative, multi-stakeholder approach, working from local to regional scales to ensure that scientific advances can inform policy and practice to deliver C, ecosystem service and poverty alleviation benefits. Dougill *et al.*, (2012) have also shown that strong existing local institutions, clear land tenure, community control over land management decision-making and up-front, flexible payment schemes are vital if payment schemes are to contribute to poverty alleviation and sustainable rangeland management.

11.5 Conclusions

Pastoral farming on unfenced communal areas is the principal livelihood activity across most of the Kalahari. The grazing land available to farmers without tenure is, however, shrinking as more is sold to private owners. This increases livestock stocking densities and communal grazing lands in Botswana typically exceed recommended carrying capacities. This leads to detrimental changes in soil properties, at least in the short term, and longer-term changes in the balance between shrubs and grasses.

Our research has shown that intensive grazing damages the biologically crusted sand surface resulting in a reduction in the amount of C fixed by autotrophic organisms. Light disturbance of the soil, however, does not appear to have an adverse effect on the ability of the organisms to photosynthesize and add C to the soil. Thus, managed grazing, where soils are only lightly disturbed, could help maintain a positive C balance in African drylands. However, the situation is complicated by longer-term changes in vegetation structure resulting from shrub encroachment. This alters the micro-climate and leads to a significant increase in soil C and N. There is a growing desire to manage land for C capture and storage, which will reward sustainable land management practices that increase the capture of photosynthetically fixed C in soils. However, what is regarded as degraded land by pastoral farmers in the Kalahari has the potential to be viewed as beneficial by C capture schemes because shrub encroachment increases the photosynthetically derived input to the soil. Thus, it is vital that land management decisions take a holistic view that includes consideration of how land use changes may affect the behaviour of human populations that rely directly on the managed land for their livelihoods.

Acknowledgements

Research was funded by a Leverhulme Trust Research Fellowship and a Leverhulme Trust research grant and conducted under Government of Botswana permit EWT8/36/4 VIII(4). We are indebted to Jill Thomas, who provided access to her land and logistical support for which we are very grateful. We thank Steve Hoon for providing the respiration chambers and the spreadsheets for soil CO_2 efflux determination. We thank two anonymous referees whose comments improved an earlier draft.

References

Behnke, R.H., Scoones, I. and Kerven, C. (eds) (1993) *Range Ecology at Disequilibrium: New Models of Natural Variability and Pastoral Adaptation in African Savannas*. Overseas Development Institute, International Institute of Environment and Development, Commonwealth Secretariat, London.

Belnap, J. and Lange, O.L. (2003) Structure and functioning of biological soil crusts: A synthesis. In: Belnap, J. and Lange, O.L. (eds) *Biological Soil Crusts: Structure, Function and Management*. Springer-Verlag, Berlin, Germany, pp. 471–479.

Berkeley, A., Thomas, A.D. and Dougill, A.J. (2005) Cyanobacterial soil crusts and woody shrub canopies in Kalahari rangelands. *African Journal of Ecology* 43, 137–145.

Campbell, A. (1990) *The Nature of Botswana. A Guide to Conservation and Development*. International Union for Conservation of Nature and Natural Resources (IUCN), Gland, Switzerland.

De Leeuw, P.N. and Tothill, J.C. (1990) *The Concept of Rangeland Carrying Capacity in Sub-Saharan Africa – Myth or Reality*. Overseas Development Institute, Pastoral Development Network, London Network Paper 29b. 20pp.

Dean, W.R.J., Milton, S.J. and Jeltsch, F. (1999) Large trees, fertile islands, and birds in arid savanna. *Journal of Arid Environments* 41, 61–78.

Dougill, A.J. and Thomas, A.D. (2004) Kalahari Sand soils: Spatial heterogeneity and land degradation. *Land Degradation and Development*. 15, 233–242.

Dougill, A.J., Thomas, D.S.G. and Heathwaite, A.L. (1999) Environmental change in the Kalahari: Integrated land degradation studies for non-equilibrium dryland environments. *Annals of the Association of American Geographers* 89, 420–442.

Dougill, A.J., Stringer, L.C., Leventon, J., Riddell, M., Rueff, H., Spracklen, D.V. and Butt, E. (2012) Lessons from community-based payment for ecosystem service projects: From forests to rangelands. *Philosophical Transactions of the Royal Society: B*, 367, 3178–3190.

Elbert, W., Weber, B., Büdel, B., Andreae, M.O. and Pöschl, U. (2009 Microbiotic crusts on soil, rock and plants: Neglected major players in the global cycles of carbon and nitrogen? *Biogeoscience Discussions* 6, 6983–7015.

Eldridge, D.J., Bowker, M.A., Maestre, F.T., Roger, E., Reynolds, J.F. and Whitford, W.G. (2011) Impacts of shrub encroachment on ecosystem structure and functioning: Towards a global synthesis. *Ecology Letters*, 14, 709–722.

FAO (1988) Guidelines: Land evaluation for extensive grazing. *FAO Soil Bulletin* 58. FAO, Rome.

FAO/IIASA/ISRIC/ISS-CAS/JRC (2009) Harmonized World Soil Database (version 1.0). FAO, Rome and IIASA, Laxenburg, Austria.

Gil-Romera, G., Lamb, H.F., Turton, D., Sevilla-Callejo, M. and Umer, M. (2010) Long-term resilience, bush encroachment patterns and local knowledge in a northeast African savannah. *Global Environmental Change* 20 (4), 612–626.

Henry, M., Valentini, R. and Benroux, M. (2009) Soil carbon stocks in ecoregions of Africa. *Biogeosciences* 6, 797–823.

Inter-Governmental Panel on Climate Change (IPCC) (2007) *Climate Change 2007: The Physical Science Basis*. Cambridge University Press, Cambridge, UK.

Langston, G. and McKenna-Neuman, C. (2005) An experimental study on the susceptibility of crusted surfaces to wind erosion: A comparison of the strength properties of biotic and salt crusts. *Geomorphology* 72, 40–53.

Ludwig, J.A. and Tongway, D.J. (1996) Rehabilitation of semi-arid landscapes in Australia. II. Restoring productive soil patches. *Restoration Ecology* 4, 398–406.

McKenna-Neuman, C. and Maxwell, C.D. (2002) Temporal aspects of the abrasion of microphytic crusts under grain impact. *Earth Surface Processes and Landforms* 27, 891–908.

Millennium Ecosystem Assessment (MEA) (2005) *Ecosystems and Human Well-Being: Synthesis.* Island Press, Washington, DC.

Moleele, N.M. and Mainah, J. (2003) Resource use conflicts: The future of the Kalahari ecosystem. *Journal of Arid Environments* 54, 405–423.

Noy-Meir, I. (1973) Desert ecosystems: Environment and producers. *Annual Reviews of Ecological Systems* 4, 25–51.

Noy-Meir, I. (1985) Desert ecosystem structure and function. In: Evenari, M., Noy-Meir, I. and Goodall, D.W. (eds) *Hot Deserts and Arid Shrublands.* Elsevier, Amsterdam.

Perkins, J.S. and Thomas, D.S.G. (1993) Spreading deserts or spatially confined environmental impacts? Land degradation and cattle ranching in the Kalahari desert of Botswana. *Land Degradation and Rehabilitation* 4, 179–194.

Porra, R.J. (2006) Spectrometric assays for plant, algal and bacterial chlorophylls. In: Grimm, B., Porra, R.J., Rüdiger, W. and Scheer, H. (eds) *Advances in Photosynthesis and Respiration. Chlorophylls and Bacteriochlorophylls: Biochemistry, Biophysics, Function and Application.* Springer-Verlag, New York.

Rohde, R.F., Moleele, N.M., Mphale, M., Allsopp, N., Chanda, R., Hoffman, M.T. *et al.* (2006) Dynamics of grazing policy and practice: Environmental and social impacts in three communal areas of southern Africa. *Environmental Science and Policy* 9, 302–316.

Schlesinger, W.H., Reynolds, J.F., Cunningham, G.L., Huenneke, L.F., Jarrell, W.M., Virginia, R.A. and Whitford, W.G. (1990) Biological feedbacks in global desertification. *Science* 247, 1043–1048.

Stocking, M.A. and Murnaghan, N. (2001) *Handbook for the Field Assessment of Land Degradation.* Earthscan, London.

Stringer, L.C., Dougill, A.J., Thomas, A.D., Spracklen, D.V., Chesterman, S., Ifejika Speranza, C. *et al.* (2012) Challenges and opportunities in linking carbon sequestration, livelihoods and ecosystem service provision in drylands, *Environmental Science and Policy* 19–20, 121–135.

Sullivan, S. and Rohde, R.F. (2002) On non-equilibrium in arid and semi-arid grazing systems. *Journal of Biogeography* 29, 1–26.

Thomas, A.D. (2012) Impact of grazing intensity on seasonal variations of soil organic carbon and soil CO_2 efflux in two semi-arid grasslands in southern Botswana. *Philosophical Transactions of the Royal Society: B* 367, 3076–3086.

Thomas, A.D. and Dougill, A.J. (2007) Spatial and temporal distribution of cyanobacterial soil crusts in the Kalahari: Implications for soil surface properties. *Geomorphology* 85, 17–29.

Thomas, A.D., Hoon, S.R. and Dougill, A.J. (2011) Soil respiration at five sites along the Kalahari transect: Effects of temperature, precipitation pulses and biological soil crust cover. *Geoderma* 167–168, 284–294.

Thomas, A.D., Hoon, S.R., Dougill, A.J. and Mairs, H. (2012) Soil organic carbon in deserts: Examples from the Kalahari. In: Mol, L. and Sternberg, T. (eds) *Changing Deserts: Integrating Environments, People and Challenges.* White Horse Press, Cambridge, UK, pp. 40–60.

Thomas, D.S.G., Sporton, D. and Perkins, J. (2000) The environmental impact of livestock ranches in the Kalahari, Botswana: Natural resource use, ecological change and human response in a dynamic dryland system. *Land Degradation and Development* 11, 327–341.

Tongway, D.J. and Ludwig, J.A. (1994) Small-scale resource heterogeneity in semi-arid landscapes. *Pacific Conservation Biology* 1, 201–208.

United Nations REDD Programme (2011) *The UN REDD Programme Strategy 2011–2015.* UN-REDD Programme Secretariat, Geneva.

Viles, H.A. (2008) Understanding dryland landscape dynamics: Do biological crusts hold the key? *Geography Compass* 2(3), 899–919.

Walkley, A. and Black. I.A. (1934) An examination of the Degtjareff method for determining organic carbon in soils: Effect of variations in digestion conditions and of inorganic soil constituents. *Soil Science* 63, 251–263.

Wang, L., Okin, G.S., Caylor, K. K. and Macko, S.A. (2009) Spatial heterogeneity and sources of soil carbon in southern African savannas. *Geoderma* 149, 402–408.

White, R. (1993) *Livestock Development and Pastoral Production and Communal Rangeland in Botswana.* Botswana Society, Gaborone, Botswana.

12 Changes in Soil Properties with Sugarcane Cropping in Mauritius

L. Ronald Ng Cheong* and Gunshiam Umrit
Mauritius Sugarcane Industry Research Institute, Moka Road, Réduit, Mauritius

12.1 Introduction and Review of Literature

Sugarcane (*Saccharum officinarum*; Poaceae) was first introduced into Mauritius in 1639 by Dutch settlers (North-Coombes, 1993) and today it is the major crop on the island, occupying some 64,000 ha, or one-third of the land area (MSIRI, 2011). However, crop yields are declining, a phenomenon that has also been observed in Australia (Garside *et al.*, 2001), which can be attributed to biological causes, such as a reduction in soil microbial biomass (SMB; Holt and Mayer, 1998) or the presence of soil organisms such as fungal root pathogens or the lesion nematode (Pankhurst *et al.*, 2003). Other possible causes have been identified, including a combination of increased soil acidification, loss of soil organic matter (SOM) and accumulation of deleterious soil organisms (Meyer and Van Antwerpen, 2001). More soil degradation is expected to occur with the increasing use of machinery, combined with large-scale rock removal and land preparation. Therefore, understanding the impacts of sugarcane cropping on soil processes is important to avoid further soil degradation. The effects of sugarcane cropping on soil have been mostly inferred from comparisons between soils under sugarcane and soils from adjacent uncultivated zones (e.g. Garside *et al.*, 1997) where it has been shown that the crop has mostly negative effects on soils (Haynes and Hamilton, 1999). Planting sugarcane on previously uncultivated land in countries as diverse as Australia, Fiji, Papua New Guinea, South Africa and Swaziland has led to a progressive degradation in soil properties, essentially through reduced SOM, acidification and compaction. Some studies have, however, shown no negative effect, e.g. Bramley *et al.* (1996) in Australia. Some positive effects have even been noted in certain instances, e.g. in Australia, McGarry *et al.* (1996) found that cropping reduced bulk density and increased SOM content.

Nevertheless, SOM tends to decline when virgin land is converted to sugarcane, this effect being most marked in the topsoil (Henry and Ellis, 1996; Deng *et al.*, 2009). This decline is a two-stepped process, with a fast initial reduction, followed by a slower process, before stabilizing at a constant equilibrium value (Dominy *et al.*, 2001). Rates of SOM decline range from 0.01% per annum (Qongqo and Van Antwerpen, 2000) to 0.14% per annum (Hartemink, 1998a). Sugarcane cropping may also lead to accumulation of inert SOM such as charcoal (Skjemstad *et al.*, 1999) when the cane is burnt, at the expense of more labile forms of carbon (C). Total SOM content would then remain similar, but as the relative

* Corresponding author: ronald.ngcheong@msiri.mu

proportion of different C pools changes, certain properties such as soil structure are affected. There is also evidence to suggest that SOM from surface horizons migrates to deeper horizons (Masilaca *et al.*, 1986) as a result of soil tillage when virgin land is cultivated to sugarcane (Basile-Doelsch *et al.*, 2009). Dominy *et al.* (2001) have identified three main causes of SOM decline with cropping, namely a lower return of organic residues to the soil, exposure of hitherto protected SOM to microbial action following aggregate disruption and more rapid SOM decomposition rates under more favourable conditions. However, this decline can be reversed by changing practices. For instance, changing from cane burning to green harvesting (where cane is not burnt prior to harvest) and trash (dry cane leaves) retention has been shown to increase SOM (Pinheiro *et al.*, 2010; La Scala Júnior *et al.*, 2012).

Concomitant with the reduction in SOM, SMB has also been found to decrease with sugarcane cropping (McGarry *et al.*, 1996). It tends to decrease rapidly at first, before stabilizing at a new equilibrium level (Dominy *et al.*, 2001). However, it has also been shown that this decline can be reversed with conservation practices (Graham *et al.*, 1999), whereby the trash is retained instead of being burnt at harvest, with an increased return of SOM to the soil from material such as decaying roots, trash and cane tops, which stimulated the activities of the microbial populations.

Soil acidification also often occurs with sugarcane cropping (Haynes and Hamilton, 1999). Meyer *et al.* (1998) have identified three main causes of acidification in sugarcane soils, namely oxidation of ammoniacal fertilizers to nitric acid, accelerated mineralization of SOM and increased leaching of basic cations from the profile. However, the one cause that is considered most important is the oxidation of ammoniacal fertilizers, particularly ammonium sulfate (McLean, 1975; Wood, 1985). Qongqo and Van Antwerpen (2000) estimated that pH declined at about 0.025 pH units per year, whereas Hartemink (1998b) calculated a rate of 0.047 pH units per year for the topsoil.

With respect to plant available nutrients, studies indicate that these are not necessarily lower in sugarcane soils compared to virgin soils. Indeed, the nutrient status depends on the nutrient balance, i.e. whether the amounts of nutrients removed in the harvested crop were higher or lower than fertilizer inputs, with other losses such as leaching being also taken into consideration (Haynes and Hamilton, 1999). Thus, calcium (Ca) and magnesium (Mg) have been found to decrease in several studies (Masilaca *et al.*, 1986; Meyer *et al.*, 1998; Qongqo and Van Antwerpen, 2000), with an average Ca concentration of 23.0 $mmol_c\ kg^{-1}$ being measured in virgin soil compared to 16.4 $mmol_c\ kg^{-1}$ in an adjacent sugarcane soil (recalculated from Wood, 1965) and potassium (K) decreasing in newly planted soil from 14.2 to 8.5 $mmol_c\ kg^{-1}$ between 1979 and 1996 (Hartemink, 1998a). Henry and Ellis (1996) also noted a decline in K concentrations, a situation further aggravated by increasing imbalances between basic cations as shown by the increases in (Ca+Mg) to K ratios. However, there were also cases where K concentrations actually increased in sugarcane soils, a situation attributed to an excessive use of K fertilizers by growers (Meyer *et al.*, 1998). In most cases, however, there were not enough bases added to compensate for the amounts that were being used by the crop and lost to leaching by rainfall or deep percolation (Masilaca *et al.*, 1986).

The increased use of heavy machinery has resulted in compaction, as shown by the significant increase in topsoil bulk density of sugarcane croplands (Wood, 1985; Van Antwerpen and Meyer, 1996). As expected, this compaction occurred mostly in the inter-row as it is traffic-induced. Qongqo and Van Antwerpen (2000) calculated that the average annual rate of bulk density increase varied between 0.004 and 0.005 $g\ cm^{-3}$. Several cultural operations contribute, with different degrees, to the compaction process, either short term from mechanized harvesting, or long term from cultivation during crop growth (McLean, 1975). The compaction process also results in lower water infiltration rates (Garside *et al.*, 1997; Hartemink, 1998a).

In addition to traffic-induced compaction, other phenomena such as surface crusting occur, which affect soil physical properties, particularly its capacity to absorb water. A lower stabilized water infiltration rate can be taken as a signal of deteriorating soil physical condition and this has been regularly observed; soils under sugarcane had a lower stabilized infiltration rate compared to those under native vegetation (Garside *et al.*, 1997). These lower rates were associated with increased soil strengths and were therefore caused by compaction, with a potentially negative effect on soil rooting depth. Hartemink (1998a) measured a low average infiltration rate of 16 mm h^{-1} in the sugarcane inter-rows compared to 259 mm h^{-1} for grassland and 290 mm h^{-1} for sugarcane rows. Since no significant difference was found between the data for grassland and sugarcane rows, it appears that infiltration rate was not affected by the crop *per se*, but most likely by its associated traffic.

Wong You Cheong and Chan (1977) were the first to study the effects of continuous sugarcane cultivation on soil properties in Mauritius. They showed that monoculture under sugarcane had a number of effects on soils, but that these effects differed depending on prevailing conditions. Thus, they demonstrated that the dry soils had acidified, probably due to the long-term use of ammonium sulfate fertilizer, but the wet soils were less acidic under sugarcane, due to liming. With acidification, exchangeable Ca and percentage base saturation decreased in the dry soils, while exchangeable aluminium levels increased. On the other hand, they found a high loss of SOM mainly in dry soils, whereas wet soils were subject to losses only in the topsoil, which they believed to be due to higher SOM turnover under sugarcane, mainly from an accelerated decomposition of acid-stable organic materials. They also found a general decrease in aggregate stability that they related to the reduction in SOM, but could find no difference in bulk density or porosity with cultivation, presumably because of limited use of heavy machinery in sugarcane cultivation in the 1970s.

The literature thus indicates that biological, chemical and physical properties of sugarcane soils have degraded over time, eventually putting at risk the long-term sustainability of sugarcane production in a number of countries. A number of cultural practices contribute to this degradation, with soil tillage, fertilization and mechanization being identified as the main factors. Research has tended to concentrate more on SOM and chemical properties, while changes in physical properties have been less studied. It has been well established that sugarcane cropping leads to a decline in SOM and SMB, coupled with acidification and soil compaction, but changes in soil chemical status are more variable as these depend on fertilization practices. However, we do not know much about the nature of the SOM in the sugarcane soils, nor do we have much information about how physical properties such as soil structure have been affected by sugarcane cropping. Furthermore, most of the research work has been undertaken in large sugarcane producing countries such as Australia and South Africa, whereas there is a paucity of data from small tropical islands such as Mauritius. In spite of its small size, Mauritius comprises a variety of climates and soil conditions that sets it apart from other sugarcane producing countries where conditions are more uniform. Its sugar industry also uses one cultural practice that does not exist elsewhere, namely 'derocking', i.e. large scale removal of stones and rocks from sugarcane fields for mechanized production (Fig. 12.1). Changes in soil properties could therefore be quite different in Mauritius compared to other countries.

The aim of the studies described in this chapter was to determine the effects of sugarcane production practices on soil properties using pristine soils as a reference, and to establish whether any such effect differed between climatic zone and soil type. It was conducted in response to apprehensions from sugar producers in Mauritius that the productive capacity of their soils could be declining because of long-term monoculture, in a manner similar to that which has been observed in Australia by Garside *et al.* (2001). The main hypothesis behind the studies was that sugarcane cropping, inclusive of practices such as mechanization, has had a degrading effect on soil properties and that soil quality has therefore been significantly affected.

12.2 Materials and Methods

Here we present the results of a study conducted in Mauritius, an island of volcanic origin located between the latitudes of 19° 58′ S and 20° 32′ S, and the longitudes of 57° 17′ E and 57° 46′ E, in the south-west Indian Ocean. The total land area is only 1860 km^2 but there is a great diversity of climatic and soil types. As it is a small, relatively isolated land mass, it is almost free from continental influences and has a maritime climate, with hot summer months from November to April and relatively cool winter months from May to October. Mean annual rainfall is around 2100 mm, but its distribution is greatly affected

Fig. 12.1. Derocking, the large-scale removal of stones and rocks from sugarcane fields, which is unique to Mauritius sugarcane production.

by orography and prevailing winds. Thus, the windward eastern coast receives 1600 mm annually, compared to almost 4000 mm in the central region and only about 1000 mm in the leeward western coast. The mean maximum temperature is around 30°C at most locations. The temperature range is of the order of 6°C in the cooler central region, and of about 8°C on the hotter coastal belt (Padya, 1984).

Soils of Mauritius are of volcanic origin and have developed from basalt as the parent material. Major soils are classified as either Latosol or Latosolic according to the soil classification system of Mauritius (Parish and Feillafé, 1965), which is based on the system adopted by Cline (1955) for Hawai'i. These are equivalent, respectively, to the Oxisol and Inceptisol in the United States Department of Agriculture (USDA) classification (Soil Survey Staff, 1999). The main soil groups within the Latosol are the Low Humic Latosol (L) and Humic Ferruginous Latosol (F), whereas the Latosolic Reddish Prairie (P) and Latosolic Brown Forest (B) soils are the main groups within the Latosolic soils (Table 12.1). The L and F soils have mostly developed on old parent material and are therefore relatively rock-free, but the L soils are found in the dry region, whereas the F soils are in the wet region. Similarly, the P and B soils have developed in the dry and the wet regions, respectively, but from younger parent material and therefore have a greater number of rocks within their profile.

Soil samples were taken from fields that were either under pristine uncropped conditions, or under long-term sugarcane cropping. The L, P, F and B soils were studied at Richeterre (20° 07′ S, 57° 31′ E), Médine (20° 16′ S, 57° 23′ E), Mon Désert Alma (20° 14′ S, 57° 33′ E) and Savannah (20° 26′ S, 57° 36′ E), respectively. Four samples were taken from three fields for both cropped and uncropped conditions, giving a total of 12 samples for each condition. Soil pits were dug for this purpose and samples were taken at different depths; 0–15 cm was considered the topsoil and 15–50 cm the subsoil. While no specific trenching direction was applied to the pristine uncropped sites, pits were dug across the cane row in the sugarcane fields and the soil was sampled across the face of the pit and mixed to produce a composite sample. Sub-samples were also taken in the uncropped sites and mixed to give composite samples that were subsequently analysed.

The soil organic carbon (SOC) concentration was determined by partial oxidation using the modified Walkley–Black procedure (Anderson and Ingram, 1993), whereas the total nitrogen (N) concentration was determined by Kjeldahl digestion followed by steam distillation and titration

Table 12.1. Characteristics of four major soil groups of Mauritius. (After Parish and Feillafé, 1965).

Soil group	Mauritius classification	USDA classification	Climatic zone
Low Humic Latosol (L)	Latosol	Oxisol	Sub-humid
Latosolic Reddish Prairie (P)	Latosolic	Inceptisol	Sub-humid
Humic Ferruginous Latosol (F)	Latosol	Oxisol	Super-humid
Latosolic Brown Forest (B)	Latosolic	Inceptisol	Super-humid

against sodium borate (Bremner and Mulvaney, 1982). Stable C isotopic composition ($\delta^{13}C$) was determined on an elemental analyser coupled to a continuous flow isotope ratio mass spectrometer with a triple collector (Europa Scientific, Crewe, UK) and results expressed in standard delta notation relative to a Pee Dee Belemnite (PDB) standard (Peterson and Fry, 1987). The fraction of total SOC derived from sugarcane was calculated using the isotope mass balance equation described by Balesdent and Mariotti (1996). Total SOC stocks (Mg ha^{-1}) were calculated using the following equation:

$$\text{Total SOC stocks (Mg ha}^{-1}) = \%C \times Z \times B \times (100 - S) \qquad (12.1)$$

Where $\%C$ is the C concentration of the sample (percent), Z is the layer thickness (m), B is the bulk density (Mg m^{-3}) and S is the stone content of soil (%). Loss of original SOC as a result of sugarcane cultivation was calculated using the following equation:

$$\%\ \text{Loss of original SOC} = 100 - \%\ C_{dfnv} \times \left[\frac{C_{\text{cane}}}{C_{\text{nv}}}\right] \qquad (12.2)$$

Where C_{dfnv} is the percentage of total SOC derived from natural vegetation, C_{cane} and C_{nv} denote the SOC stocks under sugarcane and natural vegetation, respectively. Labile C was determined by oxidation with 333 mM $KMnO_4$ according to the method described by Blair *et al.* (1995). Soil microbial biomass was measured via the chloroform fumigation–extraction method of Voroney *et al.* (1993), followed by dichromate and Kjeldahl digestion for microbial biomass C and N, respectively, using a k_{EC} value of 0.27 and a k_{EN} value of 0.53 as suggested by Anderson and Ingram (1993). Microbial biomasses of C and N were calculated by the difference between the values of fumigated samples and unfumigated ones.

Soil pH was measured electrochemically using a pH meter in a suspension of 20 g soil in 50 ml distilled water (STASM, 2003). Concentrations of individual exchangeable bases were determined using the leachate produced when 1 M unbuffered ammonium acetate solution was used to displace the exchangeable bases (Peech, 1945). The concentrations of Ca and Mg in the solution were then measured using an atomic absorption spectrophotometer, whereas that of K was determined using a flame photometer.

Bulk density and stabilized infiltration rate were both measured *in situ* using the excavation method with a sand-funnel apparatus (Blake and Hartge, 1986) and the CSIRO disc permeameter (Perroux and White, 1988), respectively. Aggregate water stability was determined from soil moisture characteristics of weathered and unweathered soil samples using the technique described by Haines (1930).

For each parameter and both conditions (uncropped versus cropped), 12 data points were obtained and statistically analysed by comparing means and standard errors. Comparisons were made between paired uncropped and cropped sites to determine whether or not differences were significant. The difference between the two cropping situations was considered significant when the standard errors did not overlap, i.e. the lower limit of the higher value is higher than the upper limit of the smaller value.

12.3 Results

12.3.1 Soil biological properties

Both SOC and total N tended to decrease in the topsoil and increase in the subsoil with sugarcane cropping (Table 12.2). With cropping, topsoil SOC and total N values became either significantly lower or were not significantly different for all four soils. The opposite was true for subsoil SOC and total N. SOM was translocated from topsoil to subsoil; as a result, there may have

Table 12.2. Effect of sugarcane cropping on soil organic matter in four major soil groups (L, P, F and B; see Table 12.1) of Mauritius.

Parameter		L U	L C	P U	P C	F U	F C	B U	B C
SOC concentration (%)	Top	2.7± 0.2	2.4 ± 0.1	3.4 ± 0.3	3.9 ± 0.2	3.6 ± 0.2	2.9 ± 0.1	3.5 ± 0.2	3.4 ± 0.1
			(–)		(0)		(–)		(0)
	Sub	1.4 ± 0.1	1.6 ± 0.1	1.8 ± 0.2	1.8 ± 0.1	1.8 ± 0.1	2.2 ± 0.1	2.3 ± 0.2	2.7 ± 0.1
			(+)		(0)		(+)		(+)
Total N concentration (%)	Top	0.29 ± 0.02	0.23 ± 0.01	0.35 ± 0.03	0.29 ± 0.02	0.26 ± 0.01	0.24 ± 0.01	0.30 ± 0.02	0.33 ± 0.01
			(–)		(–)		(0)		(0)
	Sub	0.17 ± 0.01	0.17 ± 0.01	0.20 ± 0.02	0.18 ± 0.01	0.13 ± 0.01	0.18 ± 0.01	0.20 ± 0.02	0.29 ± 0.01
			(0)		(0)		(+)		(+)
Profile SOC content (Mg ha^{-1})		89.9 ± 6.5	92.4 ± 3.8	100.6 ± 6.2	89.3 ± 4.2	99.0 ± 4.3	109.2 ± 6.4	84.8 ± 5.0	118.8 ± 4.8
			(0)		(–)		(0)		(+)
Profile total N content (Mg ha^{-1})		10.4 ± 0.6	9.3 ± 0.4	10.8 ± 0.6	9.5 ± 0.4	7.1 ± 0.4	9.1 ± 0.5	7.7 ± 0.5	12.5 ± 0.5
			(–)		(–)		(+)		(+)
C:N ratio	Top	9.3 ± 0.8	10.4 ± 0.6	9.7 ± 0.8	13.4 ± 0.6	13.8 ± 1.0	12.1 ± 1.0	11.7 ± 1.0	10.3 ± 0.5
			(0)		(+)		(0)		(0)

Note: U = uncropped, C = cropped; mean of 12 samples ± SE.
Top = topsoil (0–15 cm); Sub = subsoil (15–50 cm); (+) = significant increase with cropping; (0) = no significant difference with cropping; (–) significant decrease with cropping.

been gains or losses in the total amounts of SOC and total N within the whole soil profile (0–50 cm), depending on the extent to which these changes occurred within the upper and lower soil layers (Table 12.2). The topsoil C:N ratio increased in the dry zone, but there was no significant change in the wet zone.

12.3.2 Loss of native SOC and its replacement by sugarcane-derived C

In the Oxisols, long-term sugarcane cropping decreased SOC stocks in the surface 0–15 cm layer by 4.9 to 6.3 Mg C ha^{-1} compared to the uncropped soil. Moreover, the decline in topsoil SOC was accompanied by an increase in subsoil SOC stocks by 10.5 to 20.8 Mg C ha^{-1}. In the Inceptisols, however, changes in SOC stocks in topsoil and subsoil resulting from long-term cropping were not significant. Although long-term sugarcane production resulted in a decline of SOC in the top 15 cm layer, the total SOC stock of the top 50 cm profile did not change at any of the sites studied (Fig. 12.2).

Measurements of the $\delta^{13}C$ natural abundance showed marked depletion of the native SOC at all the study sites and its replacement by C derived from sugarcane origin. However, the amount of native C lost ranged from 27% to 70% of the original C stock in the corresponding virgin soils (Table 12.3) and represented between 26 and 71 Mg C ha^{-1}. The loss of native SOC was fully compensated by C inputs from sugarcane at all sites studied (Fig. 12.3). Thus, after more than 50 years of sugarcane cropping, sugarcane-derived C constituted between 31% and 56% of the existing C in topsoil of the Oxisols, the remaining 44% to 69% being still attributable to native C. In the Inceptisols, sugarcane-derived C contributed about 41% to 69% of the total C in the topsoil, the remaining 31% to 59% being attributed to original C derived from native vegetation.

The labile C fraction ($KMnO_4$-oxidizable; Blair *et al.*, 1995) in the upper 50 cm of the profile ranged between 21.9 and 40.5 Mg C ha^{-1} representing about 25% to 34% of the total C stock across the study sites (Table 12.4). In the P soil, labile C declined significantly by 32% as a result of sugarcane cultivation. In the Oxisols, the impact of sugarcane cropping on labile C stocks was not significant. When the labile C fraction was expressed as a proportion of non-labile C, the resulting C lability showed a more

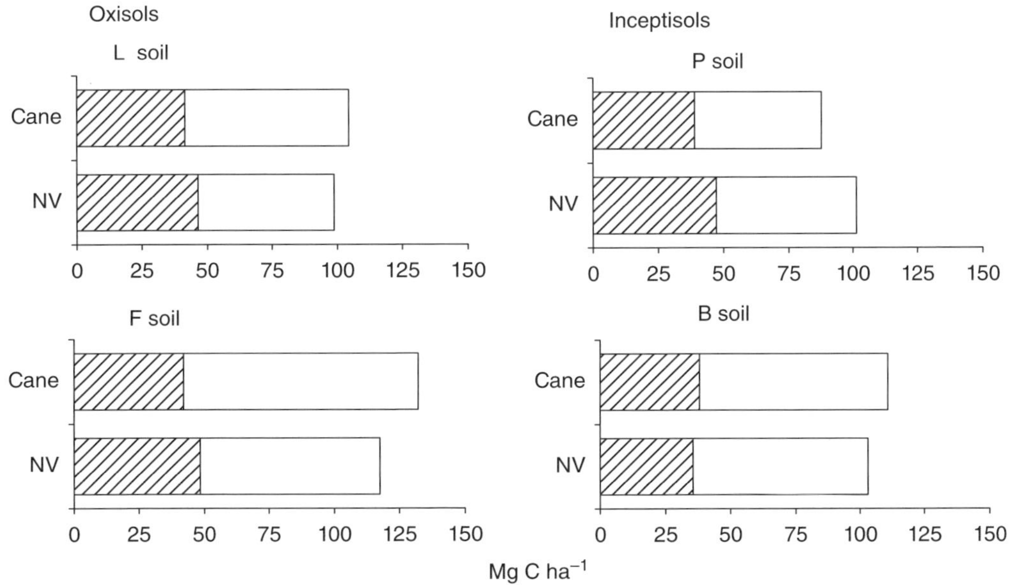

Fig. 12.2. Soil organic carbon stocks (Mg C ha^{-1}) in 0–15 cm (shaded portion), and 15–50 cm (unshaded portion) depths under long-term sugarcane cultivation (Cane) and natural vegetation (NV) in Mauritius.

Table 12.3. Loss of original C and proportion of sugarcane-derived C in two soil layers in four major soil groups (L, P, F and B; see Table 12.1) after long-term sugarcane cropping in Mauritius.

		Soil group			
Parameter	Depth (cm)	L	P	F	B
% loss of original C	0–15	38.3 ± 7.1[a]	74.1 ± 3.8[a]	62.1 ± 3.6[a]	36.5 ± 5.9[a]
	15–50	16.1 ± 3.6[b]	66.4 ± 5.9[a]	20.0 ± 4.6[b]	22.2 ± 4.7[b]
	Profile	26.5 ± 4.2	70.0 ± 4.2	37.3 ± 4.0	27.1 ± 5.3
% C derived from sugarcane	0–15	31.0 ± 4.7[a]	68.6 ± 5.4[a]	56.4 ± 1.3[a]	40.9 ± 3.4[a]
	15–50	35.9 ± 2.9[a]	62.8 ± 5.6[a]	38.4b ± 2.4[b]	27.6 ± 2.2[a]
	Profile	30.5 ± 3.4	65.3 ± 5.0	44.2 ± 1.9	32.2 ± 1.9

Note: Mean of 12 samples ± SE. For each parameter within one soil group, values followed by the same letter were not significantly different (P <0.05).

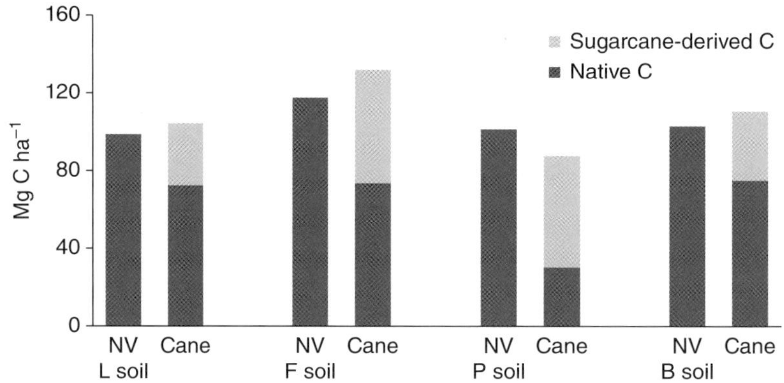

Fig. 12.3. Soil carbon stocks in the surface 50 cm of soil under natural vegetation (NV) and after long-term sugarcane cultivation (Cane) in Mauritius.

Table 12.4 Labile carbon (C) and C lability in four major soil groups (L, P, F and B; see Table 12.1) as influenced by long-term sugarcane cropping in Mauritius.

		Soil group			
Parameter		L	P	F	B
Labile C (Mg ha^{-1})	Uncropped	30.9 ± 2.0[a]	32.3 ± 2.9[a]	40.5 ± 1.9[a]	32.3 ± 2.1[a]
	Cropped	27.3 ± 2.4[a]	21.9 ± 3.3[b]	40.1 ± 2.1[a]	28.8 ± 2.0[a]
Carbon lability*	Uncropped	0.46 ± 0.02[a]	0.47 ± 0.03[a]	0.53 ± 0.02[a]	0.46 ± 0.03[a]
	Cropped	0.35 ± 0.02[b]	0.33 ± 0.02[b]	0.44 ± 0.03[b]	0.35 ± 0.03[b]

Note: Mean of 12 samples ± SE. For each parameter and soil group, values followed by the same letter were not significantly different (P < 0.05).

definitive picture of the soil's capacity to maintain the labile C fraction under sugarcane cropping. The C lability under long-term sugarcane was signficantly lower than under natural vegetation at all sites.

$$\text{*Lability} = \frac{\text{labile C stock}}{\text{non-labile C stock}}$$

With respect to SMB, sugarcane cropping either had no effect, or decreased the topsoil SMB-C and -N, except for microbial biomass C in the P soil, which increased with cropping (Table 12.5). Microbial biomass C (0–15 cm) had a mean between 419 and 648 kg ha^{-1} for uncropped conditions, compared to 280–764 kg ha^{-1} when cropped, whereas the mean SMB-N ranged from

Table 12.5. Effect of sugarcane cropping on topsoil (0–15 cm) microbial biomass in four major soil groups (L, P, F and B; see Table 12.1) of Mauritius.

	Microbial biomass C (kg ha^{-1})			Microbial biomass N (kg ha^{-1})			Microbial biomass C:N ratio		
Soil group	U		C	U		C	U		C
L	621 ± 139		653 ± 70	56 ± 8		54 ± 8	11.1 ± 2.8		12.1 ± 1.1
		(0)			(0)			(0)	
P	419 ± 97		764 ± 77	65 ± 14		79 ± 9	6.4 ± 5.8		9.7 ± 2.0
		(+)			(0)			(0)	
F	648 ± 84		280 ± 31	97 ± 22		39 ± 5	6.7 ± 1.5		7.2 ± 1.5
		(–)			(–)			(0)	
B	466 ± 82		466 ± 63	32 ± 6		36 ± 6	14.6 ± 3.5		12.9 ± 5.1
		(0)			(0)			(0)	

Note: U = uncropped; C = cropped; mean of 12 samples ± SE; (+) = significant increase with cropping; (0) = no significant difference with cropping; (–) = significant decrease with cropping.

32 to 97 kg ha^{-1} when uncropped and from 36 to 79 kg ha^{-1} when cropped. The soil type most affected was the wet, rock-free F soil where both SMB-C and SMB-N were reduced by over 50% with sugarcane cropping.

12.3.3 Soil chemical properties

Soil pH generally increased with cropping in the dry L and P soils, but decreased in the wet F and B soils (Table 12.6); mean pH values being about 5.5 in the wet soils and 7.0 in the drier ones. In general, changes with cropping in the exchangeable bases K, Ca and Mg were variable. K declined in the L soil, but mostly increased in the other soils. Ca concentrations decreased markedly in the B soil, but increased in the L and F sub-soil. Mg concentration increased with cropping in the L and P soils and was mostly unchanged in the other soils, except for the F topsoil where it declined.

12.3.4 Soil physical properties

Little change in bulk density was observed, even in the topsoil where compaction is expected to occur (Table 12.7), with only one exception for the rocky wet B soil. In fact, mean topsoil bulk density remained within the range of 0.95–1.15 g cm^{-3} and subsoil mean bulk density within the range of 1.13–1.19 g cm^{-3}. No difference was noted in terms of the aggregate stability of the four topsoils, but differing patterns were seen in the subsoils with increases in L soil, decreases in F soil and no change in the other soils. The stabilized infiltration rate was reduced significantly with mechanized sugarcane production in the wet F and B soils, falling from 190 and 216 mm h^{-1}, respectively, under native vegetation to 66 and 105 mm h^{-1} with sugarcane as well as in the dry L soils where it fell from 105 mm h^{-1} to 78 mm h^{-1}.

12.4 Discussion

The changes occurring in SOC concentration within the topsoil following sugarcane cultivation in Mauritius were reflected, to a certain extent, by changes in SMB. The reduction in SMB in the F soil follows the reduction in SOC within its topsoil, while there was no change in the B soil for both SOC and SMB. In the L topsoil, the reduction in SOC was not matched by a similar decrease in SMB, while SMB increased in the P topsoil, following the trend in its SOC concentration. Since microbial biomass C is an indicator of potential microbial activity (Rice *et al.*, 1996), sugarcane cropping tends to reduce the activity of the microbial population in the soil, presumably because of the lower SOM content, but possibly also because the remaining SOM is more recalcitrant.

The observed decline in topsoil SOC and increase in subsoil SOC concur with the findings of McGarry *et al.* (1996) and Skjemstad *et al.* (1999), who reported a decline in the SOC content of

Table 12.6. Effect of sugarcane cropping on chemical properties of four major soil groups (L, P, F and B; see Table 12.1) of Mauritius.

		Soil group							
		L		P		F		B	
Parameter		U	C	U	C	U	C	U	C
pH	Top	6.5 ± 0.2	7.0 ± 0.1 (+)	7.0 ± 0.1	7.3 ± 0.2 (0)	5.3 ± 0.1	4.7 ± 0.1 (–)	6.8 ± 0.2	5.1 ± 0.1 (–)
	Sub	6.1 ± 0.2	7.0 ± 0.1 (+)	7.0 ± 0.1	7.6 ± 0.1 (+)	5.1 ± 0.1	5.2 ± 0.1 (0)	6.9 ± 0.2	5.3 ± 0.1 (–)
K concentration ($cmol_c$ kg^{-1})	Top	1.3 ± 0.1	0.7 ± 0.1 (–)	1.0 ± 0.1	2.6 ± 0.3 (+)	0.2 ± 0.0	0.3 ± 0.0 (+)	0.8 ± 0.1	0.7 ± 0.1 (0)
	Sub	0.5 ± 0.1	0.2 ± 0.0 (–)	0.3 ± 0.1	2.3 ± 0.4 (+)	0.1 ± 0.0	0.2 ± 0.0 (+)	0.6 ± 0.1	0.4 ± 0.1 (–)
Ca concentration ($cmol_c$ kg^{-1})	Top	13.4 ± 3.0	11.7 ± 2.1 (0)	19.1 ± 1.2	15.5 ± 1.6 (–)	2.6 ± 0.9	1.6 ± 0.5 (0)	15.3 ± 3.8	3.2 ± 0.3 (–)
	Sub	6.6 ± 0.5	9.8 ± 1.6 (+)	15.4 ± 1.1	15.2 ± 1.7 (0)	0.7 ± 0.2	1.8 ± 0.4 (+)	13.1 ± 3.2	3.3 ± 0.3 (–)
Mg concentration ($cmol_c$ kg^{-1})	Top	3.9 ± 0.2	5.3 ± 0.3 (+)	5.7 ± 0.4	9.3 ± 0.8 (+)	1.2 ± 0.1	0.6 ± 0.1 (–)	2.5 ± 0.3	2.6 ± 0.3 (0)
	Sub	2.2 ± 0.1	3.9 ± 0.2 (+)	4.7 ± 0.3	9.6 ± 1.0 (+)	0.5 ± 0.1	0.5 ± 0.0 (0)	1.9 ± 0.2	2.3 ± 0.2 (0)

Note: U = uncropped; C = cropped; mean of 12 samples ± SE; Top = topsoil (0–15 cm); Sub = subsoil (15–50 cm); (+) = significant increase with cropping; (0) = no significant difference with cropping; (–) significant decrease with cropping.

Table 12.7. Effect of sugarcane cropping on physical properties of four major soil (L, P, F and B; see Table 12.1) of Mauritius.

		Soil group							
		L		P		F		B	
Parameter		U	C	U	C	U	C	U	C
Bulk density (g cm^{-3})	Top	1.14 ± 0.03	1.15 ± 0.03	1.03 ± 0.03	1.05 ± 0.04	0.89 ± 0.04	0.98 ± 0.06	0.73 ± 0.04	0.95 ± 0.03
			(0)		(0)		(0)		(+)
	Sub	1.25 ± 0.05	1.19 ± 0.02	1.09 ± 0.04	1.19 ± 0.02	1.20 ± 0.03	1.13 ± 0.03	0.86 ± 0.05	1.16 ± 0.03
			(0)		(+)		(–)		(+)
Aggregate stability index	Top	0.55 ± 0.09	0.52 ± 0.05	0.51 ± 0.10	0.52 ± 0.06	0.91 ± 0.04	0.88 ± 0.04	0.85 ± 0.07	0.88 ± 0.05
			(0)		(0)		(0)		(0)
	Sub	0.18 ± 0.02	0.36 ± 0.03	0.30 ± 0.05	0.38 ± 0.08	0.90 ± 0.04	0.76 ± 0.04	0.82 ± 0.09	0.92 ± 0.06
			(+)		(0)		(–)		(0)
Stabilized infiltration rate (mm h^{-1})		105 ± 9	78 ± 11	133 ± 18	185 ± 31	190 ± 24	66 ± 16	216 ± 36	105 ± 7
			(–)		(+)		(–)		(–)

Note: U = uncropped; C = cropped; mean of 12 samples ± SE; Top = topsoil (0–15 cm); Sub = subsoil (15–50 cm); (+) = significant increase with cropping; (0) = no significant difference with cropping; (–) significant decrease with cropping.

surface horizons but an increase in that of sub-surface horizons after different periods of sugarcane cultivation in Australia. This accumulation of SOC in deeper soil layers under sugarcane has been attributed to rhizo deposition and continual turnover of root material at lower depths as evidenced by a higher proportion of sugarcane-derived C in the subsoil compared to topsoil SOM as reported by De Resende *et al.* (2006). But our data on measurements of $\delta^{13}C$ showed that the topsoil was more depleted in native SOC than the subsoil, suggesting that repeated ploughing resulted in a greater loss of original SOC and hence lower SOC stocks. It could also be due to a downward redistribution of SOM-rich topsoil caused by tillage and disk harrowing carried out at the time of replanting (Masilica *et al.*, 1986; McGarry *et al.*, 1996) or by a vertical migration of organo-mineral complexes (Basile-Doelsch *et al.*, 2009). Apart from tillage, 'derocking' is a common practice in Mauritius, particularly in the sub-humid zone. Derocking followed by tillage probably greatly reduces SOM content, and organic waste material such as mill mud (also called filter cake, a solid waste generated on clarification of cane juice before its concentration and sugar crystallization; average composition: 8% C, 0.2% N, 0.6% P_2O_5, 0.1% K_2O), stillage slop (also called vinasse, the liquid residue left after distillation of ethanol from molasses; average composition: 1% C, 0.2% N, 0.06% P_2O_5, 1.6% K_2O) or compost is routinely added as soil amendment to restore SOM levels. When substantial amounts of stillage slop and mill mud compost were applied to a P soil over a period of 3 years, SOM concentration decreased in the topsoil but increased in the subsoil (Ng Cheong *et al.*, 2009), resulting in a net increase within the profile. Green cane trash blanketing, whereby residues from the previous crop are retained *in situ* after harvest, also contributed to increased return of residues in the field.

Although long-term sugarcane cultivation in Mauritius did not have any detrimental effect on overall SOC stocks in the top 50 cm of the profile, there was a marked depletion of the native SOC stock. Most of this loss in native C occurred in the surface 15 cm of soil, indicating that repeated tillage of topsoil resulted in enhanced mineralization of native C stocks compared to deeper layers. The highest loss of 70% was measured in a P soil and could be attributed, at least partly, to a decrease in soil clay content from an initial 63% in the virgin soil to 50% (Ng Cheong, 2007) after long-term cropping. It is generally reported that clay and silt contents play an important role in the protection of SOC by stabilizing SOC against mineralization (Bationo and Buerkert, 2001; Bationo *et al.*, 2007). In addition to soil texture, factors such as climatic variations and cultural practices such as the frequency and intensity of tillage operations could have influenced the loss of original SOC.

However, the losses of native SOC were adequately compensated by C input from sugarcane crop residues and root C turnover so that the net effect of sugarcane cultivation on total SOC stocks in the 0–50 cm profile was generally not significantly different from virgin uncultivated land. The results emphasize the importance of retaining sugarcane crop residues in the maintenance of SOC under intensive sugarcane cropping.

The proportion of sugarcane-derived C tended to decrease with depth in the wet zone but was more uniformly distributed in the soil profiles of the dry zone. Roots constitute an important source of C in sugarcane farming systems: up to 3.7 Mg C ha^{-1} annually (Suman *et al.*, 2009). The observed differences in distribution of sugarcane-derived C can, therefore, be partly explained by a lower root density in the generally water saturated layer of the wet soils at lower depths, compared to a more uniform distribution of roots down to the 50 cm depth in the dry soils. Over the 50 cm profile, the contribution of sugarcane-derived C to the total SOC stock varied from 31% to 65%. Although site-to-site variation in the gross C inputs occurs, it could not account for the observed wide variations in the contribution of sugarcane-derived C to the total SOC stock. The results tend to suggest that stabilization of C derived from sugarcane was dependent on the amounts of original C lost relative to the initial SOC stock under native vegetation and tended to continue until the initial level of SOC was restored.

Although the quantity of SOC did not change as result of long-term cropping, it resulted in a significant decline in labile C ($KMnO_4$-oxidizable), indicating that the quality of sugarcane soils, in terms of C lability was degraded, a phenomenon that was exacerbated by the shift from conventional to mechanized farming (Umrit *et al.*, 2014). The results indicated that although total SOC stocks were maintained after long-term cultivation, mainly through the return of sugarcane-derived C, the frequent tillage and disc harrowing operations have favoured the mineralization of the labile components of SOC.

Previous studies have generally shown that soil became acidified with sugarcane monoculture because of N fertilization, so the absence of acidification in sub-humid soils is surprising. This effect could be attributed to pH-enhancing practices, mainly through the addition of cement. Another element that explains the pH increase in the P soil is the use of composted mill mud as a soil conditioner after rock removal. Indeed, dry composted mill mud contains calcium oxide (Paturau, 1989), which increases soil pH. On the other hand, the B soil and the F topsoil followed the expected trend and became acidified due to ammoniacal fertilizers and urea application, compounded by leaching of basic cations out of the topsoil in the wet environment.

The relatively constant level of exchangeable bases is not surprising as K, Ca and Mg are added to the soil in the fertilization and/or pH correction process. Thus, all soils receive annual dressings of K to meet the crop requirements, 2 kg of K being required to produce 1 t of sugarcane in Mauritius (MSIRI, 2005). Some K may also have been added in excess, from either fertilizers or organic wastes such as mill mud and poultry litter. Dressings of Ca are applied in fertilizers such as calcium ammonium nitrate, as soil amendments, e.g. for liming, as calcium silicate to correct for silicon deficiency and in mill mud. In this manner, relatively high amounts of Ca are regularly applied to sugarcane fields in Mauritius. Magnesium is also regularly applied to the soil, but as a component of organic amendments, mainly in mill mud and also in poultry manure and litter. Overall, there appears to be an appropriate chemical balance in the sugarcane fields.

Since the study by Wong You Cheong and Chan (1977), who found no difference in bulk density with cropping because of the limited use of heavy machinery in the 1970s, the situation has changed. In fact, sugarcane is currently produced with the use of machines at all steps of the production process. Topsoil compaction had already been noted as a result of mechanized harvesting and loading under wet conditions in Mauritius (Ng Cheong *et al.*, 1999). The more recent data confirm that topsoil compaction does occur with sugarcane cropping, but mostly in the super-humid B soils. Soil structure has not been majorly degraded as aggregate properties are not affected in the topsoil, but care needs to be exercised in this respect as these may become degraded if the SOM content declines below a certain threshold (Ng Cheong *et al.*, 2011).

12.5 Conclusions

The fact that SOM is depleted in the topsoil with sugarcane cropping is a worrying aspect for the long-term sustainability of this crop. Not only does SOM impact directly on essential physical properties such as soil aggregation, but it also supplies a substantial part of the N that the crop requires through microbial mineralization. The associated decline in the soil microbial populations also reduces the mineralization process. The SOM will thus supply less N to the crop, with a consequent adjustment needed in terms of fertilizer N to maintain productivity, thereby increasing costs and pollution risks. The reduced SOM content in the topsoil may also increase the risk of erosion, since SOM plays a role in binding soil particles. Acidification and loss of exchangeable bases has not been noted to any great extent. Nevertheless, care should be exercised in maintaining this balance, particularly since mill mud has a major influence on this aspect. Since the degradation in topsoil SOM concentration needs to be reversed to ensure the long-term environmental and economic sustainability of the sugarcane crop in Mauritius, new strategies are needed to reverse the trend and rebuild the SOM levels. These include green cane trash blanketing, already a common practice with mechanized harvesting, the addition of large amounts of organic wastes that can be produced in large quantities from both municipal and agricultural wastes, and the addition of a leguminous break crop as green manure. Sugarcane trash is a ready source of SOM and also supplies N to the soil. It also acts as mulch, thereby conserving soil water and protecting the soil against the direct impact of rainfall.

Organic waste material such as mill mud is routinely returned to the sugarcane fields as a soil conditioner. Other forms include poultry litter, stillage slop and compost. The latter holds great promise, as it can be manufactured from both agricultural and household wastes. The addition of a leguminous break in the sugarcane monoculture should also become a regular practice. A legume reduces the amount of N fertilizer applied as it fixes N from the atmosphere and the remains of the crop increase soil SOM content when incorporated into the soil.

Another issue that needs to be addressed in the long term is soil compaction, particularly in the wet zones. Compaction leads to reduced infiltration rates and therefore degraded soil hydraulic properties, higher risks of erosion under high intensity rainfall and, of course, poorer plant growth and yield as soil–plant–water relations are affected. One approach would be to prevent all in-field traffic when soil conditions are conducive to compaction, but that would mean preventing all mechanized cultural practices and harvest under wet conditions, a solution that cannot be envisaged under commercial conditions. The trend in the Mauritius sugar industry is now for the adoption of a cropping system that includes growing sugarcane on raised beds with dual row planting, with controlled traffic confined to dedicated traffic lanes in the inter-rows, associated with minimum tillage and the inclusion of a leguminous fallow. This approach has the advantage of eliminating compaction in the sugarcane rows and of adding SOM to the soil. The adoption of new technology holds great promise for the future sustainability of the sugarcane industry of Mauritius.

References

Anderson, J.M. and Ingram, J.S.I. (1993) *Tropical Soil Biology and Fertility. A Handbook of Methods* (2nd edition). CAB International, Wallingford, UK.

Balesdent, J. and Mariotti, A. (1996) Measurement of soil organic matter turnover using ^{13}C natural abundance. In: Boutton, T.W. and Yamasaki S.I. (eds) *Mass Spectrometry of Soils*. Marcel Dekker, New York, pp. 83–112.

Basile-Doelsch, I., Brun, T., Borschneck, D., Masion, A., Marol, C. and Balesdent, J. (2009) Effect of land use on organic matter stabilized in organomineral complexes: A study combining density fractionation, mineralogy and $\delta^{13}C$. *Geoderma* 151, 77–86.

Bationo, A. and Buerkert, A. (2001) Soil organic carbon management for sustainable land use in Sudano-Sahelian West Africa. *Nutrient Cycling in Agroecosystems* 61, 131–142.

Bationo, A., Kihara, J., Vanlauwe, B., Waswa, B. and Kimetu, J. (2007) Soil organic carbon dynamics, functions and management in West African agro-ecosystems. *Agricultural Systems* 94, 13–25.

Blair, G.J., Lefroy, R.D.B. and Lisle, L. (1995) Soil carbon fractions, based on their degree of oxidation, and the development of a carbon management index for agricultural systems. *Australian Journal of Agricultural Research* 46, 1459–1466.

Blake, G.R. and Hartge, K.H. (1986) Bulk density. In: Klute, A. (ed.) *Methods of Soil Analysis Part 1, Physical and Mineralogical Methods* (2nd edition). American Society of Agronomy and Soil Science Society of America, Madison, Wisconsin, pp. 363–375.

Bramley, R.G.V., Ellis, N., Nable, R.O. and Garside, A.L. (1996) Changes in soil chemical properties under long-term sugar cane monoculture and their possible role in sugar yield decline. *Australian Journal of Soil Research* 34, 967–984.

Bremner, J.M. and Mulvaney, C.S. (1982) Nitrogen – total. In: Page, A.L., Miller, R.H. and Keeney, D.R. (eds) *Methods of Soil Analysis Part 2, Chemical and Microbiological Properties* (2nd edition). American Society of Agronomy and Soil Science Society of America, Madison, Wisconsin, pp. 595–624.

Cline, M.G. (1955) Soil survey of the territory of Hawai'i. Soil Survey Series 1939, No. 25, USDA, Washington, DC.

De Resende, A.S., Xavier, R.P., de Oliveira, O.C., Urquiaga, S., Alves, B.J.R. and Boddey, R.M. (2006) Long-term effects of pre-harvest burning and nitrogen and vinasse applications on yield of sugar cane and soil carbon and nitrogen stocks on a plantation in Pernambuco, N.E. Brazil. *Plant and Soil* 281, 339–351.

Deng, W., Wu, W., Wang, H., Luo, W. and Kimberley M.O. (2009) Temporal dynamics of iron-rich, tropical soil organic carbon pools after land-use change from forest to sugarcane. *Journal of Soils and Sediments* 9(2), 112.

Dominy, C.S., Haynes, R.J. and Van Antwerpen, R. (2001) Long-term effects of sugarcane production on soil quality in the South coast and the Midlands area of Kwazulu-Natal. *Proceedings of the South African Sugar Technologists' Association* 75, 222–227.

Garside, A.L., Bramley, R.G.V., Bristow, K.L., Holt, J.A., Magarey, R.C. and Nable, R.O. (1997) Comparisons between paired old and new land sites for sugarcane growth and yield and soil chemical, physical, and biological properties. *Proceedings of the Australian Society of Sugar Cane Technologists* 19, 60–66.

Garside, A.L., Bell, M.J. and Magarey, R.C. (2001) Monoculture yield decline – Fact not fiction. *Proceedings of the International Society of Sugar Cane Technologists* 24, 16–21.

Graham, M.H., Haynes, R.J. and Meyer, J.H. (1999) Green cane harvesting promotes accumulation of soil organic matter and an improvement in soil health. *Proceedings of the South African Sugar Technologists' Association* 73, 53–57.

Haines, W.B. (1930) Studies in the physical properties of soils. V. The hysteresis effect in capillary properties, and the modes of moisture distribution associated therewith. *Journal of Agricultural Science* 20, 97–116.

Hartemink, A.E. (1998a) Soil chemical and physical properties as indicators of sustainable land management under sugar cane in Papua New Guinea. *Geoderma* 85, 283–306.

Hartemink, A.E. (1998b) Acidification and pH buffering capacity of alluvial soils under sugarcane. *Experimental Agriculture* 34, 231–243.

Haynes, R.J. and Hamilton, C.S. (1999) Effects of sugarcane production on soil quality: A synthesis of world literature. *Proceedings of the South African Sugar Technologists' Association* 73, 46–52.

Henry, P.C. and Ellis, R.D. (1996) Soil as a factor in sugarcane ratoon yield decline on an irrigated estate in Swaziland. *Proceedings of the International Society of Sugar Cane Technologists* 22, 84–91.

Holt, J.A. and Mayer, R.J. (1998) Changes in microbial biomass and protease activities of soil associated with long-term sugar cane monoculture. *Biology and Fertility of Soils* 27, 127–131.

La Scala Júnior, N., De Figueiredo, E.B. and Panosso, A.R. (2012) A review on soil carbon accumulation due to the management change of major Brazilian agricultural activities. *Brazilian Journal of Biology* 72(3) (suppl.), 775–785.

McGarry, D., Bigwood, R.C., Pillai-McGarry, U.P., Bray, S.G. and Moody, P.W. (1996) Is cane damaging soil? A comparison of a ten-year-old block with the adjoining treeline. *Proceedings of the Australian Society of Sugar Cane Technologists* 18, 189–194.

McLean, N.R. (1975) Long-term effects of sugar cane production on some physical and chemical properties of soils in the Goondi mill area. *Proceedings of the Queensland Society of Sugar Cane Technologists, 42nd Conference*, 123–126.

Masilaca, A.S., Prasad, R.A. and Morrison, R.J. (1986) The impact of sugarcane cultivation on three Oxisols from Vanua Levu, Fiji. *Tropical Agriculture (Trinidad)* 63, 325–330.

Meyer, J.H. and Van Antwerpen, R. (2001) Soil degradation as a factor in yield decline in the South African sugar industry. *Proceedings of the International Society of Sugar Cane Technologists* 24, 8–15.

Meyer, J.H., Harding, R., Rampersad, A.L. and Wood, R.A. (1998) Monitoring long term soil fertility trends in the South African sugar industry using the FAS analytical database. *Proceedings of the South African Sugar Technologists' Association* 72, 61–68.

MSIRI (2005) Mauritius Sugar Industry Research Institute annual report 2004. MSIRI, Réduit, Mauritius.

MSIRI (2011) Mauritius Sugar Industry Research Institute annual report 2010. MSIRI, Réduit, Mauritius.

Ng Cheong, L.R., Rivière, V., Jacquin, E. and Soopramanien, C. (1999) Soil compaction due to mechanized harvesting and loading. *Proceedings of the International Society of Sugar Cane Technologists* 23, 43–51.

Ng Cheong, L.R. (2007) Effects of sugar cane production practices on soil quality in Mauritius. PhD thesis, University of Free State, Bloemfontein, South Africa.

Ng Cheong, L.R., Ng Kee Kwong, K.F., Ah Koon, P.D. and Du Preez, C.C. (2009) Changes in an Inceptisol of Mauritius after rock removal for sugar cane production. *Soil and Tillage Research* 104, 88–96.

Ng Cheong, L.R., Ng Kee Kwong, K.F. and Du Preez, C.C. (2011) Soil aggregate properties and total available water as affected by sugar cane (*Saccharum hybrid* sp.) production in Mauritius. *South African Journal of Plant and Soil* 28, 34–42.

North-Coombes, A. (1993) *A History of Sugar Production in Mauritius*. Mauritius Printing Specialists (Pte.) Ltd, Port-Louis, Mauritius, 173 pp.

Padya, B.M. (1984) *The Climate of Mauritius* (2nd edition). Meteorological Office, Vacoas, Mauritius.

Pankhurst, C.E., Magarey, R.C., Stirling, G.R., Blair, B.L., Bell, M.J. and Garside, A.L. (2003) Management practices to improve soil health and reduce the effects of detrimental soil biota associated with yield decline in sugarcane in Queensland, Australia. *Soil and Tillage Research* 72, 125–137.

Parish, D.H. and Feillafé, S.M. (1965) Notes on the 1:100,000 soil map of Mauritius. MSIRI Occasional Paper No. 22, Réduit, Mauritius.

Paturau, J.M. (1989) *By-products of the Cane Sugar Industry: An Introduction to their Industrial Utilization*. Sugar Series 11 (3rd edition). Elsevier, Amsterdam.

Peech, M. (1945) Exchangeable cations and exchange capacity. *Soil Science* 59, 25–38.

Perroux, K.M. and White, I. (1988) Designs for disc permeameters. *Soil Science Society of America Journal* 52, 1205–1215.

Peterson, B.J. and Fry, B. (1987) Stable isotopes in ecosystem studies. *Annual Review of Ecology and Systematics* 18, 293–320.

Pinheiro, E.F.M., Lima, E., Ceddia, M.B., Urquiaga, S., Alves, B.J.R. and Boddey, R.M. (2010) Impact of pre-harvest burning versus trash conservation on soil carbon and nitrogen stocks on a sugarcane plantation in the Brazilian Atlantic forest region. *Plant and Soil* 333, 71–80.

Qongqo, L.I. and Van Antwerpen, R. (2000) Effect of long-term sugarcane production on physical and chemical properties of soils in Kwazulu-Natal. *Proceedings of the South African Sugar Technologists' Association* 74, 114–121.

Rice, C.W., Moorman, T.B. and Beare, M. (1996) Role of microbial biomass carbon and nitrogen in soil quality. In: Doran, J.W. and Jones, A.J. (eds) *Methods for Assessing Soil Quality*. SSSA Special Publication No. 49, American Society of Agronomy, Madison, Wisconsin, pp. 203–215.

Skjemstad, J.O., Taylor, J.A., Janik, L.J. and Marvanek, S.P. (1999) Soil organic carbon dynamics under long-term sugarcane monoculture. *Australian Journal of Soil Research* 37, 151–164.

Soil Survey Staff (1999) *Keys to Soil Taxonomy* (8th edition). Pocahontas Press, Inc., Blacksburg, Virginia.

STASM (2003) *Manual of Sugar Cane Agronomy*. Société de Technologie Agricole et Sucrière de Maurice, Réduit, Mauritius.

Suman, A., Singh, K.P., Singh, P. and Yadav, R.L. (2009) Carbon input, loss and storage in sub-tropical Indian Inceptisol under multi-ratooning sugarcane. *Soil and Tillage Research* 24, 221–226.

Umrit, G., Ng Cheong, R., Gillabel, J., and Merckx, R. (2014) Effect of conventional versus mechanized sugarcane cropping on soil organic carbon stocks and labile carbon pools in Mauritius as revealed by ^{13}C natural abundance. *Plant and Soil* 379, 177–192.

Van Antwerpen, R. and Meyer, J.H. (1996) Soil degradation under sugarcane cultivation in Northern Kwazulu-Natal. *Proceedings of the South African Sugar Technologists' Association* 70, 29–33.

Voroney, R.P., Winter, J.P. and Beyaert, R.P. (1993) Soil microbial biomass C and N. In: Carter, M.R. (ed.) *Soil Sampling and Methods of Analysis*. Canadian Society of Soil Science, Lewis Publishers, Boca Raton, Florida, pp. 277–286.

Wong You Cheong, Y. and Chan, P.Y. (1977) Preliminary indications on the effect of continuous cane cultivation on the chemical and physical properties of Mauritius soils. *Revue Agricole et Sucrière de l'Ile Maurice* 56, 70–77.

Wood, A.W. (1985) Soil degradation and management under intensive sugarcane cultivation in North Queensland. *Soil Use and Management* 1, 120–124.

Wood, R.A. (1965) Mineralisation studies on virgin and cultivated sugar belt soils. *Proceedings of the South African Sugar Technologists' Association* 39, 195–202.

13 Patterns and Drivers of Soil Carbon Stocks and Isotopic Composition in Secondary Tropical Dry Forests of Costa Rica

Jennifer S. Powers,[1,2]* David W.P. Manning[3] and Justin M. Becknell[1]

[1]*Department of Ecology, Evolution, and Behavior, University of Minnesota, St. Paul, USA;* [2]*Department of Plant Biology, University of Minnesota, St. Paul, USA;* [3]*Odum School of Ecology, University of Georgia, Athens, USA*

13.1 Introduction

Soils contain a large pool of organic carbon (C) that may increase or decrease in response to changes in land use and management (Blair and McLean, 1917; Greenland and Nye, 1959; Schlesinger, 1977; Powers *et al.*, 2011). Natural reforestation or forest regeneration on lands that were previously used for agriculture or grazing is occurring to different degrees across the tropics (Chazdon, 2008) and understanding how this affects C stored in vegetation and soils is an important question with relevance for the global C cycle. Simple conceptual models typically assume that ecosystem C stocks including biomass and soil C (throughout the text we refer to soil organic carbon as soil C) are depleted during forest-to-pasture conversion, but gradually increase as agricultural lands are abandoned and secondary forests regenerate (Fig. 13.1) (Detwiler and Hall, 1988; Melillo *et al.*, 1996). However, a recent meta-analysis of the effects of tropical land-use changes such as reforestation on soil C stocks revealed diverse responses to land-use change, which depended upon soil mineralogy and precipitation regime (Powers *et al.*, 2011).

It has been known for decades that soil-forming factors, i.e. climate, organisms, topography, parent material and weathering state or time, affect the total amount of soil C present at a site (Jenny, 1941; Jenny *et al.*, 1948; Torn *et al.*, 1997; Amundson, 2001), thus it is not surprising that these same factors might help determine response of the soil C pool to land-use change (Post and Kwon, 2000; Laganière *et al.*, 2010). There are a number of mechanisms through which soil-forming factors might affect soil C pools during forest restoration and regeneration. Changes in vegetation productivity with succession might affect soil C stocks through inputs of organic matter to the soil. Alternatively, factors that are inherited from parent materials or reflect soil weathering processes such as the quantity and mineralogical composition of soil clays can affect soil C stabilization through mechanisms such as the formation of metal–humus complexes, sorption to clay minerals or polyvalent cation bridges (Lopez-Ulloa *et al.*, 2005; Lorenz *et al.*, 2009).

In this study we used a space-for-time substitution, also known as a chronosequence, to understand how forest regeneration on previously

* Corresponding author: powers@umn.edu

 Land-use Change Impacts on Soil Processes: Tropical and Savannah Ecosystems (eds F.Q. Brearley and A.D. Thomas)

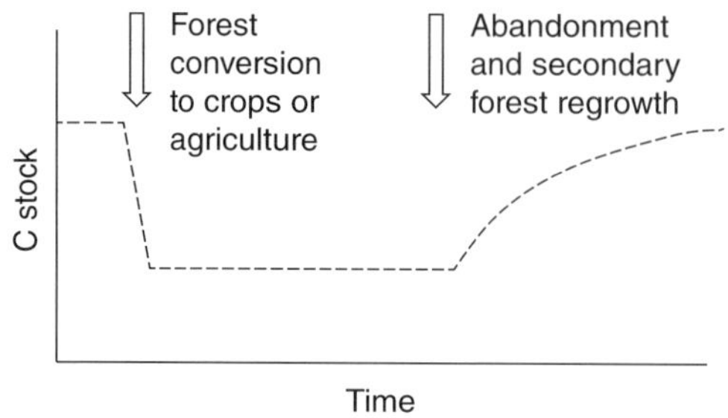

Fig. 13.1. Conceptual model of changes in biomass stocks as a function of deforestation and secondary succession. (Redrawn from Melillo *et al.*, 1996.)

grazed or cropped lands affected surface soil C in patches of seasonally dry tropical forest in north-west Costa Rica. Seasonally dry tropical forests are defined as those receiving up to 2000 mm of annual rainfall and experiencing a dry season of more than 4 months (Becknell *et al.*, 2012), and are understudied compared to wet tropical forests (Meli, 2003). The amount of dry forest in Costa Rica was severely reduced during the previous several centuries by conversion to pastures and crops (Sader and Joyce, 1988), but secondary forests are now regenerating in the Costa Rican system of protected Conservation Areas (Powers *et al.*, 2009) and on private lands (Calvo-Alvarado *et al.*, 2009). Our approach was to sample a large number of forest plots distributed across gradients of stand age and edaphic properties, and use multiple regression and correlation analysis to identify the variables that best predicted the size and isotopic composition of soil C stocks.

We focused on three response variables: (i) soil C concentrations (i.e. % C); (ii) soil C stocks, which are the product of soil C concentrations and bulk density both of which may be affected by land-use change independently; and (iii) $\delta^{13}C$ (i.e. the ratio of the two stable C compounds ^{13}C to ^{12}C), which is useful for tracking the contributions of C^4 grass-derived soil C relative to C^3 tree-derived soil C in pasture-to-forest transitions (de Koning *et al.*, 2003). The response variables we measured reflect different soil- or ecosystem-forming factors, and included forest characteristics such as stand age, species richness and aboveground biomass, and edaphic factors such as pH, clay concentrations and total elemental composition. Aboveground primary production in these forests increases with stand age (Becknell, 2012); thus, if the quantity or diversity of inputs of organic C to the soil from aboveground vegetation or roots is an important control on total soil C stocks and isotopic composition, we would expect to see strong correlations between soil C and co-varying measures of the forest community such as stand age, aboveground biomass or species richness (Maestre *et al.*, 2012). However, previous work in three forest plots in this region showed strong correlations between poorly crystallized iron (Fe) oxides and the size of stable soil C stocks (Lorenz *et al.*, 2009), suggesting that soil mineralogical composition controls soil C stocks through organic matter stabilization. Our sampling design allowed us to test the relative importance of the effect of these different factors on the total soil C stock size and its dynamics, as inferred by $\delta^{13}C$.

13.2 Methods

13.2.1 Study site

This study was conducted in two conservation areas in north-western Costa Rica, Área de Conservación Guanacaste (10° 50′ N, 85° 37′ W) and Área de Conservación Arenal-Tempisque (10° 20′ N, 85° 20′ W). Within the two conservation areas, the work was focused on the Sector Santa Rosa (formerly known as Parque Nacional Santa Rosa) in the Área de Conservación Guanacaste and Parque Nacional Palo Verde in the Área de Conservación Arenal-Tempisque (Fig. 13.2). Annual rainfall can range from 915 mm to over 3500 mm, but typically annual values are from 1500–1700 mm (Gillespie *et al.*, 2000). Mean annual temperatures at both sites are around 25°C (Gillespie *et al.*, 2000). Both parks were established in the early 1970s and include mosaics of land in various stages of abandonment from grazing, interspersed with patches of mature forests, which may have been selectively harvested or disturbed (Allen, 1988; Jiménez *et al.*, 2001). Soils at Santa Rosa have diverse parent materials, including ignimbritic pumice deposits from volcanic eruptions, which vary in composition (Leiva *et al.*, 2009), while soils derived from basalt are present at lower elevations (Hartshorn, 1983). Soils at Palo Verde include limestone-derived soils on rocky outcrops and lowland soils with high concentrations of montmorillonite

Fig. 13.2. Secondary tropical dry forests in Área de Conservación Guanacaste (a) and Arenal-Tempisque (b), Costa Rica.

clays (Hartshorn, 1983). Although detailed soil mapping and classification have not been carried out in either conservation area, the soils at Santa Rosa include Typic Ustropepts and Lithic Ustropepts (Inceptisols), while common soils at Palo Verde include Typic Pellusterts/Typic Pelluderts (Vertisols) and Typic Ustropepts (Powers *et al.*, 2009). The secondary forests we sampled are regenerating naturally following centuries of grazing and cropping (Jiménez *et al.*, 2001). Among the 84 plots we sampled, we have identified over 140 species of trees (Becknell and Powers, unpublished). We used a combination of satellite image analysis and expert opinion to assign estimates of forest age to each plot (Powers *et al.*, 2009). In general, these two sources of data were consistent and allowed us to bracket forest ages to within 5 years. Plots of mature or old growth forest were difficult to date and were assigned an arbitrary age of 100 years.

13.2.2 Plot locations and assumptions about prior land-use history

Robust statistical designs rely on random sampling schemes to make inferences about population characteristics, and also to control for factors such as the nature, intensity and duration of prior land use. Unfortunately, human use of forests (including land clearing, management and subsequent abandonment) is not random, but rather controlled by complex factors including accessibility, distance to markets, social norms and economic incentives, as well as biophysical factors (Lambin *et al.*, 2003). Our ability to locate forest patches that differed in forest age, but were otherwise similar in terms of historical land use and edaphic characteristics was limited by two factors. First, because initial conversion of unmanaged forest to small-scale cultivation began in the pre-Columbian era and continued into the Colonial period of extensive grazing (Jiménez *et al.*, 2001), it is difficult, if not impossible, to reconstruct accurate land-use histories for our sites that extend beyond 60 years or so. We assume that most, if not all, our plots were primarily grazed by cattle, which was the most common land use since colonization. Second, it was not feasible to lay out plots at random. Rather, we used a haphazard sampling approach, stratifying plots by forest cover type and aiming to obtain approximately even coverage of plots across the gradients of estimated stand age. We acknowledge this limitation of our study and discuss below how the duration and intensity of grazing prior to secondary forest regeneration is a variable that we cannot control for, but that might affect our results.

13.2.3 Soil and vegetation sampling in forests

Here we analyse soil and vegetation data from 60 plots that have previously been described (Powers *et al.*, 2009) in addition to 24 new plots. Most of the plots were on flat regions in the landscape, so we did not consider the influence of topography on soil C here. In the prior study we summarized soil C concentrations and $\delta^{13}C$, but did not analyse these data statistically as the focus of that paper was forest composition. Full methods describing plot measurements appear in Powers *et al.* (2009), and are presented here in brief. All 84 plots were 20 m × 50 m. All trees ≥10 cm diameter at breast height were identified and measured. Aboveground biomass was calculated using allometric equations that incorporated species-specific wood density values (Becknell and Powers, 2014). There are two main ways to sample soils: by fixed depth or by genetic horizons. Because we anticipated large differences among plots in soil properties and perhaps also horizons, we elected to collect two sets of soil samples using a fixed depth from the top 0–10 cm layer of mineral soil. We focused on surface soils, because they are dynamic and more likely to respond to land-use change then deeper layers. The first set of samples included three volumetric cores per plot to estimate bulk density and stone content using a rectangular turf grass sampler (2.5 cm × 9.0 cm dimensions). The second set of samples was collected by compositing 10 cores extracted with a 2.5-cm diameter punch tube corer per plot and used for chemical analysis.

13.2.4 Grasslands

While we do not have paired pasture sites that represent the pre-abandonment conditions of each secondary forest, we collected soil and vegetation samples from three grasslands that were converted from tropical dry forest decades to centuries ago, in order to have C_4-vegetation end members for stable isotope calculations. These sites are located in the Area de Conservación Guanacaste and are maintained as grasslands through annual burning and/or grazing by horses. Typical species include *Hyparrhenia rufa* and *Paspalum* species (both Poaceae). Soil samples from these sites were collected and processed using identical methods as for the forest plots. We did not quantify aboveground biomass as these sites were treeless. Because these are not currently active cattle pastures, we refer to them as grasslands.

13.2.5 Plant material for isotope analysis

In order to use simple mixing models to calculate the percentage of C^4 versus C^3-derived organic matter to the soil C pool, we measured $\delta^{13}C$ in plant tissue. We used leaf litter from a mixture of tree species collected in litter traps during the dry and wet season of 2008 to represent the isotopic composition of C^3 tree vegetation. We also collected grab samples of aboveground tissue of common C^4 grasses and C^3 sedges that were growing in the grassland sites for isotopic analysis.

13.2.6 Laboratory analyses

The laboratory methods used to characterize soil chemical properties appear elsewhere, and are presented briefly here (Powers *et al.*, 2009; Powers and Perez-Aviles, 2013). Soils were air-dried, picked free of roots and other debris, then passed through a 2-mm sieve. We measured pH in water using a 1:2.5 air-dried soil to solution ratio and particle-size distribution was determined using the hydrometer method. Unfortunately, we do not have data from selective dissolution analyses (e.g. quantities of Fe and aluminum (Al) extracted by acid ammonium oxalate, sodium pyrophosphate, etc.) for all our samples, which are useful for inferring soil mineralogy (Lorenz *et al.*, 2009; Jimenez *et al.*, 2011). Instead, we used data on total Al concentrations, which we had for all plots. Total elements including Al were measured via inductively coupled plasma emission spectroscopy on a Perkin Elmer Optima 3000 ICP Spectrometer following digestion in nitric acid in a microwave. Total Al is positively correlated with a measure of the poorly crystalline Fe oxide ferrihydrite (acid ammonium oxalate-extractable Fe minus pyrophosphate-extractable Fe = 0.40, $P < 0.01$) in a subset of our samples for which we have selective dissolution analysis data, while total Fe was not (Powers *et al.*, unpublished data). Thus, we used total Al as a proxy for the abundance of poorly crystalline minerals. Last, total soil C and its stable isotopes ($\delta^{13}C$) were measured on finely ground samples at the Stable Isotope Facility at the University of California at Davis, USA. Carbon concentrations are reported on an oven-dry weight basis and $\delta^{13}C$ is reported using standard notation in units of parts (‰), relative to the Pee Dee belemnite standard.

13.2.7 Soil carbon inventories

Soil C stocks are calculated by multiplying soil C concentrations by bulk density and sample depth; however, both soil C concentrations and bulk density may be altered by land use (Ellert and Bettany, 1995). We used two ways to calculate soil C stocks, which we refer to as the 'common depth' and 'common mass' methods. For the 'common depth' method, we multiplied each site-specific C concentration by the measured bulk density at each site, and the depth increment of 10 cm. The 'common mass' approach adjusts for possible changes in the total amount of soil sampled to a fixed depth that may be caused by differences among land uses due to compaction or expansion (Toriyama *et al.*, 2011). An implicit assumption of this approach is the untestable assumption that bulk densities were identical prior to land-use change. Thus, in chronosequence studies, comparing soils sampled under different land uses or ages on a common mass basis involves expressing soil C inventories by using a common bulk density value for a reference site that represents the 'original' land use (de Koning *et al.*, 2003). We calculated the mean bulk density of our three oldest forest sites (0.82 g cm^{-3}) and used this as a common value for expressing soil C inventories in all plots on a common mass basis.

13.2.8 Calculations of pasture- versus forest-derived C

We used a simple mixing model to estimate the contribution of C_4 (pasture) and C_3 (secondary forest) derived vegetation to the soil C in secondary forests (Balesdent and Mariotti, 1996; Lopez-Ulloa *et al.*, 2005), assuming that the dominant plant species in pastures are C^4 grasses and the dominant plant species in forests are C^3 trees. The equation is:

$$Fp = (\delta^{13}C\ soil_{sf} - \delta^{13}C\ soil_{of})/(\delta^{13}C\ detritus_{p} - \delta^{13}C\ detritus_{f}) \quad (13.1)$$

Where Fp refers to the proportion of pasture-derived soil C in the total soil C pool, $\delta^{13}C$ soil$_{sf}$ is the isotopic composition of soil C in each secondary forest plot, $\delta^{13}C$ soil$_{of}$ is the isotopic composition of soil in a reference or 'pure' C_3 or old forest soil, $\delta^{13}C$ detritus$_p$ is the isotopic signature of C^4 pasture grasses and $\delta^{13}C$ detritus$_f$ is the isotopic signature of C^3 forest vegetation. In our calculations, we set $\delta^{13}C$ soil$_{of}$ to −27.63‰, which was the lowest value for soils in our sample of secondary forests, and we assumed that this represented the end-member C^3 forest soil. The isotopic composition of pasture grasses was set to −11.24‰, which was the mean value that we measured in C^4 grass tissue, and the isotopic value of C_3 vegetation was set to −29.50‰, which is the mean value for wet and dry season litter measured in our mixed species litter traps (Table 13.1). We also ran this calculation for our three grassland samples, to check whether soil C appeared to be exclusively from C^4 vegetation, or whether it contained a mixture of C^3 and C^4 sources. Because the only term that varies in this equation for our dataset is the isotopic composition of soil C in the secondary forest or grassland sites and all other terms are fixed, this equation essentially re-scales the $\delta^{13}C$ soil values.

13.2.9 Statistical analyses

We used Spearman rank correlation coefficients as a preliminary approach to explore (non-linear) relationships among the response variables soil C, soil C stocks (Mg ha^{-1}), and $\delta^{13}C$ and the following variables: stand age, tree species richness, total aboveground biomass, soil pH, % clay and total Al. These analyses were performed for the forest plots only and excluded the three grassland sites, because we are not certain how well these three sites represent the pre-abandonment conditions of the forest sites. Next, we performed three multiple regression analyses using soil C, soil C stocks (common depth) or $\delta^{13}C$ as response variables, and stand age, species richness, aboveground tree biomass, soil pH, % clay and total Al as predictor variables. We ran all possible models and used the Akaike information criterion (AIC) to determine the model that best fit the data and minimized the AIC. We did not do these analyses on the soil C stocks calculated using common mass, as these would be identical to results for soil C. All analyses were conducted in JMP 8.0 (SAS Institute, Inc.).

13.3 Results

13.3.1 Patterns with forest age

In general, bulk density tended to decrease with increasing forest age (Fig. 13.3a), while soil C concentrations tended to increase (Fig. 13.3b) especially relative to the three grassland sites, although there was considerable variability around these mean trends. These two trends counter-balanced each other, so that soil C stocks calculated on a common depth basis using site-specific bulk density data showed little variation over succession (Fig. 13.3c). Calculating soil C stocks under the common mass assumption yields identical patterns of variation over stand age compared to soil C (Fig. 13.3d), because using a common bulk density value simply rescales the site-specific concentration data. Across the landscape, soil C stocks (0–10 cm depth) ranged from 13.1 Mg C ha^{-1} to 47.6 Mg C ha^{-1} when calculated on a common depth basis, and from 10.3 Mg C ha^{-1} to 60.5 Mg C ha^{-1} when calculated based upon a common mass with mean bulk density values from older forests (Fig. 13.3c and d). Mean soil C stocks were similar irrespective of the assumed bulk density value and were 30.3 Mg C ha^{-1} for the common depth samples and 31.0 Mg C ha^{-1} for the common mass samples.

Stable C isotope values in the grassland soils ranged from −22.25‰ to −13.37‰ (Table 13.1,

Table 13.1. Stable carbon isotope values for pasture vegetation, forest leaf litter and soil carbon (0–10 cm). Soil samples were collected from plots of regenerating tropical dry forest or three grassland sites in Costa Rica, while vegetation data are from grab samples (Poaceae) or leaf litter traps in forested plots.

Variable	Mean (‰)	Minimum (‰)	Maximum (‰)	Sample size
Vegetation				
C4 pasture plants (Poaceae)	−11.24	−12.09	−11.23	4
Forest leaf litter wet season	−29.34	−30.04	−28.31	9
Forest leaf litter dry season	−29.66	−30.02	−29.27	6
Soil				
Grasslands and pastures	−17.28	−22.25	−13.37	3
Secondary forests	−24.15	−27.63	−16.70	84

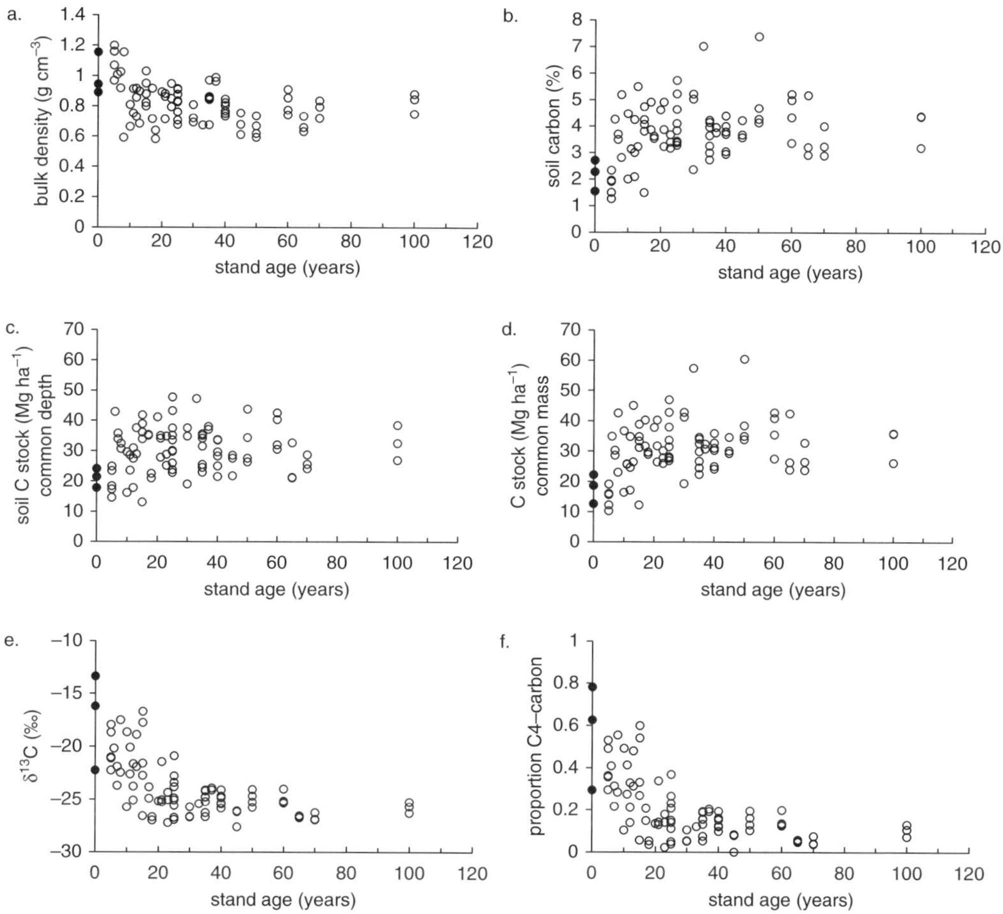

Fig. 13.3. Soil characteristics (0–10 cm sampling depth) plotted as a function of estimated forest age from 84 plots of tropical dry forest regenerating following abandonment from agriculture in Costa Rica. Response variables are as follows: (a) bulk density, (b) soil C concentrations, (c) soil C stock calculated using a common sampling depth, (d) soil C stock calculated using a common mass approach, (e) δ^{13}C and (f) proportional contribution from C^4 vegetation to total soil C. Filled symbols represent grassland sites and open symbols represent forested sites.

Fig. 13.3e), indicating that even these old grassland sites are not pure C^4 plant communities or perhaps have contributions of old, resistant soil C that persisted from the original forest vegetation. Consistent with the former explanation, the site with the lowest δ^{13}C had abundant sedges (Daniel Perez-Aviles, personal communication). Bulk δ^{13}C of surface soil C sampled in these secondary forests suggest rapid incorporation of C^3-derived material into the soil C pools (Fig. 13.3e). Within 30 years, surface soils of regenerating tropical forest had isotopic signatures similar to those of forests ≥100 years old.

As discussed above, calculating the fraction of C^4- to C^3-derived C essentially rescales the measured δ^{13}C values in each plot; however, several interesting results emerge from these data. First, soil C in both grasslands and forests consists of a mix of C^3- and C^4-derived sources. Even the grassland site that had the highest C^4 signature still consisted of almost 20% of C^3-derived C. Second, the higher values of C^4-carbon in some secondary forests compared to the three reference grassland sites suggest that when they were abandoned from pasture, they started at higher values than the current grasslands. Third, within forests of similar ages, there was a wide range of C^4-derived C. Fourth, even secondary forests of 40 years or older still show evidence of C^4-derived C.

13.3.2 Correlates of soil C and $\delta^{13}C$

The strongest correlations between predictor variables and soil C concentrations and stocks were the positive correlations with soil pH, with Spearman's correlation coefficients of 0.61 and 0.46, respectively (Table 13.2, $P < 0.0001$). Total Al was also positively correlated with soil C concentrations and % clay was positively correlated with soil C stocks (Table 13.2). Soil $\delta^{13}C$ was negatively correlated with stand age, species richness and aboveground biomass, and positively correlated with % clay (Table 13.2).

The best fit multiple regression model for soil C included tree species richness, soil pH and total Al, and had an r^2 value of 0.51 (Table 13.3). The best fit model for soil C stocks (common depth) had a much lower r^2 value and included stand age in addition to the same variables that explained variation among plots in soil C (Table 13.3). Despite the strong Spearman rank correlation between stand age and $\delta^{13}C$, the best fit regression model for $\delta^{13}C$ contained aboveground biomass, species richness, soil pH and % clay, and had an r^2 value of 0.52 (Table 13.3).

13.4 Discussion

The extent to which secondary forests in tropical landscapes provide ecosystem services such as biodiversity conservation and C storage is subject to intense debate (Wright and Muller-Landau, 2006; Laurance, 2007), and we currently lack a complete understanding of the patterns and processes of regeneration, particularly in dry tropical forests compared to wet tropical forests (Meli, 2003). The goals of this project were to determine whether simple, conceptual models of ecosystem recovery from landuse fit data from regenerating tropical dry forests and, if not, whether soil-forming factors helped to explain variation among plots in soil C and its stable C isotopic composition. In our sampling of many secondary dry forest stands in Costa Rica, we found positive but weak trends for soil C concentrations to increase with forest age, and strong and rapid decreases in C^4-derived soil C with forest age. Multiple regression models indicated that factors other than stand age such as tree species richness and soil pH explained much of the variation among sites in soil C concentrations, stocks, and isotopic composition. Below we discuss these patterns, potential mechanisms and implications for regional soil C budgets.

13.4.1 Successional patterns

Although both soil C concentrations and stocks tended to increase with secondary succession, stand age is clearly not the primary driver of soil C concentrations and stocks in our data. There are several possible explanations for our finding that a simple successional model was insufficient to account for variation across the landscape in soil C. First, it is possible that our stand age estimates were incorrect, thus obscuring the relationship with soil C. While we cannot rule this out, several lines of evidence suggest that this potential source of error is fairly small. The relationship between the fraction of C^4-derived

Table 13.2. Spearman's correlation coefficients between soil C concentrations, soil C stocks and $\delta^{13}C$ and predictor variables in plots of regenerating tropical dry forests in Costa Rica (*P* values in parentheses).

Variable	Soil % C	Soil C stock (Mg ha^{-1})	$\delta^{13}C$
Soil C stock	0.81 (<0.001)	–	0.20 (n.s.)
$\delta^{13}C$	–0.11 (n.s.)	0.20 (n.s.)	–
Stand age	0.25 (0.02)	0.08 (n.s.)	–0.61 (<0.001)
Aboveground biomass	0.11 (n.s.)	0.001 (n.s.)	–0.59 (<0.001)
Tree species richness	0.25 (<0.05)	0.18 (n.s.)	–0.50 (<0.001)
$pH_{(water)}$	0.61 (<0.001)	0.46 (<0.001)	–0.28 (<0.01)
% Clay	0.25 (<0.05)	0.28 (<0.01)	0.32 (<0.01)
Total Al	0.35 (<0.001)	0.24 (<0.05)	0.22 (<0.05)

Bonferoni-corrected $\alpha = 0.006$.

Table 13.3. Best-fitting regression models for soil % C, soil C stock and $\delta^{13}C$ in regenerating tropical dry forests in Costa Rica determined by the lowest AIC value.

Variable	Soil % C	Soil C stock (common depth)	$\delta^{13}C$
Stand age		*	
Aboveground biomass			*
Tree species richness	*	*	*
$pH_{(water)}$	*	*	*
% clay			*
Total Al	*	*	
r^2	0.51	0.33	0.52
AIC	200.49	555.53	346.39

Predictor variables included stand age, aboveground biomass, tree species richness, pH, % clay and total Al. Each response variable appears in a column and predictor variables included in the best model are denoted with an asterisk.

soil C and estimated stand age is very strong and consistent with the expectation of increasing forest-derived soil C and/or decomposition of pasture-derived soil C. Next, stocks of aboveground tree biomass and coarse woody debris in these forests increase with stand age (Kissing and Powers, 2010), and indeed, stand age explains the largest amount of variation among plots in aboveground C (Becknell and Powers, 2014). Moreover, fine root mass in a subset of the plots in which we studied soil C varies inversely with soil fertility, but not stand age, lending further support to the idea that it is possible for above- and belowground C to be controlled by different factors (Powers and Perez-Aviles, 2013). It is also possible that our sampling strategy was insufficient to characterize within-site variability of each plot or that chronosequence approaches do not account for slight variations in initial conditions among plots . However, repeated sampling of a subset of these plots using identical methods yielded the same patterns in soil C and bulk density (Powers and Nowicki, unpublished data), thus we eliminate this as a major source of variation.

A second factor that might account for the weak relationships between forest age and soil C concentrations or stocks is the possibility that the prior land use of grazing did not substantially deplete soil C stocks relative to primary or secondary forest. The fine root mass of deep-rooted pasture grasses can exceed that of forest or savanna vegetation, and add to belowground C stocks (Fisher *et al.*, 1994). Indeed, the average change in soil C stocks during forest-to-pasture conversion on low-activity clay soils in the drier tropics (1500–2500 mm mean annual rainfall) is a 26.4% increase (Powers *et al.*, 2011).

Third, it is likely that the relationship between soil C and forest age is not direct. Rather, any possible correlations are the result of factors that vary as forests age, e.g. microclimate, changes in the quantity and/or quality of organic matter inputs to the soil or changes in disturbance regimes (e.g. elimination or reduction in the number of fires). The correlation coefficients for soil C stocks (at a common depth basis) illustrate this well. Even though soil C concentrations increased slightly with forest age, bulk density also declined, leading to no significant changes in soil C stocks as a function of forest age. The possibility most supported by our data is that total soil C stocks are determined first by factors that affect the size of the stable soil C pool such as mineralogy (Lorenz *et al.*, 2009; Jimenez *et al.*, 2011), and secondarily by processes that vary with land use.

Last, we cannot rule out the possibility that the nature, duration and/or intensity of prior land use affected C stocks either during agriculture or in the post-agricultural phase of recovery (Buschbacher *et al.*, 1988), and that this accounts for some of the unexplained variation among plots. While we have confidence that most of our sites were cleared for fairly low-intensity grazing and not intensive cultivation such as rice, we do not have estimates for how long each site was subjected to grazing prior to abandonment or whether there were large differences among farms in management. Thus, unfortunately this potential source of variation cannot be addressed with this dataset. It is also possible that the uneven distribution of stand ages across soil types adds additional variation when all plots are viewed together. For example, one possible interpretation of the trend of decreasing soil % C in the oldest stands might be C saturation. However, a more likely explanation is that the oldest forests in our sample always had lower C, and some that of the high C-soils from Palo Verde are more abundant in the 30–40 year age classes but poorly represented in older age classes.

13.4.2 Soil-forming factors and soil C storage

In general, soil C concentrations were better correlated with abiotic variables (pH, % clay and total Al), while $\delta^{13}C$ (as an index of soil C turnover), was better correlated with biological variables that changed as forests aged. The multiple regression analysis revealed more complex models than the pairwise correlations, and each model included variables associated with soil C stabilization and forest characteristics (e.g. species richness), suggesting that both input processes and stabilization mechanisms are important in controlling the overall size and dynamics of the soil C pool.

One strong pattern was the positive correlation between soil pH and C concentrations. pH ranged from 5.04 to 6.74 among sites, which rules out the possibility that carbonates contributed to this pattern (note that we have indeed searched for carbonates in these soils and found them below 20 cm in certain samples, but never in the surface soils). What accounts for the positive correlations between soil C and pH? Not surprisingly, in our dataset soil pH is well correlated with the concentrations of total base cations calcium (Ca), magnesium (Mg), potassium and sodium in the soil, with Pearson correlation coefficients ranging from 0.57 to 0.79. Divalent cations like Ca can stabilize soil organic matter through the formation of bridges among organic molecules and between organic matter and clay particles (Oades, 1988). In a lab incubation study with leaf litter from two different tree species this region (the nitrogen-fixing legume *Gliricidia sepium* and an oak *Quercus oleoides*; Fagaceae), Mg addition in solution retarded decomposition, suggesting that Mg availability may play a role in determining soil C pool sizes (Powers and Salute, 2011). Total Al was also included in the regression models of soil C concentrations and stocks, but % clay was not. Again, total Al concentrations were well correlated with a measure of the poorly crystalline, Fe-containing mineral ferrihydrite. Detailed studies of soil mineralogy in four sites in the Área de Conservación Guanacaste show that ferrihydrite controls the size of the stable and passive soil C pools (Lorenz *et al.*, 2009; Jimenez *et al.*, 2011), and our results are consistent with a large role of soil mineralogy in determining the size of the total soil C pool.

However, our data also indicate a role of the forest community in soil C pool sizes and $\delta^{13}C$. Interestingly, species richness was present in all three of the regression models. Not surprisingly, species richness increases with stand age (Powers *et al.*, 2009), so it is notable that species richness adds explanatory power beyond that of stand age. Species richness is a key determinant of belowground biogeochemical function in arid ecosystems worldwide (Maestre *et al.*, 2012). In regenerating tropical dry forests, increased tree species richness might affect soil C through the quantity, quality and/or diversity of C inputs to the soil. In the forests we sampled, individual species differentially affect total and labile soil C concentrations directly under their crowns (Gei and Powers, 2013), and thus it is also possible that the identities of the species present in each stand are also important.

The actions of mycorrhizas are another possible mechanism through which species composition might affect soil C stocks and cycling, and also potentially explain the correlation between soil C and pH. Many of the trees on the poorest soils in this landscape are ectomycorrhizal (e.g. *Quercus oleoides*). A recent conceptual model posited that biogeochemical cycles in stands dominated by ectomycorrhizal trees are characterized by more organic nutrient cycling, whereas forests dominated by arbuscular mycorrhizal trees are dominated by inorganic nutrient cycling processes (Phillips *et al.*, 2013). Consistent with the hypothesis that different mycorrhizal communities yield distinct signatures on soil C is the finding that soils dominated by ectomycorrhizal plant communities store more C per unit N than plant communities that support arbuscular mycorrhizas, i.e. have greater C:N ratios (Averill *et al.*, 2014).

13.4.3 Insights from stable isotopes

The isotopic data suggest that surface soil C pools are dynamic, but with a number of caveats. The isotopic signatures in the grassland sites were not 'pure' C^4-communities, either because of the presence of C_3 species like sedges or remnant trees or residual forest-derived C from the original, primary forest (note that we have never observed charcoal or charred particles in any soil samples from this region). Our grassland site with the lowest $\delta^{13}C$ had sedges, and is likely not representative of the pre-abandonment conditions of many of our forest sites. Even so, the proportion of C^3-carbon in the grassland site with the highest $\delta^{13}C$ was 0.2. This suggests caution should be exercised in using the natural abundance isotope method as a precise tracer of soil C dynamics (Powers and Veldkamp, 2005).

Despite these caveats, our data suggest that within 30 years of regeneration the isotopic signature of surface soil C resembles that of older forests. This is similar to results from a survey of forest-to-pasture conversions in a wetter climate in north-eastern Costa Rica, where surface soil $\delta^{13}C$ values in the 0–10 cm layer of soil stabilize approximately 30 years after deforestation (Powers and Veldkamp, 2005). How can we reconcile the lack of directional changes in soil C stocks with the apparently rapid turnover of soil C? Our data suggest that the size of the total soil C pool in the topsoil is determined largely by mineralogy and secondarily by factors such as species richness, but that this carbon pool is relatively dynamic, almost completely turning over within 30 years. We also acknowledge that the r^2 values of the multiple regression equations were lowest for soil C stocks on a common depth basis and were around 0.5 for both soil C concentrations and $\delta^{13}C$, suggesting that additional factors we did not measure are also important, e.g. belowground soil C inputs from roots and microorganisms and/or duration and intensity of prior land use.

13.4.4 Methodological considerations

There is much discussion in the literature about the relative merits of sampling by fixed depth versus genetic horizon (Davidson and Ackerman, 1993) and expressing inventories on a common depth (or volume) versus common mass basis (Ellert and Bettany, 1995; Toriyama *et al.*, 2011). Although sampling by genetic horizon allows for detecting erosion, it can be difficult to accomplish in highly weathered soils that lack clearly delineated horizons and it also hinders comparisons of land-use effects among sites. Longitudinal studies where soils are repeatedly sampled over time can help resolve some of these issues and reduce errors in C accounting schemes (Toriyama *et al.*, 2011).

However, when comparing soil C stocks estimated using the common mass approach in chronosequences, adjusting samples to a common mass using a reference bulk density as we did removes any variation in soil C stocks attributed to bulk density and emphasizes the differences among sites in C concentrations. In our dataset, soil C concentrations varied more among sites than did bulk density (the coefficients of variation were 28.6% and 16.5%, respectively), and thus the choice of the common depth or common mass approach to estimate soil C stocks is not as important as understanding how soil C concentrations vary among sites.

13.5 Conclusions

A major conclusion from many recent studies of soil C responses to forest regeneration is that regional, mechanistic studies are essential for accurately understanding soil C dynamics and their consequences for CO_2 exchange with the atmosphere at coarser spatial scales (de Koning *et al.*, 2003; Lopez-Ulloa *et al.*, 2005; Powers *et al.*, 2011). We purposefully located our secondary forest plots across gradients of stand age and soil-forming factors. Our data from tropical dry forests in Costa Rica are largely consistent with results from wetter forests that show simple models of recovery of belowground C stocks with recovery from land use are insufficient to explain the patterns of soil C distribution across the landscape. Our isotopic data suggest that surface soil C pools are dynamic, but that concentrations and stocks are controlled by a number of site-specific soil-forming factors such as mineralogy, pH and species richness, in addition to forest age.

Acknowledgements

This study was supported by a McKnight Land Grant Professorship from the University of Minnesota, a NASA New Investigator Award (NS000107) and National Science Foundation CAREER Award (DEB-1053237) to J.S.P. We thank two anonymous reviewers for comments that improved this chapter.

References

Allen, W.H. (1988) Biocultural restoration of a tropical forest. *BioScience* 38, 151–161.

Amundson, R. (2001) The carbon budget of soils. *Annual Review Earth and Planetary Science* 29, 535–62.

Averill, C., Turner, B.L. and Finzi, A.C. (2014) Mycorrhiza-mediated competition between plants and decomposers drives soil carbon storage. *Nature* 505, 543–545.

Balesdent, J. and Mariotti, A. (1996) Measurement of soil organic matter turnover using 13C natural abundance. In: Boutton, T. W. and Yamasaki, S. (eds) *Mass Spectrometry of Soils*. Marcel Dekker, Inc., New York, pp. 83–111.

Becknell, J.M. (2012) Carbon cycling in secondary tropical dry forest from stands to global scales. PhD thesis, University of Minnesota, Saint Paul.

Becknell, J.M., and Powers, J.S. (2014) Stand age and soils as drivers of plant functional traits and aboveground biomass in secondary tropical dry forest. *Canadian Journal Forest Research* 44, 603–614.

Becknell, J.M., Kissing, L.B. and Powers, J.S. (2012) Aboveground biomass in mature and secondary seasonally dry tropical forests: A literature review and global synthesis. *Forest Ecology and Management* 276, 88–95.

Blair, A.W. and McLean, H.C. (1917) Total nitrogen and carbon in cultivated land and land abandoned to grass and weeds. *Soil Science* 4, 283–294.

Buschbacher, R., Uhl, C. and Serrao, E.A.S. (1988) Abandoned pastures in Eastern Amazonia. II. Nutrient stocks in the soil and vegetation. *Journal of Ecology* 76, 682–699.

Calvo-Alvarado, J., Mclennan, B., Sanchez-Azofeifa, A. and Garvin, T. (2009) Deforestation and forest restoration in Guanacaste, Costa Rica: Putting conservation policies in context. *Forest Ecology and Management* 258, 931–940.

Chazdon, R.L. (2008) Beyond deforestation: Restoring forests and ecosystem services on degraded lands. *Science* 320, 1458–1460.

Davidson, E.A. and Ackerman, I.L. (1993) Changes in soil carbon inventories following cultivation of previously untilled soils. *Biogeochemistry* 20, 161–193.

de Koning, G.H.J., Veldkamp, E. and Lopez-Ulloa, M. (2003) Quantification of carbon sequestration in soils following pasture to forest conversion in northwestern Ecuador. *Global Biogeochemical Cycles* 17, 9(1)–9(12).

Detwiler, R.P. and Hall, C.A.S. (1988) Tropical forests and the global carbon cycle. *Science* 239, 42–47.

Ellert, B.H. and Bettany, J.R. (1995) Calculation of organic matter and nutrients stored in soils under contrasting management regimes. *Canadian Journal of Soil Science* 75, 529–538.

Fisher, M.J., Rao, I.M., Ayarza, M.A., Lascano, C.E., Sanz, J.I., Thomas, R.J. and Vera, R.R. (1994) Carbon storage by introduced deep-rooted grasses in the South America savannas. *Nature* 371, 236–238.

Gei, M.G. and Powers, J.S. (2013) Do legumes and non-legumes tree species affect soil properties in unmanaged forests and plantations in Costa Rican dry forests? *Soil Biology and Biochemistry* 57, 264–272.

Gillespie, T.W., Grijalva, A. and Farris, C.N. (2000) Diversity, composition, and structure of tropical dry forests in Central America. *Plant Ecology* 147, 37–47.

Greenland, D.J. and Nye, P.H. (1959) Increases in the carbon and nitrogen contents of tropical soils under natural fallows. *Journal of Soil Science* 10, 284–299.

Hartshorn, G.S. (1983) Plants. In: Janzen, D.H. (ed.) *Costa Rican Natural History*. University of Chicago Press, Chicago, Illinois, pp. 118–157.

Hernandez, D.L., Esch, E.H., Alster, C.J., McKone, M.J. and Camill, P. (2013) Rapid accumulation of soil carbon and nitrogen in a prairie restoration chronosequence. *Soil Science Society of America Journal* 77(6), 2029–2038.

Jenny, H. (1941) *Factors of Soil Formation: A System of Quantitative Pedology*. McGraw-Hill, New York.

Jenny, H., Bingham, F. and Padilla-Saravia, B. (1948) Nitrogen and organic matter contents of equatorial soils of Columbia, South America. *Soil Science* 66, 173–186.

Jiménez, J.A., González-Jiménez, E. and Mateo-Vega, J. (2001) *Perspectives for the Integrated Management of the Tempisque River Basin*. San Jose. Organization for Tropical Studies, Costa Rica, pp. 1–24.

Jiménez, J.J., Lorenz, K. and Lal, R. (2011) Organic carbon and nitrogen in soil particle-size aggregates under dry tropical forests from Guanacaste, Costa Rica – Implications for within-site soil organic carbon stabilization. *Catena* 86, 178–191.

Kissing, L.B. and Powers, J.S. (2010) Coarse woody debris stocks as a function of forest type and stand age in Costa Rican tropical dry forest: Long-lasting legacies of previous land use. *Journal of Tropical Ecology* 26, 467–471.

Laganière, J., Angers, D.A. and Paré, D. (2010) Carbon accumulation in agricultural soils after afforestation: A meta-analysis. *Global Change Biology* 16, 439–453.

Lambin, E.F., Geist, H.J. and Lepers, E. (2003) Dynamics of land-use and land-cover change in tropical regions. *Annual Review Environment and Resources* 28, 205–241.

Laurance, W.F. (2007) Have we overstated the tropical biodiversity crisis? *Trends in Ecology and Evolution* 22, 65–70.

Leiva, J.A., Mata, R., Rocha, O.J. and Gutiérrez-Soto, M.V. (2009) Cronología de la regeneración del bosque tropical seco en Santa Rosa, Guanacaste, Costa Rica. I. Características edáficas. *Revista de Biologia Tropical* 57, 801–815.

López-Ulloa, M., Veldkamp, E. and de Koning, G.H.J. (2005) Soil carbon stabilization in converted tropical pastures and forests depends on soil type. *Soil Science Society of America Journal* 69, 1110–1117.

Lorenz, K., Lal, R. and Jiménez, J.J. (2009) Soil organic carbon stabilization in dry tropical forests of Costa Rica. *Geoderma* 152, 95–103.

Maestre, F.T., Quero, J.L., Gotelli, N.J., Escudero, A., Ochoa, V., Delgado-Baquerizo, M. *et al.* (2012) Plant species richness and ecosystem multifunctionality in global drylands. *Science* 335, 214–218.

Meli, P. (2003) Restauracion ecologica de bosques tropicales. Veinte años de investigacion academica. *Interciencia* 28, 581–589.

Melillo, J.M., Houghton, R.A., Kicklighter, D.W. and Mcguire, A.D. (1996) Tropical deforestation and the global carbon budget. *Annual Review Energy Environment* 21, 293–310.

Oades, J.M. (1988) The retention of organic matter in soils. *Biogeochemistry* 5, 35–70.

Post, W.M. and Kwon, K.C. (2000) Soil carbon sequestration and land-use change: Processes and potential. *Global Change Biology* 6, 317–327.

Powers, J.S. and Perez-Aviles, D. (2013) Edaphic factors are a more important control on surface fine roots than stand age in secondary tropical dry forests. *Biotropica* 45, 1–9.

Powers, J.S. and Salute, S. (2011) Macro- and micronutrient effects on decomposition of leaf litter from two tropical tree species: Inferences from a short-term laboratory incubation. *Plant and Soil* 346, 245–257.

Powers, J.S. and Veldkamp, E. (2005) Regional variation in soil carbon and delta^{13}C in forests and pastures of northeastern Costa Rica. *Biogeochemistry* 72, 315–336.

Powers, J.S., Becknell, J.M., Irving, J. and Perez-Aviles, D. (2009) Diversity and structure of regenerating tropical dry forests in Costa Rica: Geographic patterns and environmental drivers *Forest Ecology and Management* 258, 959–970.

Powers, J.S., Corre, M.D., Twine, T.E. and Veldkamp, E. (2011) Geographic bias of field observations of soil carbon stocks with tropical land–use changes precludes spatial extrapolation. *Proceedings National Academy of Sciences United States of America* 108, 6318–6322.

Sader, S.A. and Joyce, A.T. (1988) Deforestation rates and trends in Costa Rica, 1940–1983. *Biotropica* 20, 11–19.

Schlesinger, W.H. (1977) Carbon balance in terrestrial detritus. *Annual Review Ecology and Systematics* 8, 51–81.

Toriyama, J., Kato, T., Siregar, C.A., Siringoringo, H.H., Ohta, S. and Kiyono, Y. (2011) Comparison of depth- and mass-based approaches for estimating changes in forest soil carbon stocks: A case study in young plantations and secondary forests in West Java, Indonesia. *Forest Ecology and Management* 262, 1659–1667.

Torn, M.S., Trumbore, S.E., Chadwick, O.A., Vitousek, P.M. and Hendricks, D.M. (1997) Mineral control of soil organic carbon storage and turnover. *Nature* 389, 170–173.

Wright, S.J. and Muller-Landau, H.C. (2006) The future of tropical forest species. *Biotropica* 38, 287–301.

14 Conversion of Pastures into *Tectona grandis* Plantations in Western Panamá: Effects on Soil Properties and the Mechanisms Underlying these Changes

Luitgard Schwendenmann[1]* and Simon Kaiser[2]
[1]*School of Environment, The University of Auckland, Auckland, New Zealand;*
[2]*Tropical Silviculture and Forest Ecology, Georg-August University Göttingen, Göttingen, Germany*

14.1 Introduction

The establishment of forest plantations on pastures and cropland has the potential to lead to carbon (C) sequestration and may contribute to the restoration of ecosystem services (Fisher, 1995; Montagnini and Jordan, 2005). Where trees have rapid growth rates, particularly in the tropics and sub-tropics, the uptake of carbon dioxide from the atmosphere and subsequent storage in aboveground biomass through naturally regenerating secondary succession, afforestation (i.e. planting in an area where the previous vegetation was not forest) and reforestation (i.e. planting of forests on lands that were forested but that have been converted to non-forested land; CBD, 2003) is large (Silver *et al.*, 2000; Lal *et al.*, 2005). In this chapter, reforestation will be used to cover both reforestation and afforestation activities unless specified otherwise.

Aboveground–belowground linkages drive ecosystem processes including the fluxes of C and nutrients (Bardgett and Wardle, 2010). Whether higher standing biomass leads to enhanced soil C storage depends upon a variety of factors such as the amount of above- and belowground biomass returned to the soil as necromass, the rate of decomposition, soil moisture and temperature, soil fertility and land-use management (Gupta and Rao, 1994; Bardgett *et al.*, 2009). A number of studies and meta-analyses have shown that reforestation of pastures and cropland also leads to an increase in soil organic carbon (SOC) stocks (Paul *et al.*, 2002; Laganière *et al.*, 2010; Powers *et al.*, 2011). Other studies have observed losses in soil C following reforestation (de Koning *et al.*, 2003; López-Ulloa *et al.*, 2005; Guo *et al.*, 2008) and yet others have reported no net change (Bashkin and Binkley, 1998; Davis *et al.*, 2007). Globally, soils store three times as much C as aboveground biomass (Raich and Schlesinger, 1992), it is therefore important to quantify changes in SOC stocks following land-use conversion accurately. Research has shown that the direction and magnitude of changes in SOC following reforestation depend on a wide range of factors. These include tree species effects (Guo and Gifford, 2002; Russell *et al.*, 2007, 2010), the legacy effects of previous land use (de Koning et al., 2003), plantation age (Laganière *et al.*, 2010),

* Corresponding author: l.schwendenmann@auckland.ac.nz

soil mineralogy (Powers *et al.*, 2011) and climatic factors (Powers *et al.*, 2011).

Besides SOC, live and dead roots contribute to belowground C. Fine root biomass, root turnover and rooting depth determine the fate of root C and thus have implications for belowground C sequestration and storage (Lugo and Brown, 1993). A number of studies have examined fine root biomass following the conversion of tropical forests into other land uses (Trumbore *et al.*, 2006; Harteveld *et al.*, 2007) and during secondary succession (Hertel *et al.*, 2003; Brearley, 2011). However, the effect of reforestation on fine root biomass and fine root C and nitrogen (N) stocks is not well investigated.

Reforestation may not only lead to changes in SOC and fine roots, but may also alter the physical and chemical properties of soils due to plant–soil interactions. For example, some studies have shown that bulk density (Fisher, 1995; Salimon *et al.*, 2009) and soil pH (Berthrong *et al.*, 2009) under plantations is lower compared with croplands and pastures. However, the results are inconsistent and depend on tree species, stand age, land-use history and biophysical conditions in the study regions (Powers *et al.*, 1997; Montagnini, 2000).

One of the tree species with high growth rates and thus a high C sequestration potential is teak (*Tectona grandis*; Lamiaceae). Teak is native to south and south-east Asia, in particular India, Indonesia, Malaysia and Myanmar (Pandey and Brown, 2000), but is a widely planted hardwood species across the tropics, within and outside its natural distribution range. Of the plantations established throughout the tropics, over 2.3 million ha have been planted with teak (FAO, 2010). In Panamá, where this study was conducted, plantation forestry is rapidly growing with the addition of teak. From 1992 to 2004, an area of 46,000 ha was reforested, with approximately 75% of this area having been planted with teak (ANAM, 2006).

The aim of this study was to understand how soil C and fine roots change in response to pasture conversion and teak plantation establishment in western Panamá. We also evaluated potential positive or negative effects of teak plantations on physical (bulk density) and chemical soil properties (pH, cation exchange capacity and base saturation). A chronosequence of teak stands spanning 10 years of development on former pastures was used for this purpose. We hypothesized that pastures and teak plantations differ in their soil properties and we expected that effects, including increased soil C storage and soil fertility, are achieved to a greater extent in older teak plantations. We assumed that land-use history, site management, and species effects are the mechanisms driving these changes.

14.2 Materials and Methods

14.2.1 Study site

The study was conducted in the San Lorenzo district (8° 19′ N, 82° 9′ W) of the Chiriquí Province in western Panamá. The study area is delimited by the Pacific Ocean in the south and the Pan-American Highway in the north. The elevation of the study area ranges between 5 m and 35 m above sea level. The mean annual rainfall recorded at the nearest weather station in David, approximately 40 km west of the study area, was 2750 mm for the years 1996–2007 (ETESA, 2008). The region is characterized by distinct wet and dry seasons with the majority of rain falling between the months of June and December. The mean annual temperature is 26.7°C (ETESA, 2008).

Soils in the study area are formed by alluvial processes and are classified as Ultisols and Alfisols (B. Name, Panamá City, 2005, personal communication). The natural vegetation in the area was an evergreen seasonal lowland forest (ANAM, 2000) before deforestation in the 1950s opened the area to livestock farming. Since the 1990s, an increasing area of pasture in the study area has been converted into teak plantations. As well as teak, a number of *Acacia mangium* (Leguminosae) and *Khaya senegalensis* (Meliaceae) plantations exist in the area (R. Rojas, San Lorenzo, 2005, personal communication).

14.2.2 Land-use type, site and plot selection

The chronosequence approach (Penman *et al.*, 2003) was used to investigate whether the establishment of teak plantations on Improved Pastures and Chumico Pastures had an effect on soil properties, soil C and nitrogen (N), and fine root biomass. We selected teak plantations aged 1, 2 and 10 years old at the time of this study (Teak-1, Teak-2, Teak-10). For each land-use type (Chumico Pasture, Improved Pasture, Teak-1, Teak-2 and Teak-10) three to five sites of similar environmental conditions were selected using a stratified random design. The main characteristics of each land-use type are described below and summarized in Table 14.1.

- Chumico Pastures: Characterized by extensive grazing and a high number of native grass species. The distinct feature of this land-use type is the abundance of *Curatella americana* (Dilleniaceae) (in Spanish: *Chumico*) trees. This silvopastoral system has been in place for decades.
- Improved Pastures: The so-called Improved Pastures (in Spanish: *pasto mejorado*) consist of the introduced deep-rooted C_4 grass *Brachiaria humidicola* (Poaceae) and are intensively grazed by cattle (up to three head of cattle per hectare).
- Teak plantations: The plantations were established on both Chumico and Improved Pasture sites. Site preparation before planting included the cutting of existing trees and shrubs and the removal of stumps and coarse roots. The residual grass and herbaceous vegetation was ploughed to a depth of up to 30 cm. The soil remained bare for approximately 3 months prior to tree planting. Saplings were planted at 3 m spacing with an initial tree density of 1111 trees ha^{-1}. The area was drained at locations with high soil water content (R. Rojas, San Lorenzo, 2005, personal communication). After planting, grass, herbs and shrubs were removed by regular cutting and application of herbicides. The plantations were fertilized with 4 lbs (1 lb = 453.6 g) of organic fertilizer (in Spanish: *Gallinaza*) and 1 lb of lime per plant annually (R. Rojas, San Lorenzo, 2005, personal communication). Teak-10 was thinned and at the time of this study in 2005, 55–70% of the originally planted trees remained. Re-sprouting from cut stumps was common and these sprouts were up to 5 m high. The canopy was closed. No other tree species were found in the canopy and understorey. Grass and shrub abundance in Teak-10 was low.

Within each site, three sampling plots (21 m × 21 m) were established. In total, 57 plots across 19 sites were investigated. All plots, except for Teak-10, were located in the sub-district of San Lorenzo. The Teak-10 plantation was situated in the sub-district of Boca del Monte, 15 km to the north-west of San Lorenzo. A high percentage of clay and silt (>80%) was found across the soil profiles at all sites in San Lorenzo (Chumico Pasture, Improved Pasture, Teak-1, Teak-2). These soils were classified as clay, whereas the soils under Teak-10 were classified as loamy clay.

14.2.3 Soil and root sampling

All samples were taken between May and July 2005. For chemical analysis, three samples were taken in each plot at depths of 0–10 cm, 10–20 cm and 30–40 cm with a soil auger (2 cm in diameter). A soil pit was dug in one of the three plots per site and a sample from 90–100 cm depth was also collected. In total, 225 soil samples were collected. The samples were oven-dried (60°C, 48 h) and

Table 14.1. Characteristics of the five land-use types studied for soil properties, Chiriquí, Panamá. (Modified from Derwisch *et al.*, 2009.)

Land-use type	Year sites established	Tree characteristics (mean ± SD)			Aboveground carbon stock (kg m^{-2})	
		Density (n ha^{-1})	Height (m)	Diameter (cm)	Trees	Grass/litter
Chumico Pasture	1950–1960					5.1 ± 0.3
Improved Pasture	1950–1960					3 – 7[1]
Teak-1	2004	1079	2.1 ± 1.6	3.7 ± 1.5	0.6 ± 1.4	2.3 ± 0.1
Teak-2	2003	990	2.8 ± 2.0	3.6 ± 2.0	0.8 ± 2.1	5.8 ± 0.3
Teak-10	1995	383	20.5 ± 3.0	22.5 ± 3.6	38.0 ± 3.4	2.7 ± 0.1

The values are means ± standard deviation (*n* = 3–5 sites). [1]Pedro *et al.* (2007)

sieved (2 mm). Roots and organic material were removed before further analysis.

Volumetric samples (257.65 cm^3) were taken from the soil pit at 0–5 cm, 10–15 cm, 30–35 cm, and 90–95 cm. One sample was collected for bulk density (in total 76 samples) and a second sample was taken for fine root analysis (in total 76 samples). For fine root analysis, the samples were covered with water, gently shaken and poured on to a 1-mm mesh sieve. The procedure was repeated until all roots were on the sieve. All live roots (<2 mm diameter) were visually separated into teak and grass roots based on root colour and structure. The roots (bulked per plot) were washed again, dried at 60°C for 48 h and weighed.

14.2.4 Laboratory analyses

Soil texture was determined using a modified pipette method after treating the samples with H_2O_2 and sodium pyrophosphate (Moschrefi, 1983). Bulk density was calculated as the quotient of soil dry weight (105°C, 48 h) and fresh soil volume.

The pH in water was measured using a 1:2.5 soil-to-solution ratio (GPHR 1400A; Greisinger Electronic GmbH, Regenstauf, Germany). The effective cation exchange capacity (CEC_e) was estimated as the sum of exchangeable cations at the pH of the soil (Sumner and Miller, 1996). Dried and sieved soil samples were extracted with 1 N NH_4Cl (Meiwes *et al.*, 1984). The elements were determined using atomic absorption spectroscopy (Spectra AA.300; Varian Pty Ltd, Clayton, South Australia). Base saturation was calculated by summing the base forming cations (calcium (Ca), magnesium (Mg), potassium (K) and sodium) and then dividing this sum by the CEC_e value. The CEC_e was only determined in one soil profile under Improved Pasture, Chumico Pasture and Teak-1.

Soil and root sub-samples were ground to a fine powder using a ball mill (Fritsch GmbH, Idar-Oberstein, Germany) and then analysed for C and N with an elemental analyser (Carlo Erba NC 2500; Thermo Quest, Milan, Italy).

14.2.5 Calculation of carbon and nitrogen stocks in mineral soil and roots

Soil organic C and N stocks (kg m^2) were calculated for each soil layer separately by multiplying C and N concentration by the soil bulk density of the soil layer and by soil depth. Bulk density did not differ significantly among land-use types for a given depth (Table 14.2). However, we accounted for changes in bulk density following the approach by Solomon *et al.* (2002) by dividing the bulk density of Improved Pasture, Chumico Pasture, Teak-1 and Teak-2 by the bulk density of Teak-10 and then multiplying these values by the fixed soil depth (10 cm). Fine root C stock (g m^{-2}) for each layer was calculated as the product of root C concentration and root biomass.

14.2.6 Statistical analyses

Differences among land-use types were evaluated by Kruskal–Wallis tests. For each soil layer, the mean of three plots within each site was used in the analysis of $n = 5$ land-use types with three to five replicates per type, for a total of 19 sites analysed. If a significant effect was found, pairwise comparisons were conducted using the Mann–Whitney U test to determine which land uses were different from others. Spearman Rank correlation was applied to explore relationships among soil parameters. Significant differences were stated at the level of $P < 0.05$. Statistical analysis was conducted using SPSS 18.0.2; SPSS Inc., Chicago, Illinois, USA).

14.3 Results

14.3.1 Physical and chemical soil properties

Mean bulk density in the upper 0–10 cm ranged from 0.85 Mg m^{-3} to 0.98 Mg m^{-3} (Table 14.2). Bulk density across all land-use types tended to increase slightly with depth (Table 14.2). Bulk density values in Teak-10 were consistently lower compared to the other land-use types across the soil profile; however, differences in bulk density among land-use types were not statistically significant (Table 14.2).

The soils in all land-use types were acidic, with pH values in the topsoil between 4.7 and 5.3 (Table 14.2). In the top 10 cm, no significant differences in pH were found among land-use types (Table 14.2). In contrast, land-use had an effect on pH in deeper layers (10–20 cm, $P < 0.05$;

Table 14.2. Bulk density, pH (in water), effective cation exchange capacity (CEC_e) and base saturation of mineral soil for four depth intervals across five land-use types, Chiriquí, Panamá.

Soil properties, depth (cm)	Chumico Pasture	Improved Pasture	Teak-1	Teak-2	Teak-10	P (Kruskal Wallis Test)
Bulk density (g cm^{-3})						
0–10	0.92 ± 0.24	0.98 ± 0.09	0.92 ± 0.04	0.90 ± 0.03	0.85 ± 0.07	n.s.
10–20	1.04 ± 0.35	1.11 ± 0.10	1.04 ± 0.06	1.04 ± 0.09	0.96 ± 0.03	n.s.
30–40	1.06 ± 0.15	1.19 ± 0.10	1.10 ± 0.10	1.11 ± 0.13	0.96 ± 0.05	n.s.
90–100	1.29 ± 0.10[a]	1.14 ± 0.07[a]	0.95 ± 0.12[b]	1.19 ± 0.19[a]	0.98 ± 0.11[a]	<0.05
pH (in water)						
0–10	5.2 ± 0.3	4.9 ± 0.2	4.7 ± 0.1	4.7 ± 0.3	5.2 ± 0.3	0.05
10–20	5.3 ± 0.1[a]	4.9 ± 0.2[a]	4.6 ± 0.3[b]	4.8 ± 0.2[b]	5.1 ± 0.1[a]	<0.05
30–40	5.5 ± 0.2[a]	5.2 ± 0.1[a]	4.8 ± 0.1[b]	4.9 ± 0.1[b]	5.4 ± 0.2[a]	<0.05
90–100	5.7 ± 0.4	5.5 ± 0.2	5.2 ± 0.2	5.5 ± 0.8	5.2 ± 0.3	n.s.
CEC_e (cmol kg^{-1})*						
0–10	8.0	6.8	12.0	7.7 ± 4.0	30.2 ± 8.6	
10–20	n.d.	n.d.	n.d.	n.d.	n.d.	
30–40	9.5	2.1	10.4	5.5 ± 4.4	32.3 ± 4.1	
90–100	16.0	4.2	9.1	6.5 ± 6.1	38.7 ± 18.4	
Base saturation (%)*						
0–10	12	47	26	32 ± 24	87 ± 13	
10–20	n.d.	n.d.	n.d.	n.d.	n.d.	
30–40	23	32	9	31 ± 26	74 ± 29	
90–100	99	24	6	25 ±3 2	61 ± 46	

The values are means ± standard deviation (n = 3–5 sites). n.d = not determined. Different letters indicate a significant difference among land-use types (P < 0.05). * CEC_e and base saturation were only measured in one soil profile under Chumico Pasture, Improved Pasture and Teak-1.

30–40 cm, P < 0.05), with Teak-1 and Teak-2 having more acidic pH values. The CEC_e was 7–30 cmol kg^{-1} and the base saturation was 12–78% in the top 10 cm (Table 14.2). The CEC_e and base saturation was higher in Teak-10 than in any other land-use type at all depths.

14.3.2 Soil carbon and nitrogen

In the top 10 cm, soil C and N concentrations across land-use types ranged from 2.3% to 3.0% and from 0.21% to 0.29%, respectively (Table 14.3). Land use had a significant effect on C concentration (P < 0.05), N concentration (P < 0.05) and C:N ratio (P < 0.01) in the uppermost soil layer (0–10 cm) (Table 14.3; Fig. 14.1a). The C:N ratio at 10–20 cm (P < 0.05) and 30–40 cm depth (P < 0.05) differed among land-use types, with Teak-10 having the lowest C:N ratio (Table 14.3).

Between 2.0 kg m^{-2} and 2.6 kg m^{-2} C and 0.17 kg m^{-2} and 0.27 kg m^{-2} N, respectively, were stored in the top 10 cm of the soil profile (Table 14.3). The differences in stocks among land-use types were only significant (P < 0.05) at 0–10 cm depth, with Teak-2 having the least and Improved Pasture having the greatest C and N stocks (Table 14.3, Fig. 14.1b).

14.3.3 Fine root biomass and carbon stocks

Fine root biomass (live roots only) in the top 10 cm was 1.5 g m^{-1} C to 35.5 g m^{-1} C. Teak-1 and Teak-2 had significantly lower fine root C stocks compared with Teak-10 and the pasture sites (Table 14.3, Fig. 14.1c). A positive, but non-significant, correlation was found between fine root and soil C stocks in the uppermost soil layer (r = 0.80; P = 0.10).

Fine root C stocks decreased with depth across all the land-use types (Table 14.3). The steepest decline in fine root C stock with depth was measured under Improved Pasture, followed by Chumico Pasture and Teak-10 (Table 14.3).

Table 14.3. Soil organic carbon and nitrogen concentration, C:N ratio, soil organic carbon stocks and fine root carbon stocks in mineral soil for four depth intervals across five land-use types, Chiriquí, Panamá.

Soil properties, depth (cm)	Chumico Pasture	Improved Pasture	Teak-1	Teak-2	Teak-10	*P* (Kruskal–Wallis test)
Carbon concentration (%)						
0–10	2.87 ± 0.41[a]	3.01 ± 0.39[a]	2.39 ± 0.12[b]	2.31 ± 0.25[b]	2.86 ± 0.50[a,b]	<0.05
10–20	1.51 ± 0.13	1.59 ± 0.27	1.68 ± 0.18	1.25 ± 0.39	1.31 ± 0.29	n.s.
30–40	0.75 ± 0.15	0.91 ± 0.23	0.91 ± 0.06	0.78 ± 0.18	0.66 ± 0.18	n.s.
90–100	0.24 ± 0.02	0.37 ± 0.14	0.40 ± 0.14	0.29 ± 0.08	0.38 ± 0.20	n.s.
Nitrogen concentration (%)						
0–10	0.24 ± 0.05[a]	0.29 ± 0.04[b]	0.21 ± 0.01[a]	0.22 ± 0.04[a]	0.32 ± 0.05[b]	<0.05
10–20	0.11 ± 0.02	0.13 ± 0.03	0.13 ± 0.01	0.11 ± 0.04	0.15 ± 0.03	n.s.
30–40	0.06 ± 0.00	0.07 ± 0.02	0.08 ± 0.00	0.07 ± 0.02	0.08 ± 0.02	n.s.
90–100	0.03 ± 0.00	0.04 ± 0.01	0.04 ± 0.01	0.04 ± 0.00	0.05 ± 0.02	n.s.
C:N ratio						
0–10	11.8 ± 0.7[a]	10.2 ± 0.3[b]	11.5 ± 0.2[a]	10.5 ± 1.0[a,b]	8.8 ± 0.2[c]	<0.01
10–20	13.4 ± 1.6[a]	12.0 ± 0.7[a]	13.0 ± 0.3[a]	12.1 ± 1.6[a]	8.4 ± 0.2[b]	<0.05
30–40	12.4 ± 2.0[a]	12.9 ± 1.2[a]	11.9 ± 0.3[a]	11.6 ± 1.1[a]	7.8 ± 0.3[b]	<0.05
90–100	8.4 ± 1.4	9.3 ± 2.2	10.5 ± 1.1	8.7 ± 1.6	7.1 ± 0.6	n.s.
Soil carbon stocks (kg m^{-2})						
0–10	2.45 ± 0.35[a]	2.57 ± 0.33[a]	2.04 ± 0.10[b]	1.98 ± 0.21[b]	2.43 ± 0.36[a]	<0.05
10–20	1.45 ± 0.12	1.53 ± 0.26	1.61 ± 0.17	1.20 ± 0.38	1.26 ± 0.29	n.s.
30–40	0.71 ± 0.15	0.87 ± 0.22	0.87 ± 0.06	0.75 ± 0.18	0.62 ± 0.15	n.s.
90–100	0.24 ± 0.02	0.37 ± 0.14	0.39 ± 0.14	0.29 ± 0.08	0.37 ± 0.19	n.s.
Fine root carbon stocks (g m^{-2})						
0–5 cm	19.3 ± 10.6[a]	35.5 ± 1.6[b]	1.5 ± 0.2[c]	12.1 ± 8.7[a]	29.1 ± 12.1[a,b]	<0.05
10–15 cm	8.2 ± 7.9	5.1 ± 1.4	1.5 ± 1.5	2.9 ± 2.6	8.7 ± 5.1	n.s.
30–35 cm	3.1 ± 3.2	5.2 ± 2.8	0.4 ± 0.6	2.4 ± 2.9	4.4 ± 2.2	n.s.
90–95 cm	0.1 ± 0.1	3.1 ± 2.1	0.0 ± 0.0	0.0 ± 0.0	0.9 ± 1.2	<0.05

The values are means ± standard deviation (n = 3–5 sites). Different letters indicate a significant difference among land-use types ($P < 0.05$).

14.4 Discussion

As hypothesized, pastures and teak plantations differed in a number of soil properties including pH, CEC_e, soil C and N, and fine root biomass. Most of the differences between land uses and age of conversion were limited to the top layer of soil (0–10 cm). In contrast to our expectations, teak plantation development did not lead to a substantial increase in soil C storage. Pastures had higher soil C and N stocks and fine root biomass than the plantations. The main mechanisms underlying the observed changes following pasture conversion are discussed below.

14.4.1 Land-use history, soil mineralogy, site preparation and management

A trend for lower bulk density with increasing plantation age was found. This is in line with other studies where lower bulk densities were found in tree plantations or secondary forests compared to pastures and grasslands (Fisher, 1995; Salimon *et al.*, 2009; Don *et al.*, 2011). Fisher (1995) explained the decrease in bulk density in 4-year-old native and exotic tree plantations established on an abandoned pasture with the input of leaf litter and soil organic matter (SOM) and high biotic activity in the litter such as earthworm and arthropods. In contrast, trampling by cattle and the use of heavy machinery before and during the establishment of plantations was likely to have contributed to the compaction of soil resulting in higher bulk densities in the young teak plantations (Reiners *et al.*, 1994).

The amount of C stored (around 2.5 kg m^{-2}) in the surface layer (0–10 cm) under pastures in San Lorenzo is similar to soil C stocks in pasture sites in central Panamá (Schwendenmann *et al.*, 2007; Neumann-Cosel *et al.*, 2011). In contrast, pastures on old volcanic soils in Costa Rica stored more C in the top 10 cm of soil (around

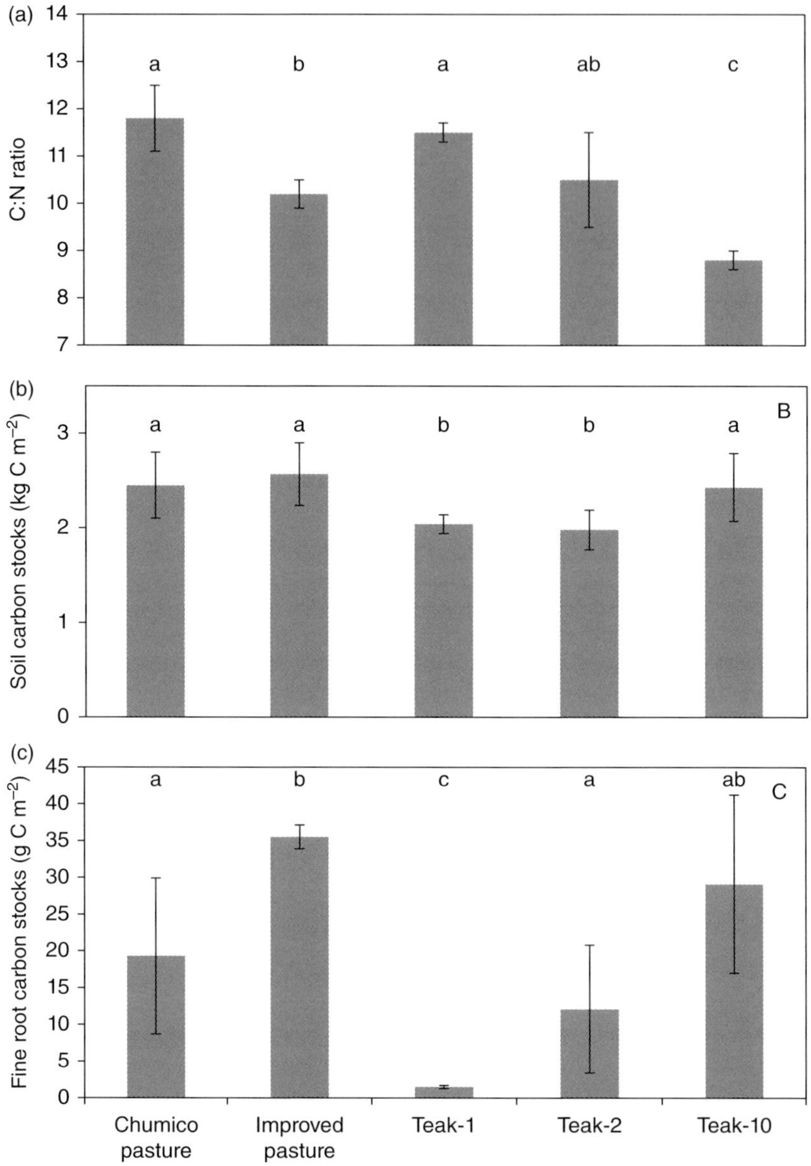

Fig. 14.1. (a) C:N ratio (0–10 cm depth), (b) soil carbon stocks (0–10 cm depth) and (c) fine root carbon stocks (0–5 cm depth) in each of five land-use types, Chiriquí, Panamá. All values are mean ± standard deviation (n = 3–5 sites). Different letters indicate a significant difference among land-use types ($P < 0.05$).

4.5 kg m^{-2}) (Schedlbauer and Kavanagh, 2008). Soil mineralogy and pasture management may explain the differences among study locations (Powers and Schlesinger, 2002).

A reduction in C concentration and SOC stocks after the conversion of pastures and cropland into plantations is frequently observed (Guo and Gifford, 2002; Lagnière *et al.*, 2010). Measuring SOC concentrations in the O/A horizons of active pastures and teak plantations in Costa Rica, Boley *et al.* (2009), found that soils under 8–10-year-old teak plantations had significantly lower C concentrations than active pastures. The establishment of native tree (*Vochysia guatemalensis*; Voychysiaceae) plantations on pastures in Costa Rica reduced SOC stocks in the top 30 cm for the first 10 years (Powers, 2004), suggesting that soil C loss is not limited to the conversion of pastures into exotic species.

Soil disturbance during preparation of the plantation is often cited as a reason for SOC loss in young tree plantations established on pastures and grasslands (Paul *et al.*, 2002; Langière *et al.*, 2010). Prior to tree planting at San Lorenzo, the sites were ploughed to a depth of up to 30 cm. Ploughing and tilling operations can cause loss of substantial amounts of SOC due to accelerated decomposition rates and erosion (McLauchlan, 2006). Low net primary productivity of newly planted trees and the elimination of the grass and shrub vegetation in the first few years after plantation establishment, as shown by low aboveground and fine roots biomass in Teak-1 and Teak-2, is another reason explaining the SOC losses (Turner and Lambert, 2000).

The teak plantations were fertilized, which is likely to have implications for soil microbial communities and soil C storage. The fertilization of *Eucalyptus saligna* (Myrtaceae) seedlings in Hawai'i increased plant productivity and aboveground C storage but resulted in lower fine root biomass and had a negative effect on soil C storage (Giardina *et al.*, 2004).

Soil C gains or losses following land-use conversion also depends on the C stock of the original land-use type. The comparatively high SOC measured in Improved Pasture may be associated with high fine root biomass as roots have been found to be a more effective pathway to build up SOC stocks than foliar litter (Lugo and Brown, 1993). A decline in fine root biomass may be characteristic of reforestation of active pastures where fine root productivity can be high due to frequent grazing (Pucheta *et al.*, 2004).

Statistically significant differences in soil C and N storage between pastures and teak plantations were only evident in the surface soil layer (0–10 cm). This has been reported by others and indicates that surface soils are more sensitive to changes in vegetation and/or site preparation (Don *et al.*, 2011). However, it is interesting to note that SOC stocks in Teak-1 at 10–20 cm tended to be higher compared to the other land-use systems. This observation is in agreement with findings from studies conducted in croplands where the subsoil C stock was higher due to mixing with C-rich topsoil through tillage. This highlights that sampling depth has to be as deep as the tillage depth to quantify land-use change effects on SOC stocks accurately (Don *et al.*, 2011).

14.4.2 Species effects and soil–plant interactions

The highest concentrations of the base-forming cations Ca and Mg across the soil profile were found in Teak-10. A growing body of evidence suggests that tree species can alter soil properties (Binkley and Ryan, 1998; Russell *et al.*, 2004, 2007). For example, significantly higher CEC_e and Ca concentrations were measured in the topsoil of 8-year-old and 12-year-old teak plantations in Costa Rica compared to active pasture, undisturbed forest and native tree plantations (Boley *et al.*, 2009). Increased concentrations of exchangeable Ca and percentage base saturation have also been found under teak plantations in India (Singh *et al.*, 1985). Similarly, Ca enrichment under teak plantations was observed in Myanmar (Suzuki *et al.*, 2007). Higher CEC_e and Ca in soil under teak are attributed to a high leaf litterfall and higher concentrations of Ca and Mg in teak leaves (Singh *et al.*, 1985). We measured a significantly higher pH under Teak-10 compared to the younger plantations despite liming. Salifu and Meyer (1998) also reported high pH values under teak in Ghana, which was attributed to higher Ca concentrations and teak's ability to act as a cation pump from deeper soil layers as shown in a temperate forest study by Dijkstra and Smits (2002).

Species effects and land-use management may also explain the slightly higher pH and CEC_e under Chumico Pasture compared to Improved Pasture. Although we did not find any charcoal, Chumico Pastures might have been burned in the past, affecting soil cation concentration and pH. Further, the mineralization of N, SOM, exchangeable Ca, Mg and K, and total soil N were found to be higher under evergreen trees (among others *Curatella americana*) than in soils under savanna grasses (García Miragaya *et al.*, 1994).

Surface soil erosion, low fine root densities and organic matter quality seem to be responsible for impeding SOC accumulation in the teak plantations investigated. Canopy closure has been reached in Teak-10, limiting the amount of light reaching the forest floor. This may explain the low abundance of understorey vegetation, as observed in Teak-10. Besides increasing plant diversity in teak plantations, understorey vegetation may also contribute to overall ecosystem C storage (Powers *et al.*, 2013). Further, more rapid teak litter decomposition leads to an exposed soil surface that is prone to erosion (Tangtham, 1992; Boley *et al.*, 2009). The risk of soil erosion in teak plantations, caused by large raindrops falling from broad and large teak leaves, has been noted by Calder (2001). In turn, the high level of erosion may create an additional barrier to plant recruitment as seed resources are washed away. In addition to topsoil erosion, root litter deposition is not likely to contribute large amounts to SOC as teak fine-root densities in our study plots were comparatively low. The low soil C:N ratio in Teak-10 suggests high N mineralization rates. It could also suggest N fixation by free living N fixers (Russell and Raich, 2012). This may result in enhanced nutrient and organic matter return to the soil, but also a greater C loss from decomposition (Fisher and Binkley, 2000).

Live fine root biomass and C stocks (0–10 cm) were highest in the top 10 cm under Chumico Pasture and Improved Pasture. This is a species-related effect, as studies have shown that grasslands allocate most of their biomass to the root system (Jackson *et al.*, 1996; Poorter *et al.*, 2012). Further, the turnover of root biomass is much faster than in forest environments (Guo *et al.*, 2007). This may result in higher root C inputs in herbaceous than in forest ecosystems. The high root C stock at 90 cm depth under Improved Pasture is considerable compared to the other land-use systems. The potential of the deep-rooted grass *Brachiaria humidicola* to sequester significant amounts of C deep in the soil was also found in South American savannas (Fisher *et al.*, 1994).

14.4.3 Plantation age

Soil organic C losses were partly reversible as indicated by higher C stocks at 0–10 cm in Teak-10 compared to Teak-1 and Teak-2. According to the outcome of a recent meta-analysis, changes in SOC storage following afforestation of grassland and cropland are closely linked to plantation age. While a loss in SOC stocks of 5.6% was found in young plantations (<10 years), SOC stocks in medium aged (10–30 years) and older (>30 years) plantations increased by 6.1% and 18.5%, respectively (Lagnière *et al.*, 2010). Enhanced SOC accumulation over time after planting trees has been associated with an increase in C input from litterfall and the turnover of dead roots (Richter *et al.*, 1999) and improved SOM protection (Six *et al.*, 2002; Del Galdo *et al.*, 2003).

Low fine root biomass, especially in Teak-1 and Teak-2, is likely to be an effect of tree age and soil disturbance. A positive effect of tree age on fine root biomass has also been reported by Hertel *et al.* (2003). Overall, fine root biomass in our study was low when compared with other teak plantations (Takahashi *et al.*, 2012) and other tropical primary and secondary forest datasets (Brearley, 2011). Besides unsuitable soil conditions (see below), low fine root biomass and soil C stocks at our sites might be related to the low tree density. Roots intensively colonize the soil and add SOM if trees are planted in high density (Rao *et al.*, 2004). Previous research has also shown that fine root density in seasonal tropical forests is strongly influenced by soil moisture, with 1.5 times higher root biomass in the wet season compared to the dry season (Cuevas, 1995). Our root biomass measurements were conducted at the beginning of the dry season. Thus, root biomass during this period may have been lower than that in the rainy season.

Fine root biomass and fine root C stocks were concentrated in the surface soil layer (0–10 cm) independent of land-use type. This is in agreement with the meta-analysis by Jackson *et al.* (1996), who found that the vertical distribution of roots across many ecosystems can be described with an asymptotic equation. Teak-10 showed a deeper root penetration compared to Teak-1 and Teak-2, but we assume that in Teak-10 the maximum depth of fine root penetration has not been reached yet.

14.4.4 Consequences of teak establishment on soil quality, soil C storage and plant diversity

Teak occurs naturally on fertile soils (Tanaka *et al.*, 1998). Teak is a ‘calcareous’ species and requires relatively large amounts of Ca for its growth (Singh *et al.*, 1985; Zech and Drechsel, 1991). Higher uptake and less return during the early period of plantation establishment may result in the depletion of soil nutrients. The lower concentration of exchangeable cations in Teak-1 and Teak-2 may be explained by nutrient depletion despite fertilizing the plots. It is interesting to note that the plantation established in 2004 (Teak-1) was later abandoned as teak grew poorly on these soils (R. Rojas, San Lorenzo, 2012, personal communication). Several authors also highlight that the removal of teak biomass through harvesting will result in a considerable loss of Ca from the system (Hase and Fölster, 1983; Nwoboshi, 1984).

Soil organic C storage under teak increased with plantation age. This is in accordance with other studies where initial loss in SOC that occurred during the first few years after afforestation was followed by a gradual return of C stocks to levels comparable to those in the control agricultural soil, and then increased to generate net soil C gains (Paul *et al.*, 2002; Davis *et al.*, 2007). Our study did not represent the effects of a full teak rotation period of 25 years, thus we may have underestimated the soil C sequestration potential of these plantations. However, the soil C sequestration potential of teak in a separate study in Panamá seemed to be limited, as 20-year-old teak plantations did not accumulate more soil C (between 2.0 and 2.4 kg m^{-2}) in the top 10 cm than Teak-10 (Kraenzel *et al.*, 2003). Soil properties (low pH, limited Ca, aluminium toxicity) might not be suitable for teak establishment in certain areas in Panamá (Craven *et al.*, 2007). Lack of proper silvicultural management (e.g. pruning, thinning) might also result in poor tree growth and C sequestration rates. As described above, soil erosion and rapid decomposition may also hamper the soil C sequestration potential under teak (Takahashi *et al.*, 2012).

The conversion of traditional pasture systems such as the Chumico Pasture into Improved Pasture and teak plantations also has implications for the diversity of plant species. The Chumico Pasture has a diversity of native herbaceous species. In contrast, Improved Pasture is dominated by an introduced grass (*Brachiaria humidicola*) and the plantations are dominated by an exotic tree. Mono-specific and intensively managed teak plantations can improve soil conditions and produce high yields. Negative effects of mono-specific teak plantations are loss of biodiversity, soil erosion and pests (Hallet *et al.*, 2011). To ensure long-term sustainability, potential positive and negative effects on ecosystem health have to be evaluated before establishing new teak plantations. To improve ecosystem function and biodiversity, planting teak with other species is suggested (Brockerhoff *et al.*, 2008; Takahashi *et al.*, 2012).

14.5 Conclusions

Conversion of pastures into teak plantations in Panamá altered soil properties and fine root biomass. Differences were strongest in the surface layer. In contrast to our expectations, soil C stocks under young teak plantations did not exceed the soil C stock under pastures. Still, soil C is an important C stock as more C was stored in the topsoil (0–30 cm) than in the aboveground teak biomass. Despite the pressure to plant fast-growing tree species such as teak for timber and C sequestration, the potential negative effects on ecosystem health have to be evaluated before establishing plantations. Based on our findings, landowners and decision makers should consider the following aspects when establishing and managing teak plantations in order to maintain SOC concentrations at pre-plantation levels and enhance other ecosystem services: planting teak in mixture, minimizing soil disturbance during site preparation and planting, promoting the development of an understorey and limiting activities that restrict or undermine its development.

Acknowledgements

We thank Prime Forestry Group Panamá S.A. (now Forests for Friends) and Tree Gold Cooperation for granting access to their plantations and providing logistical support. Laboratory assistance was provided by Ann-Kathrin Bauer and Helge Dickau. S.K. was supported by a scholarship from the German Academic Exchange Service (DAAD).

References

ANAM (2000) Mapa de vegetación de Panamá. Informe final. Autoridad Nacional del Ambiente (ANAM), Panama. Balboa-Ancón, Panamá, República de Panamá.

ANAM (2006) Indicatores Ambientales de la Republica de Panama 2006. Autoridad Nacional del Ambiente (ANAM), Panama. Balboa-Ancón, Panamá, República de Panamá.

Bardgett, R.D. and Wardle, D.A. (2010) *Aboveground-Belowground Linkages: Biotic Interactions, Ecosystem Processes, and Global Change*. Oxford University Press, Oxford, UK.

Bardgett, R.D., De Deyn, G.B and Ostle, N.J. (2009) Plant–soil interactions and the carbon cycle. *Journal of Ecology* 97, 838–839.

Bashkin, M.A. and Binkley, D. (1998) Changes in soil carbon following afforestation in Hawaii. *Ecology* 79, 828–833.

Berthrong, S.T., Jobbàgy, E.G. and Jackson, R.B. (2009) A global meta-analysis of soil exchangeable cations, pH, carbon and nitrogen with afforestation. *Ecological Applications* 19, 2228–2241.

Binkley, D. and Ryan, M. (1998) Net primary production and nutrient cycling in replicated stands of *Eucalyptus saligna* and *Albizia falcataria*. *Forest Ecology and Management* 112, 79–85.

Boley, J.D., Drew, A.P. and Andrus, R.E. (2009) Effects of active pasture, teak (*Tectona grandis*) and mixed native plantations on soil chemistry in Costa Rica. *Forest Ecology and Management* 257, 2254–2261.

Brearley, F.Q. (2011) Below-ground secondary succession in tropical forests of Borneo. *Journal of Tropical Ecology* 27, 413–420.

Brockerhoff, E.G., Jactel, H., Parrotta, J.A., Quine, C.P. and Sayer, J. (2008) Plantation forests and biodiversity: Oxymoron or opportunity? *Biodiversity and Conservation* 17, 925–951.

Calder, I.R. (2001) Canopy processes: Implications for transpiration, interception and splash induced erosion, ultimately for forest management and water resources. *Plant Ecology* 153, 203–214.

CBD Secretariat of the Convention on Biological Diversity (2003) Interlinkages between biological diversity and climate change. Advice on the integration of biodiversity considerations into the implementation of the United Nations Framework Convention on Climate Change and its Kyoto protocol. Montreal, SCBD, 154pp. (CBD Technical Series no. 10).

Craven, D., Braden, D., Ashton, M.S., Berlyn, G.P., Wishnie, M. and Dent, D. (2007) Between and within-site comparisons of structural and physiological characteristics and foliar nutrient content of 14 tree species at a wet, fertile site and a dry, infertile site in Panama. *Forest Ecology and Management* 238, 335–346.

Cuevas, E. (1995) Biology of the belowground system of tropical dry forests. In: Bullock, S.H., Mooney, H.A. and Medina, E. (eds) *Seasonally Dry Tropical Forests*, Cambridge University Press, New York, pp. 362–383.

Davis, M., Nordmeyer, A., Henley, D. and Watt, M. (2007) Ecosystem carbon accretion 10 years after afforestation of depleted subhumid grassland planted with three densities of *Pinus nigra*. *Global Change Biology* 13, 1414–1422.

de Koning, G.H.J., Veldkamp, E. and López-Ulloa, M. (2003) Quantification of carbon sequestration in soils following pasture to forest conversion in northwestern Ecuador. *Global Biogeochemical Cycles* 17, 1098.

Del Galdo, I., Six, J., Peressotti, A. and Cotrufo, M.F. (2003) Assessing the impact of land-use change on soil sequestration in agriculture soils by means of organic matter fraction and stable C isotopes. *Global Change Biology* 9, 1204–1213.

Derwisch, S., Schwendenmann, L., Olschewski, R. and Hölscher, D. (2009) Estimation and economic evaluation of aboveground carbon storage of *Tectona grandis* plantations in Western Panama. *New Forests* 37, 227–240.

Dijkstra, F.A. and Smits, M.M. (2002) Tree species effects on calcium cycling: The role of calcium uptake in deep soils. *Ecosystems* 5, 385–398.

Don, A., Schumacher, J. and Freibauer, A. (2011) Impact of tropical land-use change on soil organic carbon stocks – A meta-analysis. *Global Change Biology* 17, 1658–1670.

ETESA (2008) *Total mensual de precipitation*, Santa Fe. Gerencia de hidrometerologica y Estudios – Empresa de transmision electrica S.A., Panama City, Panama.

FAO (2010) *Global Forest Resources Assessment 2010: Main Report*. FAO Forestry Paper No. 163. FAO, Rome. Available at: www.fao.org/docrep/013/i1757e/i1757e.pdf (accessed 2 April 2015).

Fisher, R.F. (1995) Amelioration of degraded rain forest soils by plantations of native trees. *Soil Science Society of America Journal* 59, 544–549.

Fisher, R.F. and Binkley, D. (2000) *Ecology and Management of Forest Soils* (3rd edition). Wiley, New York.

Fisher, M.J., Rao, I.M., Ayarra, M.A., Lascano, C.E., Sanz, J.I., Thomas, R.J. and Vera, R.R. (1994) Carbon storage by introduced deep-rooted grasses in the South American savannas. *Nature* 371, 236–239.

García Miragaya, J., Flores, S. and Chacón. N (1994) Soil chemical properties under individual evergreen and deciduous trees in a protected Venezuelan savanna. *Acta Oecologica/International Journal of Ecology* 15, 477–484.

Giardina, C.P., Binkley, D., Ryan, M.G., Fownes, J.H. and Senock, R.S. (2004) Belowground carbon cycling in a humid tropical forest decreases with fertilization. *Oecologia* 139, 545–550.

Guo, L.B. and Gifford, R.M. (2002) Soil carbon stocks and land use change: A meta analysis. *Global Change Biology* 8, 345–360.

Guo, L.B., Wang, M.B. and Gifford, R.M. (2007) The change of soil carbon stocks and fine root dynamics after land use change from a native pasture to a pine plantation. *Plant and Soil* 299, 251–262.

Guo, L.B., Cowie, A.L., Montagu, K.D. and Gifford, R.M. (2008) Carbon and nitrogen stocks in a native pasture and an adjacent 16-year-old *Pinus radiata* D. Don. plantation in Australia. *Agriculture, Ecosystems and Environment* 124, 205–218.

Gupta, R.K. and Rao, D.L.N. (1994) Potential of wastelands for sequestering carbon by reforestation. *Current Science* 66, 378–380.

Hallett, J.T., Diaz-Calvo, J., Villa-Castillo, J. and Wagner, M.R. (2011) Teak plantations: Economic bonanza or environmental disaster? *Journal of Forestry* 109, 288–292.

Hase, H. and Foelster, H. (1983). Impact of plantation forestry with teak (*Tectona grandis*) on the nutrient status of young alluvial soils in west Venezuela. *Forest Ecology and Management* 6, 33–57.

Harteveld, M., Hertel, D., Wiens, M. and Leuschner, C. (2007) Spatial and temporal variability of fine root abundance and growth in tropical moist forests and agroforestry systems (Sulawesi, Indonesia). *Ecotropica* 13, 111–120.

Hertel, D., Leuschner, C. and Hölscher, D. (2003) Size and structure of fine root systems in old-growth and secondary tropical montane forests (Costa Rica). *Biotropica* 35, 143–153.

Jackson, R.B., Canadell, J., Ehleringer, J.R., Mooney, H.A., Sala, O.E. and Schulze, E.D. (1996) A global analysis of root distributions for terrestrial biomes. *Oecologia* 108, 389–411.

Kraenzel, M., Castillo, A., Moore, T. and Potvin, C. (2003) Carbon storage of harvest-age teak (*Tectona grandis*) plantations, Panama. *Forest Ecology and Management* 173, 213–225.

Laganière, J., Angers, D.A. and Paré, D. (2010) Carbon accumulation in agricultural soils after afforestation: A meta-analysis. *Global Change Biology* 16, 439–453.

Lal, R. (2005) Soil carbon sequestration in natural and managed tropical forest ecosystems. *Journal of Sustainable Forestry* 21, 1–30.

López-Ulloa, M., Veldkamp, E. and de Koning, G.H.J. (2005) Soil carbon stabilization in converted tropical pastures and forests depends on soil type. *Soil Science Society of America Journal* 69, 1110–1117.

Lugo, A.E. and Brown, S. (1993) Management of tropical soils as sinks or sources for atmospheric carbon. *Plant and Soil* 149, 27–41.

McLauchlan, K. (2006) The nature and longevity of agricultural impacts on soil carbon and nutrients: A review. *Ecosystems* 9, 1364–1382.

Meiwes, K.-J., König, N., Khanna, P.K. and Ulrich, B. (1984) Chemische Untersuchungsverfahren für Mineralböden, Auflagehumus und Wurzeln zur Charakterisierung und Bewertung der Versauerung in Waldböden. Bericht des Forschungszentrums Waldökosysteme, Göttingen, Germany.

Montagnini, F. (2000) Accumulation in aboveground biomass and soil storage of mineral nutrients in pure and mixed plantations in a humid tropical lowland. *Forest Ecology and Management* 134, 257–270.

Montagnini, F. and Jordan, C.F. (2005) *Tropical Forest Ecology. The Basis for Conservation and Management*. Springer-Verlag, Berlin/New York.

Moschrefi, N. (1983) Ein neues Verfahren der Schlämmanalyse für die Bestimmung der Korngrößenzusammensetzung. *Mitteilungen der Deutschen Bodenkundlichen Gesellschaft* 38, 115–118.

Neumann-Cosel, L., Zimmermann, B., Hall, J.S., van Breugel, M. and Elsenbeer, H. (2011) Soil carbon dynamics under young tropical secondary forests on former pastures – A case study from Panama. *Forest Ecology and Management* 261, 1625–1633.

Nwoboshi, L.C. (1984) Growth and nutrient requirements in teak plantation age series in Nigeria. II. Nutrient accumulation and minimum annual requirements. *Forest Science* 30, 35–40.

Pandey, D. and Brown, C. (2000) Teak: A global overview. *Unasylva* 51, 3–13.

Paul, K.I, Polglase. P.J, Nyakuengama, J.G. and Khanna, P.K. (2002) Change in soil carbon following afforestation. *Forest Ecology and Management* 168, 241–257.

Pedro, J.A., Miles, J.W., Guiot, D.J., Cuadrado, H. and Lascano C.E. (2007) Cultivar Mulato II (*Brachiaria* hybrid CIAT 36087): A high-quality forage grass, resistant to spittlebugs and adapted to well-drained, acid tropical soils. CIAT, Cali, Colombia.

Penman, J., Gytarsky, M., Hiraishi, T., Krug, T., Kruger, D., Pipatti, R. *et al.* (2003) Good practice guidance for land use, land-use change and forestry. IPCC National Greenhouse Gas Inventories Programme and Institute for Global Environmental Strategies, Kanagawa, Japan.

Poorter, H., Niklas, K.J., Reich, P.B., Oleksyn, J., Poot, P. and Mommer L. (2012) Biomass allocation to leaves, stems and roots: meta-analyses of interspecific variation and environmental control. *New Phytologist* 193, 30–50.

Powers, J.S. (2004) Changes in soil carbon and nitrogen after contrasting land-use transitions in northeastern Costa Rica. *Ecosystems* 7, 134–146.

Powers, J.S., and Schlesinger, W.H. (2002) Relationship among soil carbon distributions and biophysical factors at nested spatial scales in rain forests of northeastern Costa Rica. *Geoderma* 109, 165–190.

Powers, J.S, Haggar, J.P. and Fisher, R.F. (1997) The effect of overstory composition on understory woody regeneration and species richness in 7-year-old plantations in Costa Rica. *Forest Ecology and Management* 99, 43–54.

Powers, J.S., Corre, M.D., Twine, T.E. and Veldkamp E. (2011) Geographic bias of field observations of soil carbon stocks with tropical land-use changes precludes spatial extrapolation. *Proceedings of the National Academy of Sciences of the USA* 108, 6318–6322.

Powers, R.S., Busse, M.D., McFarlane, K.J., Zhang, J. and Young, D.H. (2013) Long-term effects of silviculture on soil carbon storage: Does vegetation control make a difference? *Forestry* 86, 47–58.

Pucheta, E., Bonamici, I., Cabido, M. and Diaz, S. (2004). Below-ground biomass and productivity of a grazed site and a neighbouring ungrazed exclosure in a grassland in central Argentina. *Austral Ecology* 29, 201–208.

Raich, J.W. and Schlesinger, W.H. (1992) The global carbon dioxide flux in soil respiration and its relationship to vegetation and climate. *Tellus, Series B: Chemical and Physical Meteorology* 44, 81–89.

Rao, M.R., Schroth, G., Williams, S.E., Namirembe, S., Schaller, M. and Wilson, J. (2004) Managing below-ground interactions in agroecosystems. In: Van Noordwijk, M., Cadisch, G. and Ong, C.K. (eds) *Below-ground Interactions in Tropical Agroecosystems: Concepts and Models with Multiple Plant Components*. CAB International, Wallingford, UK, pp. 309–327.

Reiners, W.A., Bouwman, A.F., Parsons, W.F.J. and Keller, M. (1994) Tropical rainforest conversion to pasture: Changes in vegetation and soil properties. *Ecological Applications* 4, 363–377.

Richter, D.D., Markewitz, D., Trumbore, S.E. and Wells, C.G. (1999). Rapid accumulation and turnover of soil carbon in a re-establishing forest. *Nature* 400, 56–58.

Russell, A.E. and Raich, J.W. (2012) Rapidly growing tropical trees mobilize remarkable amounts of nitrogen, in ways that differ surprisingly among species. *Proceedings of the National Academy of Science of the USA* 109, 10398–10402.

Russell, A.E., Cambardella, C.A., Ewel, J.J. and Parkin, T.B. (2004) Species, rotation, and life-form diversity effects on soil carbon in experimental tropical ecosystems. *Ecological Applications* 14, 47–60.

Russell, A.E., Fisher R.F. and Raich, J.W. (2007) Tree species effects on soil properties in experimental plantations in tropical moist forest. *Soil Science Society of America Journal* 71, 1389–1397.

Russell, A.E., Raich, J.W., Bedoya, R., Valverde-Barrantes, O. and González, E. (2010) Impacts of individual tree species on carbon dynamics in a moist tropical forest environment. *Ecological Applications* 20, 1087–1100.

Salifu, K.F. and Meyer, W.L. (1998). Physico-chemical properties of soils associated with logged forest and areas converted to Teak (*Tectona grandis*) in Ghana. *Commonwealth Forestry Review* 77, 91–99.

Salimon, C.I., Davidson, E.A., Victoria, R.L., Melo, A.W.F. (2004) CO_2 flux from soil in pastures and forests in southwestern Amazonia. *Global Change Biology* 10, 833–843.

Schedlbauer, J.L. and Kavanagh, K.L. (2008) Soil carbon dynamics in a chronosequence of secondary forests in northeastern Costa Rica. *Forest Ecology and Management* 255, 1326–1335.

Schwendenmann, L., Pendall, E. and Potvin, C. (2007) Surface soil organic carbon pools, mineralization and CO_2 efflux rates under different land-use types in Central Panama. In: Tscharntke, T., Leuschner, C., Zeller, M. and Guhardja, E. (eds) *The Stability of Tropical Rainforest Margins: Linking Ecological, Economic and Social Constraints of Land Use and Conservation*. Springer-Verlag, Heidelberg, Germany, pp. 109–131.

Silver, W.L., Ostertag, R. and Lugo, A.E. (2000) The potential for carbon sequestration through reforestation of abandoned tropical agricultural and pasture lands. *Restoration Ecology* 8, 394–407.

Singh, S.B., Nath, S., Pal, D.K. and Banarjee, S.K. (1985) Changes in soil properties under different plantations of the Darjeeling Forest Division. *Indian Forester* 111, 90–98.

Six, J., Conant, R.T., Paul, E.A. and Paustian, K. (2002) Stabilization mechanisms of soil organic matter: Implications for C-saturation of soils. *Plant and Soil* 241, 155–176.

Solomon, D., Fritsche, F., Lehmann, J., Tekalign, M. and Zech, W. (2002) Soil organic matter dynamics in the subhumid agroecosystems of the Ethiopian highlands: Evidence from natural ^{13}C abundance and particle size-fractionation. *Soil Science Society of America Journal* 66, 969–978.

Sumner, M.E. and Miller W.P. (1996) Cation exchange capacity and exchange coefficients. In: Sparks, D.L., Page, A.L., Helmke, P.A., Loeppert, R.H. Soltanpour, P.N., Tabatabai, M.A., Johnston, C.T. and Sumner M.E.

(eds) *Methods of Soil Analysis. Part 3 – Chemical Methods.* SSSA Book Series No. 5. Soil Science Society of America and American Society of Agronomy, Madison, Wisconsin, pp. 1201–1230.

Suzuki, R., Takeda, S. and Thein, H. M. (2007) Chronosequence changes in soil properties of teak (*Tectona grandis*) plantations in the Bago mountains, Myanmar. *Journal of Tropical Forest Science* 19, 207–217.

Takahashi, M., Marod, D., Panuthai, S. and Hirai, K. (2012) Carbon cycling in teak plantations in comparison with seasonally dry tropical forests in Thailand. In: Blanco, J.A. and Lo, Y.H. (eds) *Forest Ecosystems – More than Just Trees.* InTech Rijeka, Croatia, pp. 209–230.

Tanaka, N., Hamazaki, T. and Vacharangkura, T. (1998) Distribution, growth and site requirements of teak. *Japan Agricultural Research Quarterly* 32, 65–77.

Tangtham, N. (1992) Soil erosion problem in teak plantation. In: *Proceeding of the Seminar on 50 Anniversary of Huay-Tak Teak Plantation: 60th Birthday Celebration of Her Majesty the Queen of Thailand,* Lampang, Thailand, 5–8 Aug 1992, pp. 247–259 (in Thai with English summary).

Trumbore, S.E., da Costa, E.S., Nepstad, D.C. *et al.* (2006) Dynamics of fine root carbon in Amazonian tropical ecosystems and the contribution of roots to soil respiration. *Global Change Biology* 12, 217–229.

Turner, J. and Lambert M. (2000) Change in organic carbon in forest plantation soils in eastern Australia. *Forest Ecology and Management* 133, 231–247.

Zech, W. and Drechsel, P. (1991) Relationships between growth, mineral nutrition and site factors of the teak (*Tectona grandis*) plantations in the rainforest zone of Liberia. *Forest Ecology and Management* 41, 221–235.

15 Land-use Change Impacts on Soil Processes in Tropical and Savannah Ecosystems: Emerging Themes and Future Research Directions

Andrew D. Thomas,[1]* Marife D. Corre,[2] Luitgard Schwendenmann,[3] Edzo Veldkamp,[2] Kazumichi Fujii,[4,5] Krista L. McGuire,[6,7] David D. Mkwambisi,[8] L. Ronald Ng Cheong,[9] Jennifer S. Powers[10,11] and Francis Q. Brearley[12]

[1]Department of Geography and Earth Sciences, Aberystwyth University, Aberystwyth, UK; [2]Büsgen Institute – Soil Science of Tropical and Subtropical Ecosystems, Georg-August University Göttingen, Göttingen, Germany; [3]School of Environment, The University of Auckland, Auckland, New Zealand; [4]Forestry and Forest Products Research Institute, Tsukuba, Japan; [5]Graduate School of Agriculture, Kyoto University, Kyoto, Japan; [6]Department of Biology, Barnard College, Columbia University, New York, USA; [7]Department of Ecology, Evolution, and Environmental Biology, Columbia University, New York, USA; [8]Lilongwe University of Agriculture and Natural Resources, Lilongwe, Malawi; [9]Mauritius Sugarcane Industry Research Institute, Réduit, Mauritius; [10]Department of Ecology, Evolution and Behavior, University of Minnesota, St. Paul, USA; [11]Department of Plant Biology, University of Minnesota, St Paul, USA; [12]School of Science and the Environment, Manchester Metropolitan University, Manchester, UK

15.1 Emerging Themes in Land-use Change Effects on Soils in the Tropics

This book contains a series of research studies and state-of-knowledge review chapters that illustrate the range of drivers of land-use change in the tropics and the ways in which these are affecting soils. There are contributions from all continents where tropical and savannah environments are found and from authors across the world. Contributions have been grouped into three broad areas covering: soil microbial populations and processes; urban soils, agriculture and soil contamination; and soil carbon (C) and soil organic matter (SOM). Although we do not claim to have covered all the factors and processes affecting soils in the tropics, we are able to see the emergence of key themes that can be used to inform future research directions. To conclude this book we have produced a synthesis that draws on findings from each of the chapters and includes the thoughts of many of

* Corresponding author: ant23@aber.ac.uk

the authors on the main challenges ahead for a sustainable future for tropical and savannah soils. These are based around where (1) and how (2, 3) we should be conducting our sampling and analyses; adopting a holistic approach to understanding soil functioning and fertility (4–6); utilizing new molecular techniques to test hypotheses relating to soil processes and the impacts of land-use change (7); and, finally, reconciling the multiple demands on soils in rural and urban settings (8, 9).

15.2 Working at Appropriate Scales to Address Relevant Questions About Land Management

1. Rates of land-use change in the tropics are high and there are a multitude of different uses after natural vegetation clearance. The implications of some of these changes for soils remain poorly understood due to a lack of research. Examples include forest clearance for oil palm and rubber plantations and savannah clearance for sugar cane. Importantly, these land-use changes are not represented in recent guidelines from the Intergovernmental Panel on Climate Change (IPCC, 2013) on national greenhouse gas inventories. Their Tier 1 method assumes soil C changes to be zero for conversions from forest to tree or rubber plantations, but only because of the limited data available (Aalde *et al.*, 2006; Lasco *et al.*, 2006). Furthermore, there is a geographical bias in soil research, notably in relation to soil C, with drier areas of the tropics (500–1500 mm annual precipitation) and areas of high-activity clay presently underrepresented (Powers *et al.*, 2011; Thomas, 2012). The importance of tropical peatland soils as major C stocks leads to a need for increased research on this soil type, as it is impossible to generalize from the more 'typical' clay-rich tropical soils. If we are to have a more comprehensive understanding of the global soil resource, and in particular on the effects of land-use change on land–atmosphere C fluxes, then addressing these gaps should be a research priority.

2. Soil sampling and the subsequent laboratory analyses can be time consuming and costly. Logistical and financial constraints on data collection are limiting our understanding of soil processes, properties and responses to land-use change in many areas within the tropics. Ideally, studies should be conducted at landscape scales and include landscape variability since this strongly affects the results (Corre *et al.*, Chapter 10, this volume). Studies on soil C stock estimation, for example, should conduct cost-optimization analyses to answer the question: how can we optimize field-based sampling with a limited amount of money? Many studies of land-use change effects on soil C come from paired plots or chronosequences; these should be complemented with longitudinal studies that are established across gradients of soil-forming factors and common land-use transitions (e.g. Thomas *et al.*, 2011). Finally, all studies should be designed with explicit consideration of how data collected at the plot or field scale can be linked to regional or pan-tropical databases, such as the Harmonized World Soil Database (Nachtergaele *et al.*, 2008).

3. Many tropical soils are old and highly weathered with deep soil profiles. Tropical trees can root very deeply and, as a consequence, add C to the deepest layers of sub-soil. Currently, most evaluations of soil C account for the top 1 m (Jobbágy and Jackson, 2000) and there is a paucity of surveys that assess deeper horizons. However, important changes may actually occur in these deeper soil horizons (Trumbore *et al.*, 1995; Veldkamp *et al.*, 2003). Studies in tropical soils have shown that decomposition rates of SOM decrease with soil depth (van Dam *et al.*, 1997). While the exact mechanisms behind this are unclear, it remains important to understand the full profile when evaluating the complete effects of land-use changes on soil C stocks (Schmidt *et al.*, 2011). The temperature and moisture regimes at depth in the soil will be different to those closer to the surface and this has implications for modelling the effects of climate on C turnover and respiration. Depth-dependent decomposition rates may be especially important in landscapes where active erosion and sedimentation are happening since they may affect landscape-scale soil C balances (Van Oost *et al.*, 2007). Therefore, we recommend additional studies that examine soil properties and dynamics at depths of greater than 1 m.

15.3 Adopting a Holistic Approach to Understanding Soil Functioning and Fertility

4. During the past decade, considerable progress has been made in advancing our understanding of biogeochemical cycles in tropical ecosystems. Many biogeochemists and soil scientists have traditionally focused on specific cycles, in particular the C or nitrogen (N) cycle. However, recent findings highlight the importance of micronutrients and metals in C, N and phosphorus (P) cycling (Bellenger *et al.*, 2011; Wurzburger *et al.*, 2012). Furthermore, hydrological processes are important drivers of soil biogeochemistry (Castellano *et al.*, 2012). As biogeochemical cycles do not exist in isolation, future research should aim to investigate a wider range of elements and determine how they, and their couplings, are affected by land use and climatic changes. Coordinated distributed experiments as described by Fraser *et al.* (2013) might be a way forward for understanding the general principles in coupled biogeochemical cycles in tropical ecosystems.

5. An improved understanding of plant–soil interactions will be critical for predicting the response of tropical ecosystems to land-use change. Soil and plant processes are intricately linked and there is a large body of literature exploring interactions at the plant–soil interface (Bardgett and Wardle, 2010). Investigating soil characteristics under different native tree species in Costa Rica revealed that soil organic C and nutrients were correlated with the amount of leaf litter input but also with litter and root chemistry (Russell *et al.*, 2004). Besides species identity, soil processes are also influenced by plant diversity (Fornara and Tilman, 2008). Soil–plant feedbacks are likely to vary over time due to changes in plant composition and resource availability (Peltzer *et al.*, 2010; Kardol *et al.*, 2013). Thus, future research should address the temporal dynamic of plant composition and related soil processes with a particular focus on the rhizosphere.

6. Soil acidification is likely to become increasingly problematic in tropical agriculture (Sumner and Noble, 2003). Soil organic matter cycling influences acidification through effects on proton generation (nitrification) and consumption (net mineralization of organic anions). Therefore, modelling of soil organic C dynamics is required for predicting soil acidification, especially under land-use change, such as shifting cultivation, deforestation and increasingly mechanized agriculture (see Fujii *et al.*, Chapter 6, this volume). In addition, the role of SOM in proton generation and consumption needs to be studied in greater depth and in more soil types. Both laboratory studies (reciprocal dynamics between SOM and acid production) and field monitoring of SOM and pH (long-term manipulation or chronosequence approaches) in tropical regions are needed (Kemmitt *et al.*, 2005; Guo *et al.*, 2010). The construction of a model of reciprocal dynamics between SOM and acidity will contribute to improved management of SOM for ameliorating acidification and ultimately in better land management strategies.

15.4 Utilizing New Molecular Techniques to Test Hypotheses Relating to Soil Processes and the Impacts of Land Use

7. Microorganisms play a vital role in the cycling of elements and nutrients within all soils. Previous research has been limited in the extent to which we can describe and analyse soil microbial communities but the advent of next generation sequencing has opened up new possibilities for research in soil microbial studies at a finer taxonomic resolution. Linked to functional gene studies (He *et al.*, 2007; Paula *et al.*, 2014), these types of investigations have the potential to unlock the composition and functions of the soil microbial community. Future research has the possibility of using microbial responses as a quantitative indicator of the impact of land-use change on tropical soils. It is currently unknown to what extent degraded and cultivated lands retain microbial communities capable of supporting regenerating forest. For example, oil palm agricultural sites labelled as 'sustainable' with the Roundtable on Sustainable Palm Oil certification have mostly been assessed using macroscopic taxon surveys (Edwards *et al.*, 2010) but incorporation of soil microbial communities into sustainability assessments will be vital for mitigating further biodiversity losses and for evaluating forest regeneration potential.

15.5 Reconciling the Multiple Demands on Soils in Rural and Urban Settings

8. There is huge potential to close nutrient gaps and reduce nutrient losses by adopting approaches such as 'regenerative cycling' (*sensu* Pearson, 2007) whereby waste becomes an agricultural input and is recycled. In the urban environment of developing countries, many cities are faced with challenges associated with unregulated waste disposal and increased waste burning leading to a loss of usable nutrients and many other environmental problems. Examples of organic material that can be added to soils include: eco-sanitation products (Mkwambisi *et al.*, Chapter 8, this volume), bones (Simons *et al.*, 2014), biochar (Abiven *et al.*, 2014) and agricultural wastes such as concentrated molasses stillage (da Silva *et al.*, 2013). There is a clear need, however, to match demand with supply and to determine the long-term use of these products on soil 'quality' to ensure that soils do not become contaminated or imbalanced in nutrients. For example, data from an irrigated trial in Zimbabwe (Booth and Lightfoot, 1990) suggest that there is cause for concern if sugarcane vinasse is applied in large quantities, as this could lead to salinity problems in the long term.

9. Soil nutrient and organic C retention can also be improved through appropriate agricultural management options. If applied correctly, they can reduce inorganic fertilizer requirements, for example, through improved understanding of soil-specific nutrient application rates to optimize soil nutrient cycling while increasing SOM levels. The major benefits of SOM, however, arise not from its accumulation but from its decomposition (Janzen, 2006) thus creating a potential conflict between decisions taken to enhance SOM storage or utilization (e.g. Thomas, 2012). Research is needed to develop, test and validate alternative agricultural management strategies that will optimize SOM storage while at the same time ensuring available nutrients are released to satisfy crop nutrient demand. As an example, in the case of sugarcane, there are many practices such as double-row planting on raised beds, minimum tillage, controlled traffic and legume fallows (Garside *et al.*, 2004) that contrast markedly when compared with a classical approach comprising continuous mono-cropping under single row planting and normal tillage operations. Research is needed to develop reliable management techniques to ensure soil C and key plant nutrients are retained and recycled appropriately to avoid soil degradation and sustain plant productivity (Lal, 2000).

15.6 A Way Forward

In concluding this book we take guidance and inspiration from the United Nation's Food and Agricultural Organization Global Soil Partnership, which raises awareness of soils as a key component of the environment and celebrates their importance as a vital contributor to human well-being (refer to the extensive material on the FAO website for more information: www.fao.org/globalsoilpartnership/en). The lack of awareness of the critical role of soils in the environment amongst the public and policy makers reflects poorly on our collective effort to communicate our expertise outside the narrow confines of academia. This is a vitally important task at a time when soil degradation is increasing in extent and severity, our food, fuel and fibre demands are increasing and yet the political will to cherish and protect soil remains weak. There is no doubt this will be challenging, particularly as the processes we describe are often subtle, complex and hidden below ground. Compared to some of the other environmental issues we face today, soils are rarely considered newsworthy, particularly on television media. As a coalition of soil scientists, ecologists, biologists and geographers we must rise to this challenge and drastically improve the level of engagement with key stakeholders and the way in which soils are taught about at all levels of education.

The Global Soil Partnership have set out 'Five Pillars of Action' that we summarize here as a guide to the way forward:

1. Promote sustainable management of soil resources for soil protection, conservation and sustainable productivity.
2. Encourage investment, technical cooperation, policy, education, awareness and extension in soil.
3. Promote targeted soil research and development focusing on identified gaps, priorities and

synergies with related productive, environmental and social development actions.
4. Enhance the quantity and quality of soil data through improved data generation, analysis, validation, reporting, monitoring and integration with other disciplines.
5. Harmonize methods, measurements and indicators for the sustainable management and protection of soil resources.

The United Nations has declared 2015 to be the International Year of Soils, which the Global Soil Partnership, in conjunction with the Food and Agricultural Organization, will implement. This is a timely opportunity to raise awareness of our soils and to push for more effective protection for this vital and fragile resource. If the five pillars were widely adopted by policy makers, not just in the tropics, but globally, then the benefits would be widespread. Healthier soils would lead to healthier people in an environment that would be more sustainably managed. A few years ago this may have been seen as a rather naïve Utopian vision, held by a committed but eccentric few. Today, we believe it needs to be an integral part of all agricultural and natural resources policy if we are to ensure a viable future for all.

References

Aalde, H., Gonzalez, P., Gytarsky, M., Krug, T., Kurz, W.A., Ogle, S. *et al.* (2006) Forest land. In: *Intergovernmental Panel on Climate Change. Guidelines for National Greenhouse Gas Inventories*. Institute for Global Environmental Strategies Hayama, Japan, pp. 4.1–4.83.

Abiven, S., Schmidt, M.W.I. and Lehmann, J. (2014) Biochar by design. *Nature Geoscience* 7, 326–327.

Bardgett, R.D. and Wardle, D.A. (2010) *Aboveground-Belowground Linkages: Biotic Interactions, Ecosystem Processes, and Global Change*. Oxford University Press, Oxford, UK.

Bellenger, J.-P., Wichard, T., Xu, Y. and Kraepiel, A.M.L. (2011) Essential metals for nitrogen fixation in a free-living N_2-fixing bacterium: Chelation, homeostasis and high use efficiency. *Environmental Microbiology* 13, 1395–1411.

Booth, R.J. and Lightfoot, C.J. (1990) The reticulation of ethanol stillage through irrigation systems and its use for fertilization of sugarcane in Zimbabwe. *Agricultural Water Management* 17, 49–58.

Castellano, M.J., Lewis. D.B., Andrews, D.M. and McDaniel, M.D. (2012) Coupling biogeochemistry and hydropedology to advance carbon and nitrogen cycling science. In: Lin, H. (ed.) *Hydropedology: Synergistic Integration of Soil Science and Hydrology*. Academic Press, Waltham, Massachusetts, pp. 711–735.

da Silva, A., Rossetto, R., Bonnecine, J., Piemonte, M. and Muraoka, T. (2013) Net and potential nitrogen mineralization in soil with sugarcane vinasse. *Sugar Tech* 15, 159–164.

Edwards, D.P., Hodgson, J.A., Hamer, K.C., Mitchel, S.L., Ahmad, A.H., Cornell, S.J. and Wilcove, D.S. (2010) Wildlife-friendly oil palm plantations fail to protect biodiversity effectively. *Conservation Letters* 3, 236–242.

Fornara, D.A. and Tilman, D. (2008) Plant functional composition influences rates of soil and nitrogen accumulation. *Journal of Ecology* 96, 314–322.

Fraser, L.H., Henry, H.A.L., Carlyle, C.N., White, S.R., Beierkuhnlein, C., Cahill Jr, J.F. *et al.* (2013) Coordinated distributed experiments: An emerging tool for testing global hypotheses in ecology and environmental science. *Frontiers in Ecology and the Environment* 11, 147–155.

Garside, A.L., Watters, T.S., Berthelsen, J.E., Sing, N.J., Robotham, B.G. and Bell, M.J. (2004) Comparisons between conventional and alternative sugarcane farming systems which incorporate permanent beds, minimum tillage, controlled traffic and legume fallows. Available at: www.sugarresearch.com.au/icms_docs/157327_Comparisons_between_conventional_and_alternative_sugarcane_farming_systems.pdf (accessed 2 April 2015).

Guo, J.H., Liu, X.J., Zhang, Y., Shen, J.L., Han, W.X., Zhang, W.F. *et al.* (2010) Significant acidification in major Chinese croplands. *Science* 327, 1008–1010.

He, Z., Gentry, T.J., Schadt, C.W., Wu, L., Liebich, J., Chong, S.C. *et al.* (2007) GeoChip: A comprehensive microarray for investigating biogeochemical, ecological and environmental processes. *The ISME Journal* 1, 67–77.

IPCC (2013) *Climate Change 2013: The Physical Science Basis. Contribution of Working Group I to the Fifth Assessment Report of the Intergovernmental Panel on Climate Change*. (eds T.F. Stocker, D. Qin, G.-K. Plattner, M. Tignor, S.K. Allen, J. Boschung, A. Nauels, Y. Xia, V. Bex and P.M. Midgley). Cambridge University Press, Cambridge, UK.

Janzen, H.H. (2006) The soil carbon dilemma: Shall we hoard it or use it? *Soil Biology and Biochemistry* 38, 419–424.

Jobbágy, E.G. and Jackson, R.B. (2000) The vertical distribution of soil organic carbon and its relation to climate and vegetation. *Ecological Applications* 10, 423–436.

Kardol, P., de Deyn, G.B., Laliberté, E., Mariotte, P. and Hawkes, C.V. (2013) Biotic plant–soil feedbacks across temporal scales. *Journal of Ecology* 101, 309–315.

Kemmitt, S.J., Wright, D. and Jones, D.L. (2005) Soil acidification used as a management strategy to reduce nitrate losses from agricultural land. *Soil Biology and Biochemistry* 37, 867–875.

Lal, R. (2000) Soil management in the developing countries. *Soil Science* 165, 57–72.

Lasco, R.D., Ogle, S., Raison, J. and Verschot, L.W.R. (2006) Cropland. In: *Intergovernmental Panel on Climate Change. Guidelines for National Greenhouse Gas Inventories.* Institute for Global Environmental Strategies Hayama, Japan, pp. 5.1–4.66.

Nachtergaele, F., van Velthuizen, H., Verelst, L., Batjes, N., Dijkshoorn, K., van Engelen, V. *et al.* (2008) *Harmonized World Soil Database (version 1.0)*. FAO, Rome, Italy.

Paula, F.S., Rodrigues, J.L.M., Zhou, J., Wu, L., Mueller, R.C., Mirza, B.S. *et al.* (2014) Land use change alters functional gene diversity, composition, and abundance in Amazon forest soil microbial communties. *Molecular Ecology* 23, 2988–2999.

Pearson, C.J. (2007) Regenerative, semiclosed systems: A priority for twenty-first-century agriculture. *BioScience* 57, 409–418.

Peltzer, D.A., Wardle, D.A., Allison, V.J., Baisden, W.T., Bardgett, R.D., Chadwick, O.A. *et al.* (2010) Understanding ecosystem retrogression. *Ecological Monographs* 80, 509–529.

Powers, J.S., Corre, M.D., Twine, T.E. and Veldkamp, E. (2011) Geographic bias of field observations of soil carbon stocks with tropical land-use changes precludes spatial extrapolation. *Proceedings of the National Academy of Sciences of the USA* 108, 6318–6322.

Russell, A.E., Cambardella, C.A., Ewel, J.J. and Parkin, T.B. (2004) Species, rotation-frequency, and life-form diversity effects on soil carbon in experimental tropical systems. *Ecological Applications* 14, 47–60.

Schmidt, M.W.I., Torn, M.S., Abiven, S., Dittmar, T., Guggenberger, G., Janssens, I.A. *et al.* (2011) Persistence of soil organic matter as an ecosystem property. *Nature* 478, 49–56.

Simons, A., Solomon, D., Chibssa, W., Blalock, G. and Lehmann, J. (2014) Filling the phosphorous fertilizer gap in developing countries. *Nature Geoscience* 7, 3.

Sumner, M.E. and Noble, A.D. (2003) Soil acidification: The world story. In: Rengel, Z. (ed.) *Handbook of Soil Acidity*. Marcel Dekker, New York, pp. 1–28.

Thomas, A.D. (2012) Impact of grazing intensity on seasonal variations of soil organic carbon and soil CO_2 efflux in two semi-arid grasslands in southern Botswana. *Philosophical Transactions of the Royal Society Series B – Biological Sciences* 367, 3076–3086.

Thomas, A.D., Hoon, S.R. and Dougill, A.J. (2011) Soil respiration at five sites along the Kalahari Transect: Effects of temperature, precipitation pulses and biological soil crust cover. *Geoderma* 167–168, 284–294.

Trumbore, S.E., Davidson, E.A., de Camargo, P.B., Nepstad, D.C. and Martinelli, L.A. (1995) Belowground cycling of carbon in forests and pastures of eastern Amazonia. *Global Biogeochemical Cycles* 9, 515–528.

van Dam, D., Veldkamp, E. and van Breemen, N. (1997) Soil organic carbon dynamics: Variability with depth in forested and deforested soils under pasture in Costa Rica. *Biogeochemistry* 39, 343–375.

Van Oost, K., Quine, T.A., Govers, G., De Gryze, S., Six, J., Harden, J.W. *et al.* (2007) The impact of agricultural soil erosion on the global carbon cycle. *Science* 318, 626–629.

Veldkamp, E., Becker, A., Schwendenmann, L., Clark, D.A. and Schulte-Bisping, H. (2003) Substantial labile carbon stocks and microbial activity in deeply weathered soils below a tropical wet forest. *Global Change Biology* 9, 1171–1184.

Wurzburger, N., Bellenger, J.P., Kraepiel, A.M.L. and Hedin, L.O. (2012) Molybdenum and phosphorus interact to constrain asymbiotic nitrogen fixation in tropical forests. *PLoS ONE* 7, e33710.

Index

Page numbers in **bold** refer to illustrations and tables